BASIC ELECTRICAL ENGINEERING

RITU SAHDEV

KHANNA BOOK PUBLISHING CO. (P) LTD.
Publisher of Engineering and Computer Books
4C/4344, Ansari Road, Darya Ganj, New Delhi-110002
Phone : 011-23244447-48 **Mobile:** +91-9910909320
E-mail : contact@khannabooks.com
Website : www.khannabooks.com

Dear Readers,

To prevent the piracy, this book is secured with HIGH SECURITY HOLOGRAM on the front title cover. In case you don't find the hologram on the front cover title, please write us to at contact@khannabooks.com or whatsapp us at +91-99109 09320 and avail special gift voucher for yourself.

Specimen of Hologram on front Cover title:

Moreover, there is a SPECIAL DISCOUNT COUPON for you with EVERY HOLOGRAM.

How to avail this SPECIAL DISCOUNT:

Step 1: Scratch the hologram
Step 2: Under the scratch area, your "coupon code" is available
Step 3: Logon to www.khannabooks.com
Step 4: Use your "coupon code" in the shopping cart and get your copy at a special discount
Step 5: Enjoy your reading!

Price: ₹ 425.00

ISBN: 978-93-86173-49-2

Basic Electrical Engineering
by Ritu Sahdev

First Edition: 2018
Reprint: 2022, 2024, **2026**

To view complete list of books, Please scan the QR Code:

Published by:
Khanna Book Publishing Co. (P) Ltd.
CIN: U22110DL1998PTC095547

Visit us at: **www.khannabooks.com**
Write us at: **contact@khannabooks.com**

Printed in Bharat.

Dedication

This book is dedicated to
my **Parents**,
son – **Rohit Sahdev**,
daughter-in-law – **Robina Sahdev**
and
grandsons – **Arnav** and **Adhiraj**

PREFACE

In the present scenario, electrical energy has become an integral part of all the engineering applications. Therefore, a course of **'Basic Electrical Engineering'** has been introduced in all the engineering disciplines by all the universities in India as well as abroad. Usually, this course is taught in the 1st or 2nd semester.

It is observed that most of the students, particularly the students belonging to the disciplines of civil, mechanical, computers, textile etc. face difficulties in understanding the contents of this course. To keep this in mind, every effort has been made to make the text students' friendly by using simple and lucid language.

The book is prepared as per the **latest model curriculum of All India Council for Technical Education (AICTE)** for the Degree and Diploma courses in all branch of engineering.

Each chapter of the book contains much needed text supported by neat and self-explanatory diagrams to make the subject self-speaking to a great extent. A large number of solved and unsolved examples have been added in various chapters to enable the students to attempt different types of questions asked in the examination without any difficulty. Practice Exercises have been added in all the chapters at regular intervals to maintain the students regular in their study. At the end of each chapter Summary, Objective Type Questions, Short-Answer Questions, Test Questions and Unsolved Examples have been added to make the book complete and comprehensive unit in all respects.

The author lays no claim to original research in preparing the text. Materials available in the research work of eminent authors have been used liberally. But the author claims that she has organized the subject matter in very systematic manner. She also claims that the language of the text is lucid, direct and easy to understand.

Although, every care has been taken to eliminate errors but it is very difficult to claim perfection. I shall be very grateful to the readers (students and teachers) and users of this book if they point out any mistake that might have crept in. Suggestions for the improvement of the book shall be highly appreciated.

Ritu Sahdev
Jalandhar

ACKNOWLEDGEMENTS

There are several people to whom I would like to thank. Firstly, I would like to thank my husband Mr. S K Sahdev who has encouraged me and helped me in preparing this text in a presentable form.

Secondly, I would like to appreciate and thank all the Deans, faculty and students of various Technical Universities for giving me their cooperation, support, collaboration and friendship while compiling the work.

I would like to thank all my colleagues and friends who have given me suggestions, ideas and guidance. I owe my family members and relatives a special word of thanks for their moral support and encouragement.

I do express my gratitude to the Publishers "M/S Khanna Publishing House, New Delhi", its Director Mr. Mukul Seth and his team for guidance and bringing out the book in such a short time and reasonable price.

Ritu Sahdev

Jalandhar

CONTENTS

DC CIRCUITS

LEARNING OBJECTIVES

- What are the main elements of an electric circuit?
- What are the units of such elements?
- What is the meaning of constant voltage and constant current source?
- What is Ohm's law? What are its limitations?
- What are the effective value of resistance when number of resistors are connected in series, parallel, or series-parallel combinations?
- What are the applications of series and parallel circuits?
- What is an electric network?
- What is the behaviour of different elements employed in an electric network?
- How electric or electronic circuits are analysed and designed? How the value of different components of an electric or electronic circuit are determined by using:
  - Kichhoff's law
  - Delta-star or star-delta transformation
  - Superposition theorem
  - Thevenin's theorem
  - Norton's theorem
- What is maximum power transfer theorem and its applications?

INTRODUCTION

The path for flow of electric current in called *electric circuit.* The electric circuit in just an arrangement in wihch electrical energy sources and various elements such as resistor, inductors and capacitors are connected in series, parallel or series-parallel combinations.

For simplification or analysing these circuits, various laws and theorems have been developed. In this chapter, we shall discuss some of the important laws and theorems.

1.1 CIRCUIT ELEMENTS

In general, the circuit elements can be categorised as:

1. Active and passive elements.
2. Unilateral and bilateral elements.
3. Linear and non-linear elements.
4. Lumped and distributed elements.

1. Active and passive elements

The elements which supply energy to the circuit or network are called *active elements*. All energy sources (batteries or generators) are the active elements. The elements which receive energy are called *passive elements* (such as resistors, inductors and capacitors).

2. Unilateral and Bilateral Elements

The elements which conduct current in one direction only are called *unilateral elements* such as semiconductor diodes, vacuum tubes, selenium rectifiers etc.

Bilateral elements: The elements which conduct current in both the directions are called *bilateral elements* such as resistors etc. In other words, the elements which behave similarly when current flows through them in either direction are called *bilateral elements.*

3. Linear and Non-linear Elements

Linear Elements: The elements which have V-1 characteristics as straight line are called *linear elements* such as resistors.

Non-linear Elements: The elements which do not follow V-1 characteristics as straight line are called *non-linear elements,* such as diodes, transistors etc.

4. Lumped and Distributed Elements

Lumped Elements: The elements in which action takes place simultaneously are called *lumped elements,* such as resistors, inductors and capacitors. These elements are smaller in size.

Distributed Elements: The elements in which action for a given cause is not occurring simultaneously at the same instant but it is distributed are called *distributed elements,* such as transmission line which is having distributed resistance, inductance and capacitance along its length.

1.2 RESISTORS

Resistor is a component, used to limit the amount of current or divide the voltage in an electronic circuit. The ability of a resistor to oppose the current is called *resistance*. The unit of resistance R is *ohm* for which the symbol is W (the Greek capital letter omega). The schematic symbol of R is shown in fig.1.1.

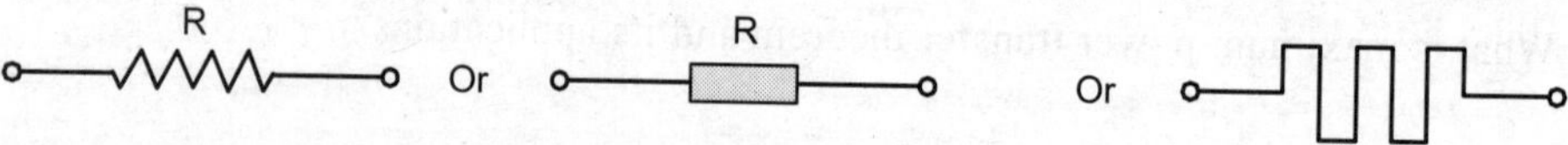

Fig.1.1: Symbol of fixed resistor

Each resistor has two main characteristics *i.e.* its resistance (R) in ohm and its power rating in watt (W). The resistors having wide range of resistance (from a fraction of an ohm to many mega ohm) are available. The power rating may be as low as 1/10 W to as high as several hundred watts. The value of R is selected to obtain a desired current I or voltage drop IR in the circuit. At the same time wattage of the resistor is selected so that it can dissipate the heat losses without overheating itself.

Classification of resistors

The resistors may be classified as fixed and variable resistors.

Fixes resistors: The resistors which have fixed value of resistance are called *fixed resistors.* These resistors may be carbon composition (shown in fig.1.2) or wire-wound resistors (shown in fig.1.3)

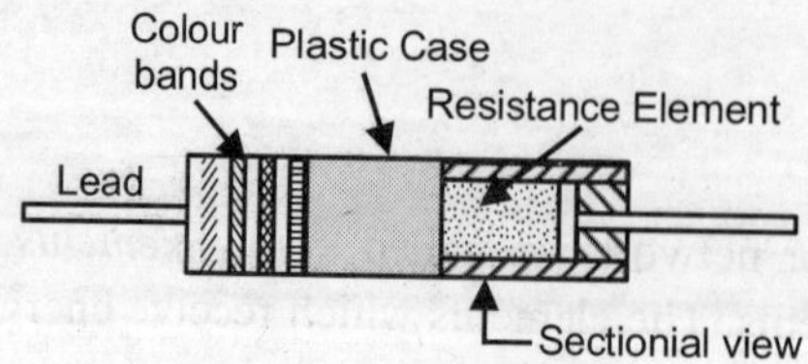

Fig.1.2: Carbon-composition resistor

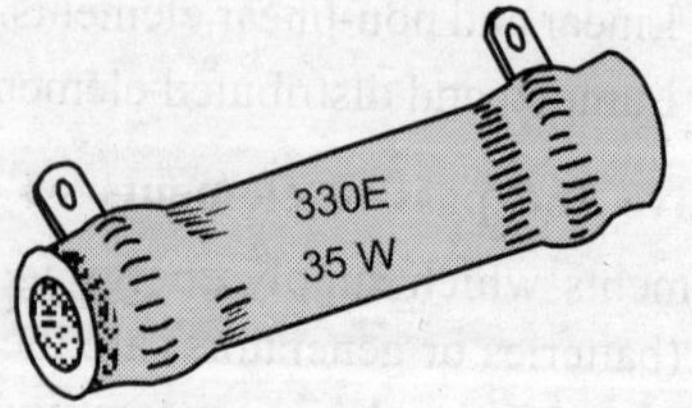

Fig.1.3: Wire-wound resistor

Variable resistors: The resistors in which value of resistance can be varied as per the need are called *variable resistors.* The variable resistors may be carbon composition resistors (shown in fig.1.4) or wire-wound resistors (show in fig.1.5)

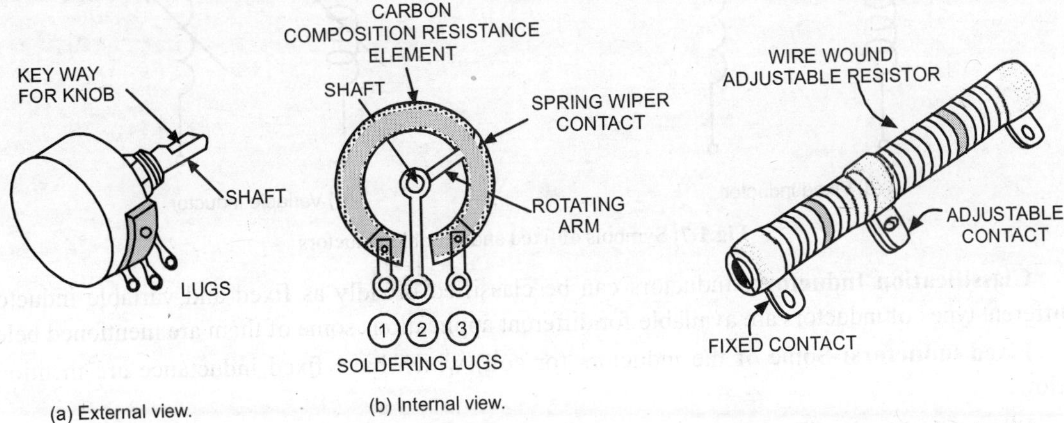

Fig.1.4: Variable carbon-composition resistor

Fig.1.5: Variable wire wound resistor.

The schematic symbol of variable resistor is shown in fig.1.6.

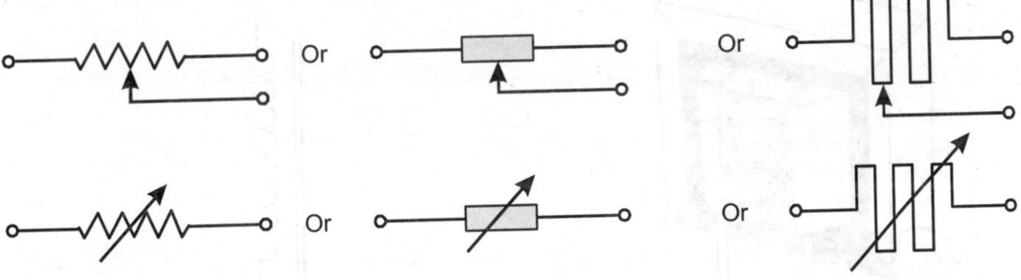

Fig.1.6: Schematic symbol of Variable resistor

The resistance of a wire depends upon:

(i) its length *i.e.* $R \alpha l$

(ii) its area of cross-section *i.e.* $R \alpha \frac{1}{a}$ and

(iii) nature of material of which the wire is made of *i.e.* $R \alpha \frac{l}{a}$ or $R = \rho \frac{l}{a}$

where r ('Rho' a Greek letter) is called resistivity of the material. Its unit is ohm-m.

1.3 INDUCTORS

A component that opposes the change of current in circuit is called an *inductor*. The ability of an inductor or coil due to which it opposes the change of current flowing through it is called its *inductance*. The *inductance* of an inductor may also be defined as the ability of an inductor (or coil) to produce voltage when the current varies through it. The unit of inductance is *henry*. The schematic symbol of a fixed and variable inductor is shown in fig.1.7.

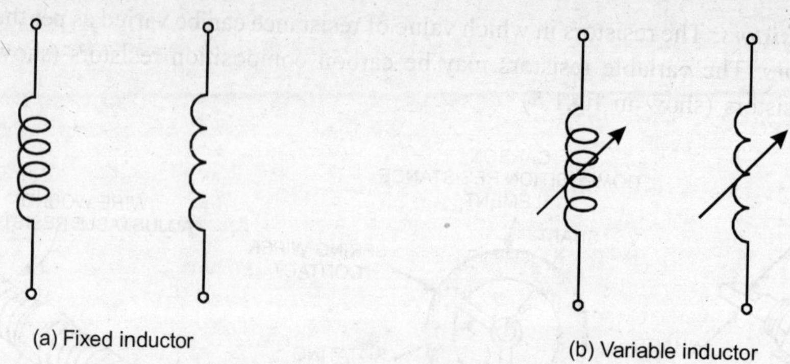

(a) Fixed inductor (b) Variable inductor

Fig.1.7: Symbols of fixed and variable inductors

Classification Inductors: Inductors can be classified broadly as fixed and variable inductors. Different types of inductors are available for different applications some of them are mentioned below.

Fixed inductors: Some of the inductors (or coil) which have fixed inductance are mentioned below:

- **Filter Chokes:** A filter choke or inductor is shown in fig.1.8. It is used to block a.c. signal (voltage) and allows the d.c. signal (voltage) to pass through it.

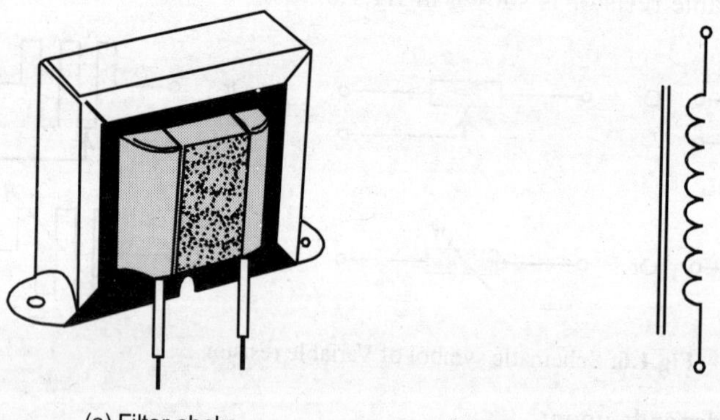

(a) Filter choke. (b) Schematic diagram (symbol) of filter choke.

Fig.1.8: Inductor with magnetic core

- **Radio-Frequency Chokes:** A typical air-core radio-frequency inductor (choke) is shown with its schematic symbol in Fig.1.9.

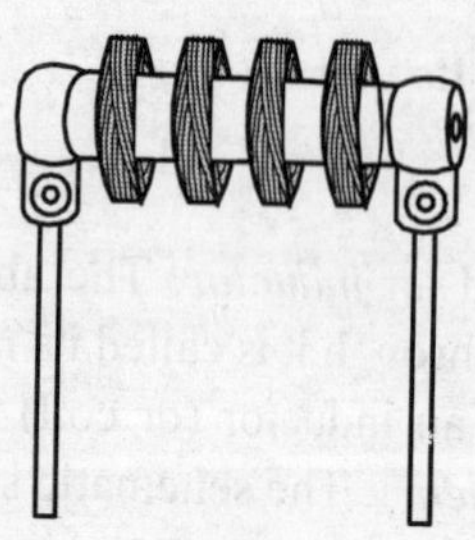

(a) Radio frequency choke. (b) Schematic diagram (symbol) of radio frequency choke.

Fig.1.9: Inductor with air core

Variable Inductors: A variable inductors (or coil) is shown in fig.1.10. The inductance of the coil is varied by varying the position of core. When current flows through a coil flux is produced by it which links with its outer turns. If current changes in the coil, the flux linking with its own turns changes and emf is induced in it called self induced emf which is proportional to rate of change of current or flux linkages, *i.e.*

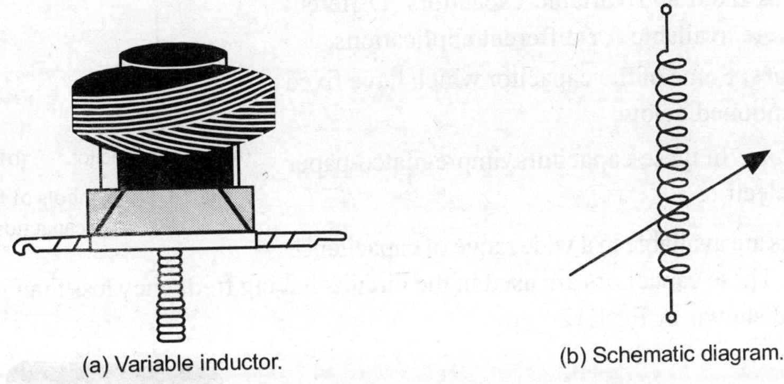
(a) Variable inductor. (b) Schematic diagram.

Fig.1.10: Variable inductor with its schematic diagram.

$$e \propto \frac{di}{dt} \quad or \quad e = L\frac{di}{dt} \quad or \quad L = \frac{e}{di/dt} \qquad \text{...............(i)}$$

Where L is the inductance of the coil.

Now $$e = N\frac{d\phi}{dt} \quad also \ e = L\frac{di}{dt} \quad \therefore \quad L\frac{di}{dt} = N\frac{d\phi}{dt}$$

Or $$L = \frac{Nd\phi}{di} = \frac{N\phi}{I} \qquad \text{................(ii)}$$

$$L = \frac{N}{I} \times \frac{NI}{l} \times a\mu_o\mu_r = \frac{N^2}{\text{Raluctance}} = \frac{N^2}{l} \times a\mu_o\mu_r \qquad \text{...................(iii)}$$

$$L = \frac{e}{di/dt} = \frac{N\phi}{I} = \frac{N^2}{l} \times a\mu_o\mu_r$$

The energy stored in the magnetic field of an inductor is given by the relation

$$W = \frac{1}{2}LI^2 \ or \ w = \frac{1}{2}Li^2 \text{ joule}$$

Note: Inductance opposes the change of current, therefore, it opposes the alternating current (ac) but does not oppose the direct current (dc)

1.4 CAPACITORS

Two conducting plates separated by an insulating material (or dielectric) forms a *capacitor*. Capacitor has the ability to store electric charge. The capacity of a capacitor to store charge per unit potential difference is called its *capacitance*. The unit of capacitance is farad (F). However, the unit farad is too large, the capacitors are specified practically in microfarad (mF) or pico-farad (pF).

A capacitor is a component which offers low impedance to a.c. but very high impedance (resistance) to d.c. The schematic symbol of fixed and variable capacitor (*C*) is shown in Fig.1.11.

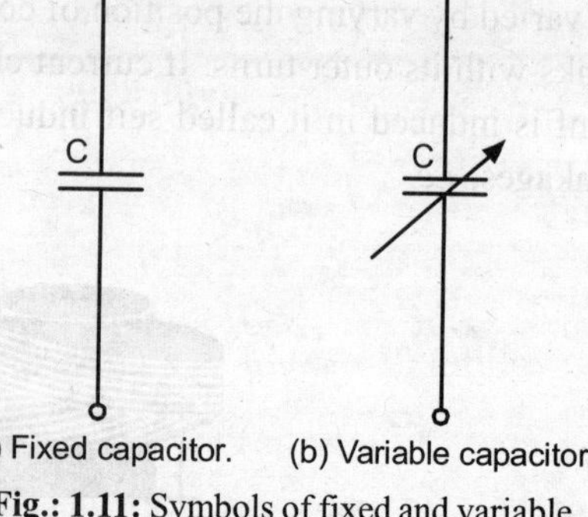

(a) Fixed capacitor. (b) Variable capacitor.

Fig.: 1.11: Symbols of fixed and variable capacitors.

Classification of capacitors: The capacitors can also be classified broadly as fixed and variable capacitors. Different types of capacitors are available for different applications.

Fixed capacitors: Some of the capacitor which have fixed capacitance are mentioned below:

- *Paper Capacitors:* In these capacitors, impregnated paper is used as a dielectric.

Paper capacitors are available in a wide range of capacitance and voltage ratings. These capacitors are used in the circuits having frequency less than *radio-frequency.* Such capacitors are shown in fig.1.12.

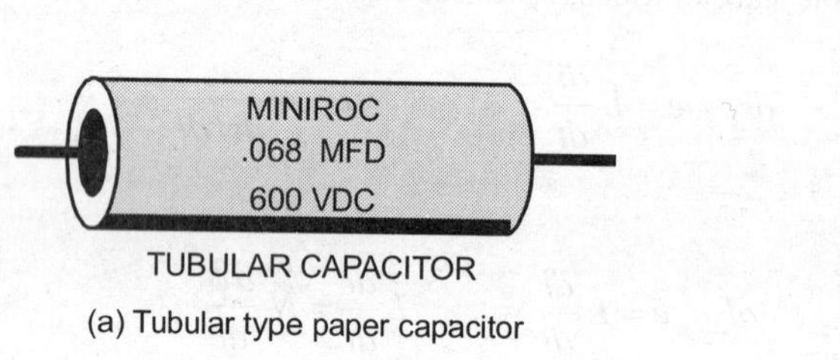

(a) Tubular type paper capacitor

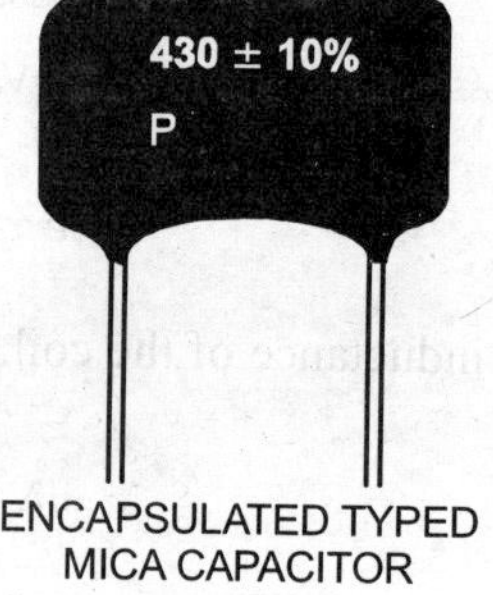

(b) Encapsulated type capacitor.

Fig.1.12: Paper Capacitors

- *Mica Capacitors:* In these capacitors, mica is used as a dielectric. These capacitors are used in the circuits operating at radio-frequency. One of such capacitor is shown in fig.1.13

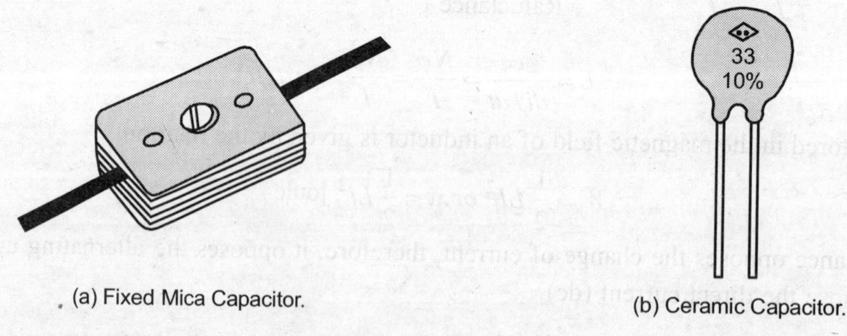

(a) Fixed Mica Capacitor.

(b) Ceramic Capacitor.

Fig.1.13: Fixes Mica capacitor

Fig.1.14: Ceramic Capacitors

- *Ceramic Capacitors:* Ceramic is used as dielectric material in these capacitors. A disc type ceramic capacitor is shown in fig.1.14.
- *Electrolytic Capacitors:* In these capacitors, an electrolyte is used as a dielectric. These are mostly used in the filter section of dc power supplies. Such capacitors are shown in fig.1.15.

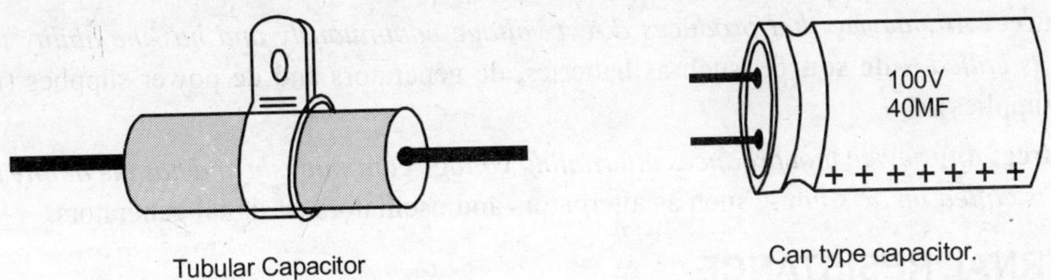

Fig.1.15: Electrolytic Capacitors.

Variable Capacitors

The electronic circuits, in which frequency is to be changed as per requirement such as tuning circuits, variable capacitors are used. *Air-gang capacitor* is, shown in fig.1.16(a). and trimmer or padder with mica sheet capacitors are shown in fig.1.16(b). These are the most common variable capacitors.

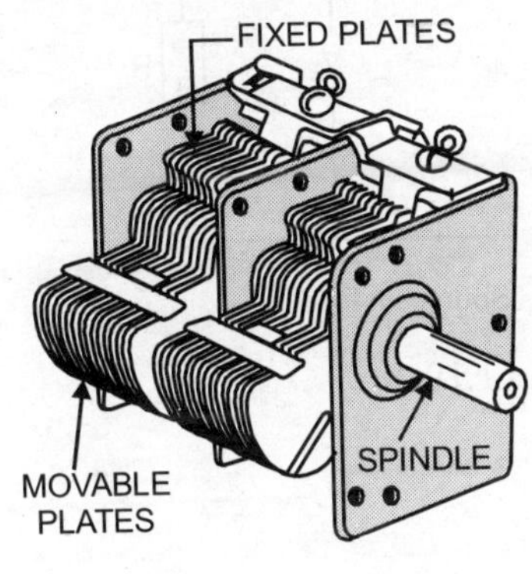

(a) Air-gang capacitor

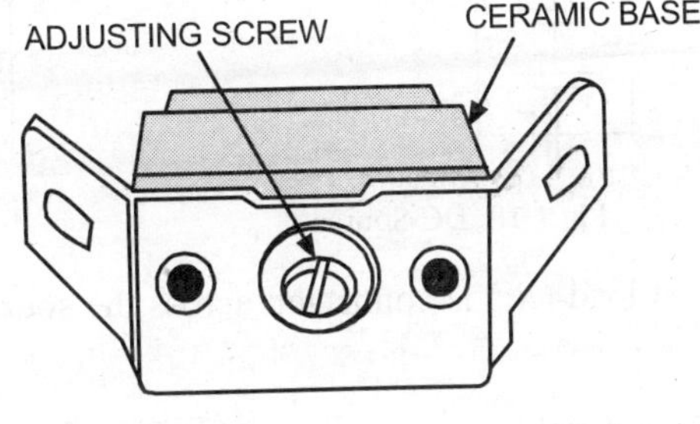

(b) Trimmer or padder capacitor

Fig.1.16: Variable capacitors

The capacitance of a capacitor is defined as charge per unit potential difference, *i.e.*

$$C = \frac{Q}{V} \text{ farad} \qquad or \qquad c = \frac{q}{v} \text{ farad}$$

The energy stored in a capacitor is given by the relation

$$W = \frac{1}{2} CV^2 \text{ joule} \qquad or \qquad w = \frac{1}{2} cv^2 \text{ joule}$$

1.5 VOLTAGE AND CURRENT SOURCES

To deliver electrical energy to electrical/electronic circuits, a source is required. A load is connected to the source as shown in Fig.1.17. The source may be either a d.c. (direct current) source or an a.c. (alternating current) source.

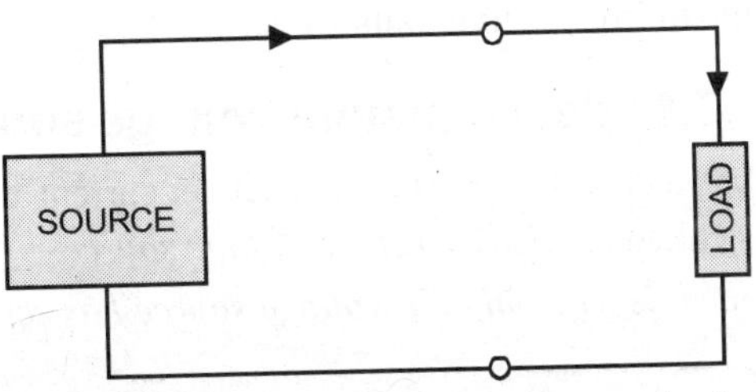

Fig.1.17: Transfer of energy from source to load

- **DC Source:** *Any device that produces direct voltage continuously and has the ability to deliver current is called a* **dc source,** such as batteries, dc generators and dc power supplies (regulated power supplies).
- **AC Source:** *Any device that produces alternating voltage continuously and has the ability is deliver current is called an ac source,* such as alternators and oscillators or signal generators.

1.6 INTERNAL RESISTANCE

All the sources (battery, dc generator or rectifier type supply) have some internal resistance R_i. The equivalent circuit of a dc source is represented by e.m.f. E in series with internal resistance R_i of the source as shown in fig.1.18.

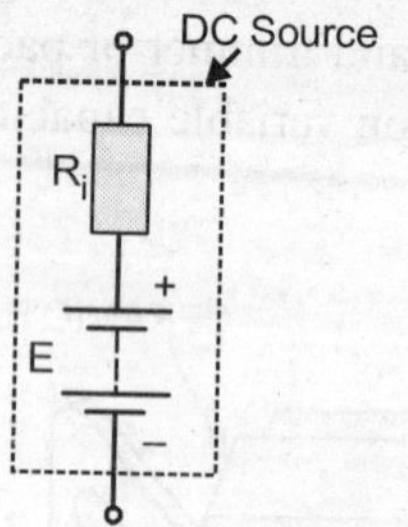

Fig.1.18: DC Source

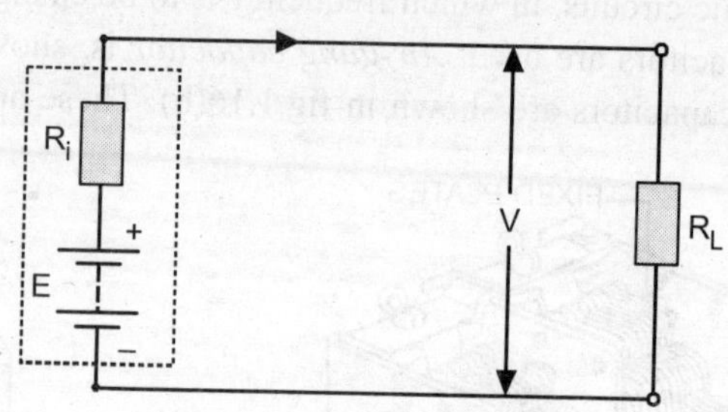

Fig.1.19: DC Source on load

When load (R_L) is connected across the source as shown in fig.1.19.

Load current, $$I_L = \frac{E}{R_L + R_i}$$

Terminal voltage, $$V = E - I_L R_i$$

or $$V = I_L R_L = \frac{E}{R_L + R_i} \times R_L = \frac{E}{1 + R_i / R_L}$$

The voltage across the load terminals is reduced because of the voltage drop in the internal resistance of the source. A source having smaller internal resistance will provide more voltage across the load terminals.

1.7 CONSTANT VOLTAGE SOURCE

An electrical source is called a constant voltage source if it supplies power at almost constant voltage to a load irrespective to its value.

1.7.1 Ideal constant voltage source

An electrical source that can deliver current at constant voltage irrespective of the value of load resistance is called an *ideal constant voltage source.*

It is true only if a voltage source has zero internal resistance. We know,

Terminal voltage, $$V = \frac{E}{1 + R_i / R_L}$$

Since R_i = zero,

$$V = E$$

An ideal constant voltage source and its characteristics are shown in fig.1.20(a) and 1.20(b) respectively..

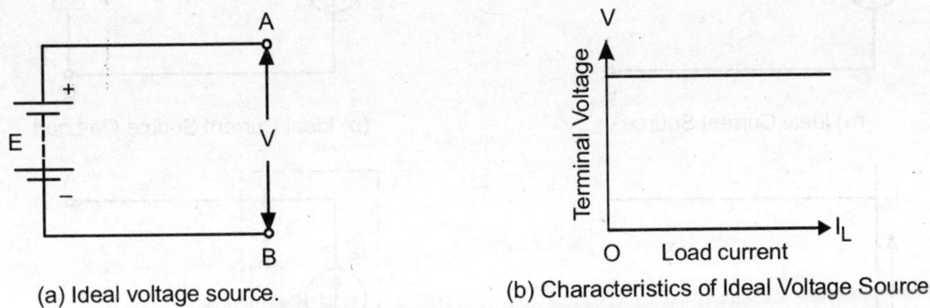

(a) Ideal voltage source. (b) Characteristics of Ideal Voltage Source.

Fig.1.20: Ideal constant voltage source

An ideal constant voltage source is just an idea, it cannot exist in nature. All the voltage sources have some internal resistance (or impedance), although its value may be very small.

1.7.2 Real Constant Voltage Source

All the voltage sources have some internal resistance which limits the current at short circuit. A real voltage source and its characteristics are shown in Fig.1.21(a) and 1.21(b) respectively.

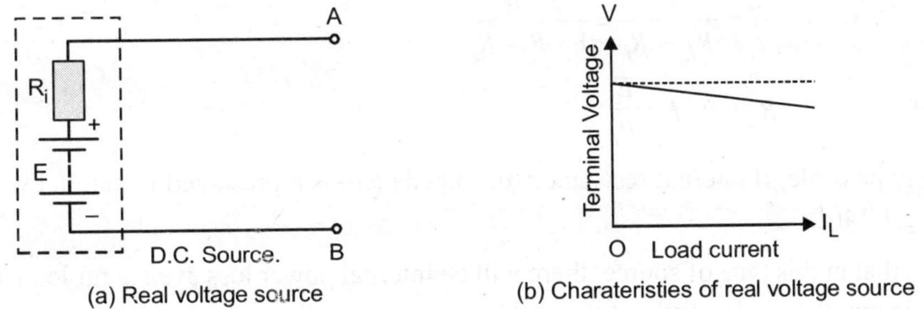

(a) Real voltage source (b) Charateristies of real voltage source

Fig.1.21: Real Voltage Source

To maintain the terminal voltage of the source almost constant from no-load to full-load, the internal resistance of the source should be very small.

A voltage source that has very low internal resistance (or impedance) as compared to load resistance (or impedance) is known as a real **constant voltage source.**

1.8 CONSTANT CURRENT SOURCE

An electrical source is called a constant current source if it supplies power at almost constant current to a load irrespective to its value.

1.8.1 Ideal constant current source

An electrical source that can deliver constant current irrespective of the value of load resistance is called *ideal constant current source.*

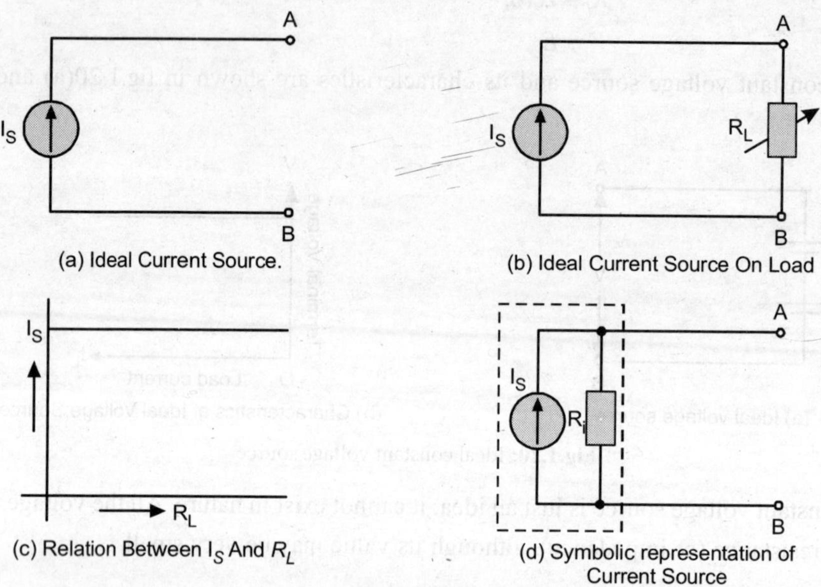

Fig.1.22: Ideal current source, its characteristics and symbol

It is true only if the source has infinite internal resistance.

Since, $$I = \frac{E}{R_L + R_i} = \frac{E/R_i}{1 + R_L/R_i}$$

i.e. even if $$R_L = 0,\ I = \frac{E}{R_i}$$

This is only possible, if internal resistance (or impedance) is represented in parallel with the load as shown in fig.1.14(d)

This shows that in this type of source, there will be internal power loss even at no-load. Hence ideal current source is merely an idea and not the real one.

1.8.2 Real constant current source

An ideal constant current source is just idea. A real constant current source is basically a voltage source which delivers almost the same current at all values of load resistance. This is only possible if the source has very high internal resistance *i.e.* $R_i >> R_L$.

A source that has very high internal resistance (or impedance) as compared to the load resistance (or impedance) is considered as a **constant current source.**

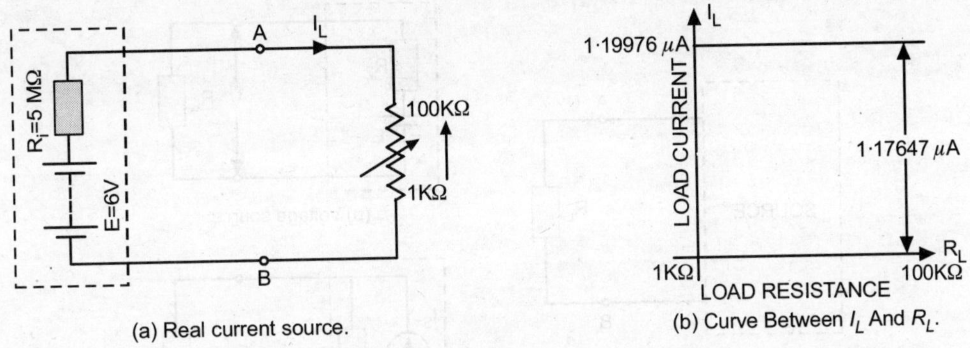

(a) Real current source. (b) Curve Between I_L And R_L.

Fig.1.23: Real current source and its characteristics

A real current source having very high internal resistance say 5 MW with a comparatively low load resistance R_L (say varying from 1 K to 100 K) is shown in Fig.1.23(*a*). Its characteristics are show in fig.1.23(b).

This shows that the load current remains almost constant and the source behaves as a constant current source irrespective of the value of load resistance.

1.9 DIFFERENCE BETWEEN VOLTAGE SOURCE AND CURRENT SOURCE

The behaviour of an electrical source depends upon its working conditions. When the value of load resistance (or impedance) is very large as compared to the internal resistance (or impedance) of the source, the source is treated as a voltage source.

Whereas, when the value of load resistance is very small as compared to the internal resistance of the source, the source is treated as a current source.

For instance, consider a 12 V battery with an internal resistance of 0.5 ohm. If the load applied on the battery varies from 1 KW to 10 KW, the terminal voltage varies from 11.994V to 11.9994 $\left[V = \frac{E}{1 + R_i / R_L}\right]$. The source is considered as constant voltage source.

If the load applied on the battery varies from 1 mW to 10 mW the current supplied by the battery varies from 23.952 A to 23.529 A $\left[I = \frac{E / R_i}{1 + R_L / R_l}\right]$. The source is considered as constant voltage source.

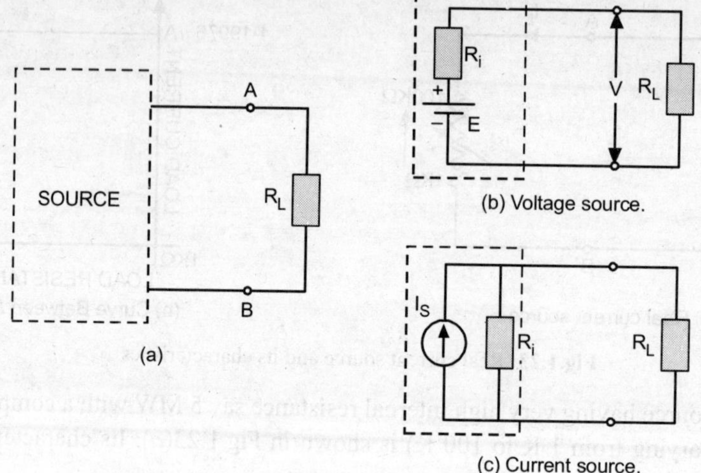

(a)

(b) Voltage source.

(c) Current source.

Fig.1.24

Conclusion: It clearly shows that the source is the same but it is considered as a constant voltage source if $R_i << R_L$ and the same source is considered as a constant current source If $R_i >> R_L$. Hence practically, there is *no difference* between constant voltage source and constant current source.

1.10 SOURCE TRANSFORMATION (Conversion of Voltage Source to Current Source and Vice-Versa)

It has been seen that, the same electrical source behaves as a constant voltage source and as a constant current source. Accordingly the same source can be represented in two different ways as shown in fig1.1.25(b) and 1.25(c) respectively.

Thus, a voltage source can be represented as a current source as shown in fig.1.25(a) and a current source can be represented as a voltage source as shown in fig.1.25(b)

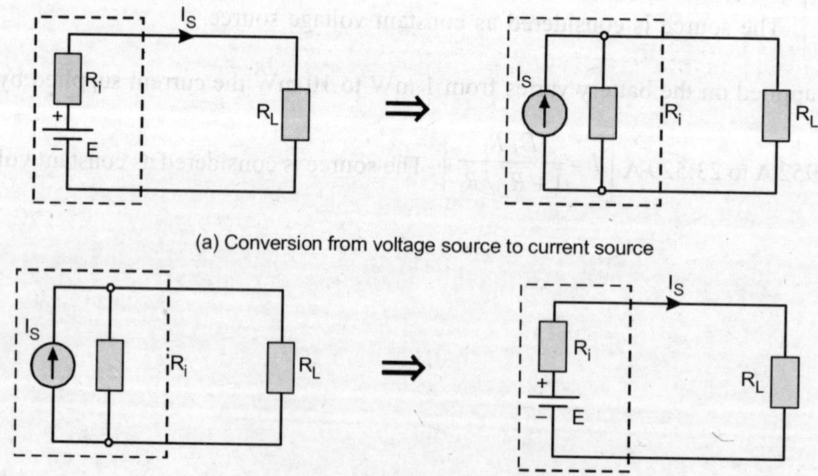

(a) Conversion from voltage source to current source

(b) Conversion from current source to voltage source

Fig.1.25: Conversion of voltage source to current source and vice-versa.

Example 1.1. *Fig.1.26 shows a d.c. voltage source having an open circuit voltage of 12 V and an internal resistance of 1 ohm. Obtain its equivalent current source representation.*

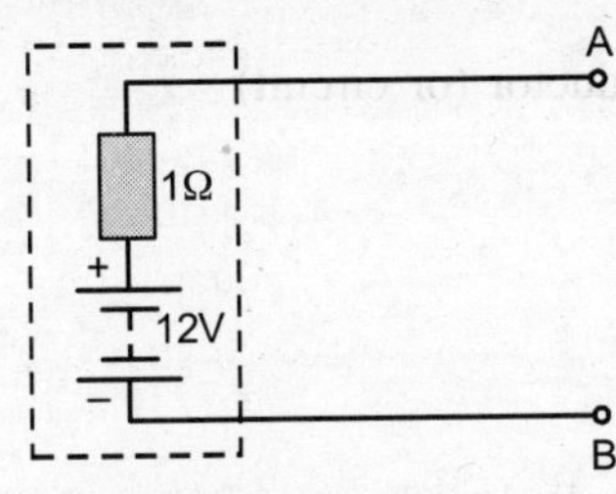

Fig.1.26: Voltage Source

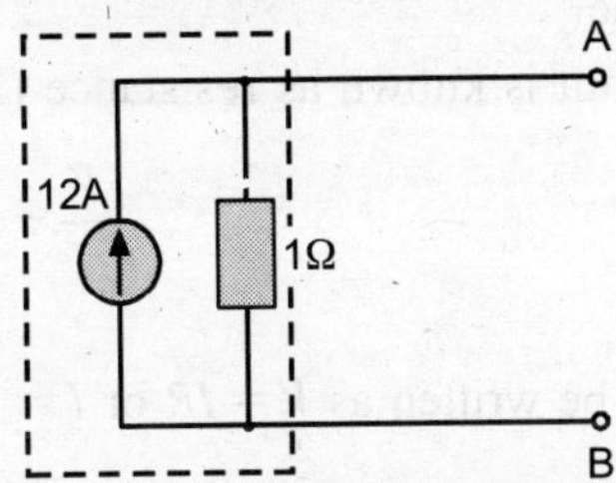

Fig.1.27: Equivalent Current source

Sol. If the terminals A and B of voltage-source are short-circuited, the current supplied by the source,

$$I_S = \frac{E}{R_i} = \frac{12}{1} = 12\ \text{A}$$

In the equivalent current-source representation, the current source is of 12A. The internal resistance of the source is represented in parallel with the current source as shown in Fig.1.27.

Example 1.2. *Fig.1.28 shows a d.c. current-source, obtain its equivalent voltage-source representation.*

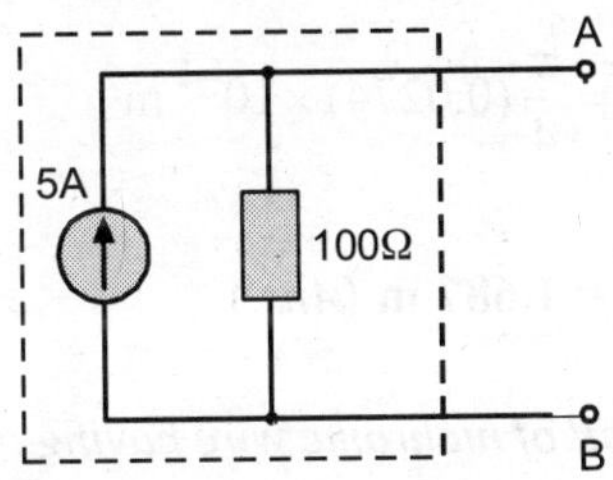

Fig.1.28: D.C Current Source

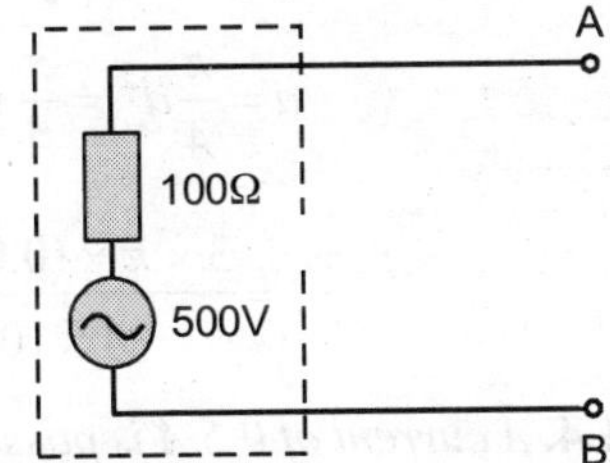

Fig.1.29: Equivalent Voltage Source

Sol. The open circuit voltage of the current source across the terminals A and B.

$$E = I_S R_i = 5 \times 100 = 500\ \text{V}$$

This will be the generated e.m.f. or ideal voltage of the equivalent voltage source representation. The internal impedance ($R_i = 100\ \Omega$) is put in series with the ideal voltage source as shown in Fig.1.29. This gives the equivalent voltage source representation of the given current source.

1.11 OHM'S LAW

This law states that the current flowing between any two points of a conductor (or circuit) is directly proportional to the potential difference across them, provided physical conditions (i.e. temperature etc.) remain the same.

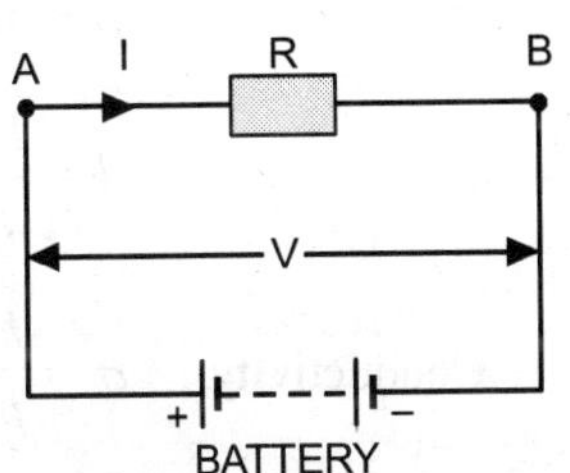

Fig.1.30: A resistor is connected across a battery

Considering circuit shown in fig.1.30.

Mathematically; $I \propto V$

or $\frac{V}{I} = Constant$ or $\frac{V_1}{I_1} = \frac{V_2}{I_2} = \ldots\ldots = \frac{V_n}{I_n} = Constant$

In other words, the ratio of potential difference across any two points of a conductor to the current flowing between them is always constant, provided the physical conditions (*i.e.* temperature etc.) remain the same

This constant is known as resistance (R) of the conductor (or circuit)

$$\therefore \qquad \frac{V}{I} = R$$

It can also be written as $V = IR$ or $I = \dfrac{V}{R}$

Limitations of Ohm's law: Ohm's law cannot be applied to the circuits:

(*i*) Consisting of electronic tubes or transistors because these eliments are not bilateral.

(*ii*) Consisting of non-linear elements such as electric arc etc.

Example 1.3. *The specific resistance of platinum at 0°C is 10.5 micro - ohm - cm. What should be the length of platinum wire of diameter 0.0274 cm to have a resistance of 3 ohm at 0°C ?*

Solution: Resistance of a wire at 0°C;

$$R_0 = \rho_0 \frac{l}{a} \quad \text{or} \quad l = \frac{R_0 a}{\rho_0}$$

where, $R_0 = 3\ \Omega$; $\rho_0 = 10{\cdot}5 \times 10^{-6}\ \Omega$ cm $= 10{\cdot}5 \times 10^{-8}\ \Omega$ m

$$a = \frac{\pi}{4} d^2 = \frac{\pi}{4}(0.0274)^2 \text{ cm}^2 = \frac{\pi}{4}(0.0274) \times 10^{-4} \text{ m}^2$$

$$\therefore \qquad l = \frac{3 \times \pi \times (0.0274)^2 \times 10^{-4}}{4 \times 10.5 \times 10^{-8}} = \mathbf{1.687\ m}\ (\textit{Ans.})$$

Example 1.4. *A current of 0·5 A is passed through a coil of nichrome wire having a cross - sectional area of 0·01 cm². If the resistivity of the nichrome is 108 × 10⁻⁶ ohm – cm and the p.d. across the ends of the coil is 54 volt, what is the length of the wire ? What is the conductivity and conductance of the wire ?*

Solution: Resistance, $R = \rho \dfrac{l}{a}$

Where, $R = \dfrac{V}{I} = \dfrac{54}{0.5} = 108\,\Omega$; $a = 0.01 \text{ cm}^2 = 0.01 \times 10^{-4} \text{ m}^2$

$$\rho = 108 \times 10^{-6}\ \Omega \text{ cm} = 108 \times 10^{-8}\ \Omega \text{ m}$$

$$\therefore \qquad l = \frac{Ra}{p} = \frac{108 \times 0.01 \times 10^{-4}}{108 \times 10^{-4}} = \mathbf{100\ m}\ (\textit{Ans.})$$

Conductivity, $\sigma = \dfrac{I}{p} = \dfrac{1}{108 \times 10^{-8}} = \mathbf{92{\cdot}59 \times 10^4\ mho/m}$ (*Ans.*)

Conductance, $G = \dfrac{I}{R} = \dfrac{1}{108} = \mathbf{9{\cdot}259 \times 10^{-3}\ mho}$ (*Ans.*)

1.12 D.C CIRCUITS

The closed path for the flow of direct current is called **d.c. circuit.**

A simple *d.c.* circuit is shown in fig.1.31. which contains a *d.c.* source (battery), a load (lamp,), a switch, connecting leads and measuring instruments like ammeter and voltmeter.

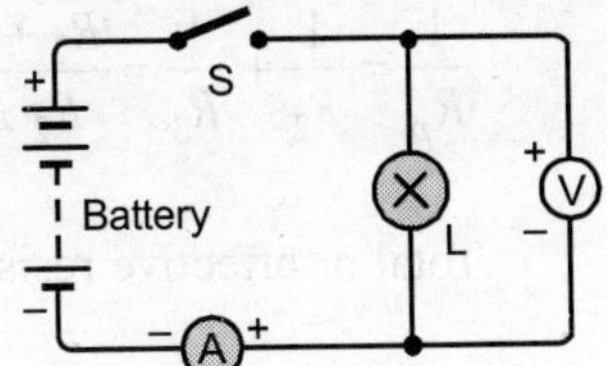

Fig.1.31: Simple dc circuit

1.13 SERIES CIRCUITS

A circuit is which number of resistors are connected such that same current flows through them is called *series circuit.*

Fig.1.32 shows a simple series circuit which carries three resistors R_1, R_2 and R_3 connected in series across a supply voltage of V volt.

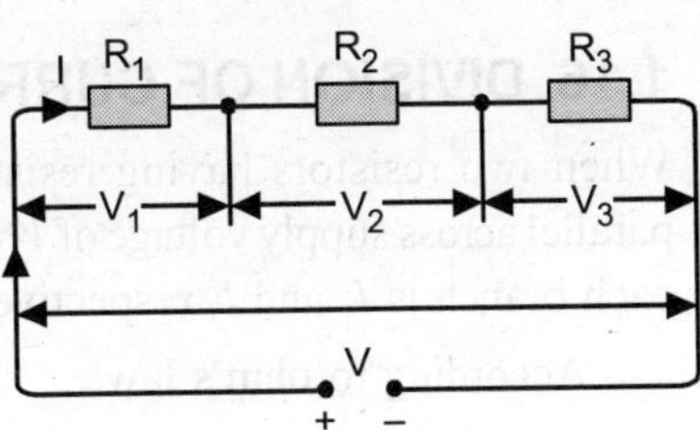

Fig.1.32: Three resistors connected in series.

Current flowing through the circuit is I ampere.

$V = V_1 + V_2 + V_3 = IR_1 + IR_2 + IR_3$ (As per Ohm's law)

If R is the effective resistance of the circuit, then

$IR = IR_1 + IR_2 + IR_3$ or $R = R_1 + R_2 + R_3$

i.e. Total resistance = Sum of the individual resistances.

This circuit is usually employed in marriages for decoration purposes.

1.14 PARALLEL CIRCUITS

A circuit is which number of resistors are connected such that different current flows through them but potential difference across them is the same as shown in fig.1.33. Current flowing through the resistors R_1, R_2 and R_3 is I_1, I_2 and I_3 respectively, then;

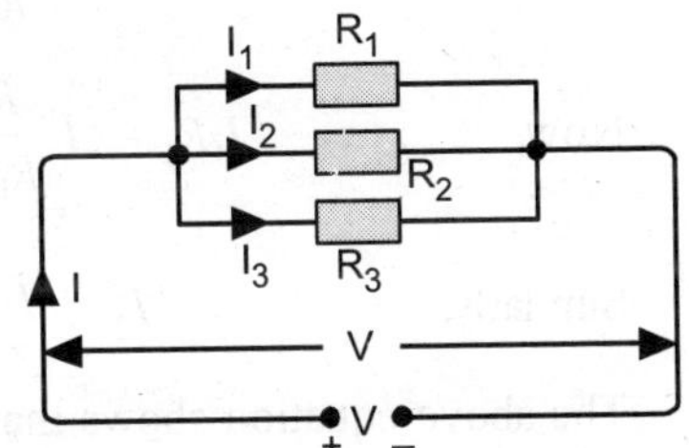

Fig.1.33: Three resistors connected in parallel

The total current drawn by the circuit,

$$I = I_1 + I_2 + I_3 = \frac{V}{R_1} + \frac{V}{R_2} + \frac{V}{R_3} \quad \text{(according to ohm's law)}$$

If R is the effective resistance of the circuit, then

$$\frac{V}{R} = \frac{V}{R_1} + \frac{V}{R_2} + \frac{V}{R_3} \quad \text{or} \quad \frac{1}{R} = \frac{1}{R_1} + \frac{1}{R_2} + \frac{1}{R_3}$$

i.e. Reciprocal of effective resistance = sum of reciprocal of the individual resistances.

All domestic appliances are connected in parallel since these are opperated at the same voltage and controlled independently.

1.15 SERIES – PARALLEL CIRCUITS

The circuit in which resistors are connected in series and parallel combination is called *series - parallel circuit,* as shown in fig.1.34.

Here, resistor R_2 and R_3 are connected in parallel, but their combination is connected in series with R_1. Say effective value of R_2 and R_3 is R_P, then

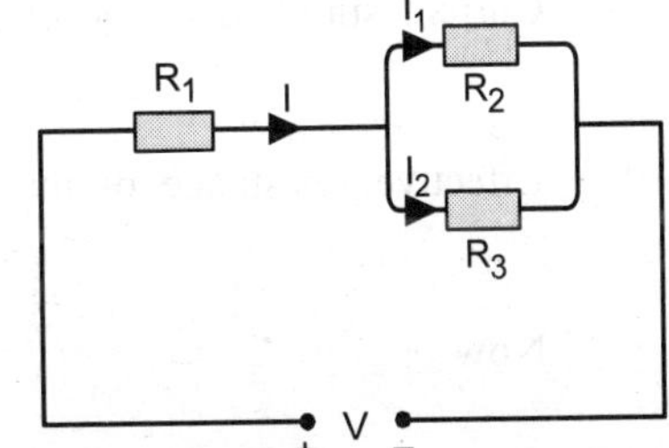

Fig.1.34: Series parallel circuit.

$$\frac{1}{R_p} = \frac{1}{R_2} + \frac{1}{R_3} = \frac{R_2 + R_3}{R_2 R_3} \quad \text{or} \quad R_p = \frac{R_2 R_3}{R_2 + R_3}$$

Total or effective resistance of the circuit, $R = R_i + R_p = R_l + \dfrac{R_2 R_3}{R_2 + R_3}$

Alternately; total or effective resistance of the circuit, $R = R_1 + (R_2 \| R_3) = R_1 + \left(\dfrac{R_2 R_3}{R_2 + R_3}\right)$

1.16 DIVISION OF CURRENT IN PARALLEL CIRCUITS

When two resistors having resistance R_1 and R_2 are connected in parallel across supply voltage of V volt, say the current flowing through each branch is I_1 and I_2 respectively.

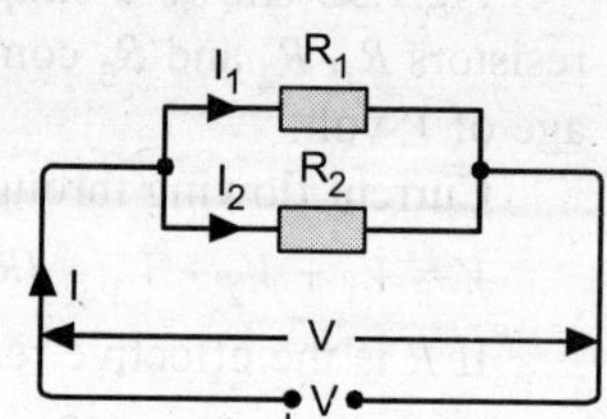

Fig.1.35: Two resistors connected in parallel

According to ohm's law;

$$I_1 R_1 = I_2 R_2 = IR = V$$

where R is total or effective resistance of the circuit and I is the total current.

$$R = \frac{R_1 R_2}{R_1 + R_2}$$

Now, $\quad I_1 R_1 = I \dfrac{R_1 R_2}{R_1 + R_2} \quad$ or $\quad I_1 = I \dfrac{R_2}{R_1 + R_2}$

Similarly, $\quad I_2 = I \dfrac{R_1}{R_1 + R_2}$

The above equation shows that

$$\text{Current in one branch} = \frac{\text{Resistance of other branch}}{\text{sum of resistance of two branches}} \times \text{total current}$$

Example 1.5. *A resistor R is connected in series with a parallel circuit comprising of two resistors having resistance of 6 and 4 ohm respectively. The total power dissipated in the circuit is 48 watt, applied voltage is 12V. Calculate the value of R .*

Solution: Total power dissipated, $P = 48$ W; Applied voltage, $V = 12$ V

The circuit is shown in Fig.1.36.

Current supplied to the circuit, $\quad I = \dfrac{P}{V} = \dfrac{48}{12} = 4$ A

Effective resistnace of the circuit, $R_{eff} = \dfrac{V}{I} = \dfrac{12}{4} = 3$ ohm

Now $\quad R_{eff} = R + \left(\dfrac{6 \times 4}{6 + 4}\right)$

Fig.1.36: Circuit diagram

Or $\quad 3 = R + 2.4$ or $R = 3 - 2.4 =$ **0.6 ohm** *(Ans.)*

Example 1.6. *Determine current I in the circuit shown in fig.1.37, all the resistors are given in ohm.*

Solution: Simplified view of the circuit is shown in fig.1.38.

∴

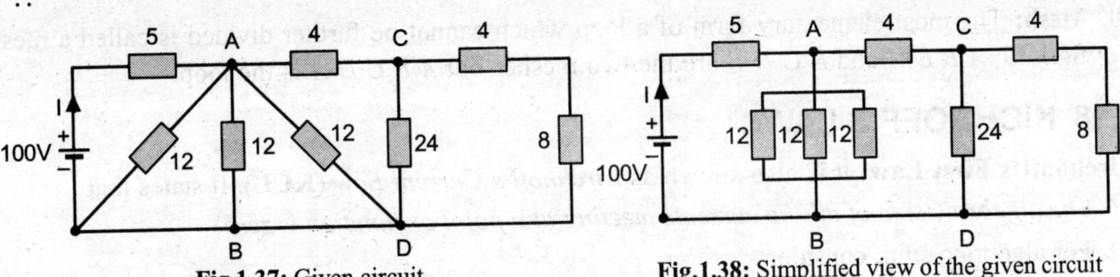

Fig.1.37: Given circuit **Fig.1.38:** Simplified view of the given circuit

The effective resistance across the supply.

$$R = [[\{(4 + 8) \parallel 24\} + 4] \parallel 12 \parallel 12 \parallel 12] + 5$$

$$= [\{(12 \parallel 24) + 4\} \parallel 12 \parallel 12 \parallel 12] + 5 = \left[\left(\frac{12 \times 24}{12 + 24} + 4\right) \parallel 12 \parallel 12 \parallel 12\right] + 5$$

$$= [(8 + 4) \parallel 12 \parallel 12 \parallel 12] + 5 = \left(\frac{12}{4}\right) + 5 = 3 + 5 = 8\ \Omega$$

$$\text{Current, } I = \frac{V}{R} = \frac{100}{8} = \mathbf{12.5\ A}\ (\textit{Ans.})$$

1.17 NETWORK TERMINOLOGY

Network theorems are applied to analyze the electrical network. While discussing these theorems one comes across various terms discussed below:

1. **Electric Network:** The combination of various electric elements, connected in any manner, is called an *electric network.*
2. **Electric circuit:** An *electric circuit* is a closed conducting path through which an electric current either flows or is intended to flow.
3. **Parameters:** Various elements of an electric circuit are called its *parameters* such as resistors, inductors, capacitors etc.
4. **Active network:** An electric network which contains one or more than one source of emf is called *active network.*
5. **Passive network:** An electric network which does not contain any source of emf is called *passive network.*
6. **Node:** A node is a point in the network where two or more circuit elements are joined. In fig.1.39, *A*, *B*, *C* and *D* are the nodes.
7. **Junction:** A junction is a point is the network where three or more circuit elements are joined. In fact, it is a point where current is divided. In fig.1.39, *B* and *D* are junctions.
8. **Branch:** The part of a network which lies between two junction points is called branch. In fig.1.39. *D A B*, *B C D* and *B D* are the three branches.

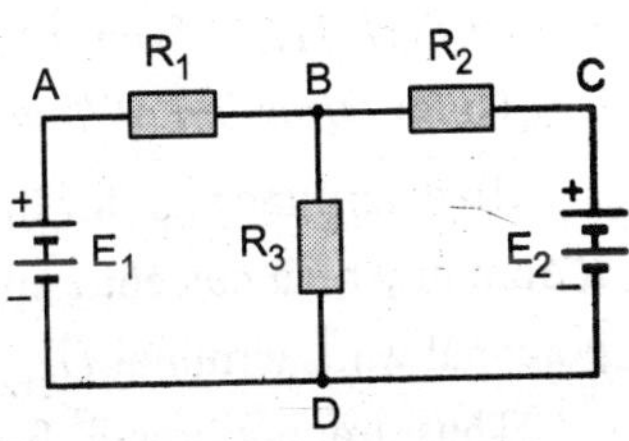

Fig.1.39: An electric nrtwork

9. **Loop:** The closed path of a network is called a loop. In fig.1.39, *A B D A*, *B C D B* and *A B C D A* are the three loops.

10. **Mesh:** The most elementary form of a loop which cannot be further divided is called a mesh. In fig.1.39, *A B D A* and *B C D B* are the two meshes but *A B C D A* is the loop.

1.18 KICHHOFF'S LAWS

Kirchhoff's First Law: It is also known as *Kirchhoff's Current Law* (**KCL**). It states that

The algebraic sum of all the currents meeting at a point or junction is zero.

For algebraic sum, consider;

Incoming currents as +ve; and Outgoing currents as – ve.

Applying Kirchhoff's current law to junction *O* in fig.1.40, we get,

$$I_1 + I_2 - I_3 + I_4 - I_5 = 0$$

or

$$I_1 + I_2 + I_4 = I_3 + I_5$$

i.e. Sum of incoming currents = sum of outgoing currents.

In other words,

At any junction of an electric network, the sum of incoming currents are equal to the sum of outgoing currents.

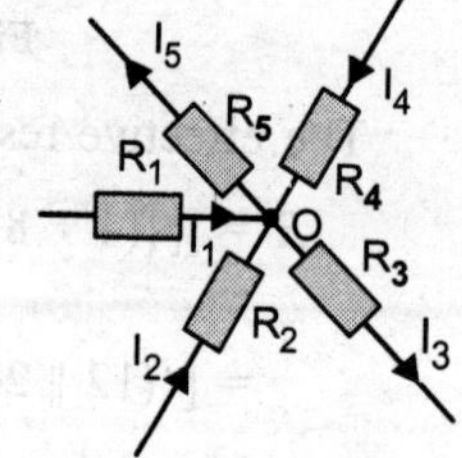

Fig.1.40: Five branches meeting at a juction

Kirchhoff's Second Law: It is also known as a *Kirchhoff's Voltage Law* (**KVL**) or *Kirchhoff's mesh law*. It states that

In a closed circuit or mesh, the algebraic sum of all the e.m.fs. plus the algebraic sum of all the voltage drops (*i.e.* product of current and resistances) *is zero.*

i.e. in a closed circuit or mesh,

algebraic sum of all the e.m.fs. + algebraic sum of all the voltage drops = 0

For algebraic sum, consider

A rise in potential as + ve; and A fall in potential as – ve.

Consider a branch *A B* containing only one source of e.m.f. (*E*) as shown in fig.1.41(*a*). If this branch is traced from *A* to *B*, there is fall in potential, therefore, *E* is taken as negative (*i.e.*–*E*).

Similarly, if the branch is traced from *A* to *B* in fig.1.41(*b*), there is rise in potential and *E* will be taken as positive (*i.e.*+*E*), *i.e.*

- *In fig.1.41 (b) from A to B, E* is negative (– *E*);
- *In fig.1.41(b), from A* to *B*, *E* is positive (+ *E*).

Note: *Direction of flow of current is neriter considered nor marked.*

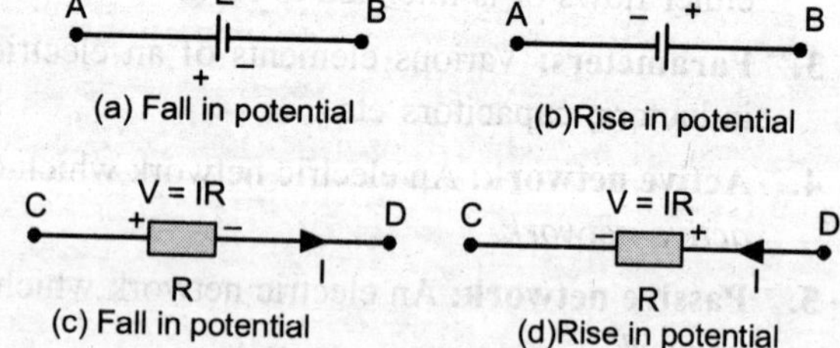

Fig.1.41: Algebraic signs for emf and voltage drop.

To determine sign for the voltage drop *V* (= *IR*), consider a branch containing resistor of resistance *R* ohm in which current *I* ampere flows from *C* to *D* as shown in fig.1.41(*c*). Therefore, *C* is at higher potential w.r.t. terminal *D*.

Thus, tracing branch from *C* to *D*, *V* is negative (*i.e.* – *V*) as there is *fall in potential.* As we are moving in the direction of flow of current.

Similarly, while tracing branch *C* to *D* in fig.1.41(*d*), *V* is positive (*i.e.* + *V*) as there is rise in potential and we are moving opposite to the flow of current.

Note: *Only the direction of flow of current determines the sign of V.*

Illustration:

Consider a network shown in fig.1.42. In this circuit, we can apply *kVL* for the closed circuit *ABEFA* and form an equation given below:

$$-R_1(I_1 + I_2) - I_2 R_3 + E_1 = 0$$

Where $R_1(I_1 + I_2)$ (voltage drop) is considered as negative, since we are tracing the circuit along the direction of flow of current.

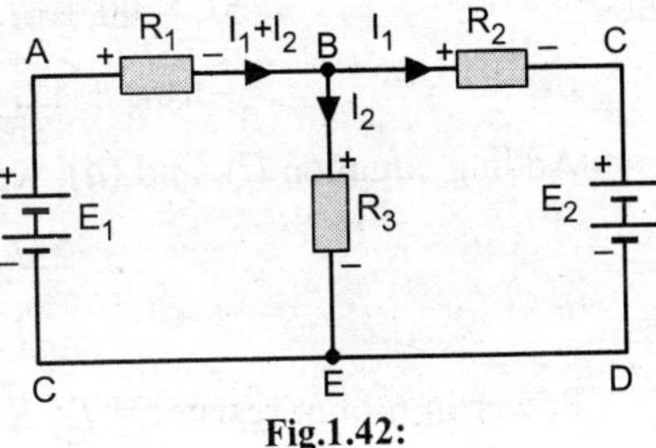

Fig.1.42:

$I_2 R_3$ (voltage drop) in taken as negative as we are moving along the direction of flow of current.

E_1 (e.m.f. of battery) in taken as positive as we move from –ve to +ve terminal at the battery and there in rise in potential (Note: The direction of flow of current is not to be considered).

Similarly, *kVL* can be applied the other closed circuit *BCDEB*, we get,

$$-I_1R_2 - E_2 + I_2R_3 = 0$$

1.19 SOLUTION OF NETWORK BY KIRCHHOFF'S LAWS

For solution of network, while applying Kirchhoff's laws, the following steps are taken:

- Name all the nodes and junctions as *A, B, C, D*, ...
- Mark the assumed direction of flow of current in all the branches of the network according to *K C L*.
- Choose as many number of closed circuits, as the number of unknown quantities.
- For the chosen closed circuits frame the equations by applying *KVL*.
- Solve the equations and determine the unknown values.

Note: *If the determined current carries the – ve sign, it shows that the actual direction of flow of current is opposite to that of the assumed direction of flow of current in the given branch.*

Example 1.7. *In the circuit shown in fig.1.43. calculate current and power in the 6 ohm resistor.*

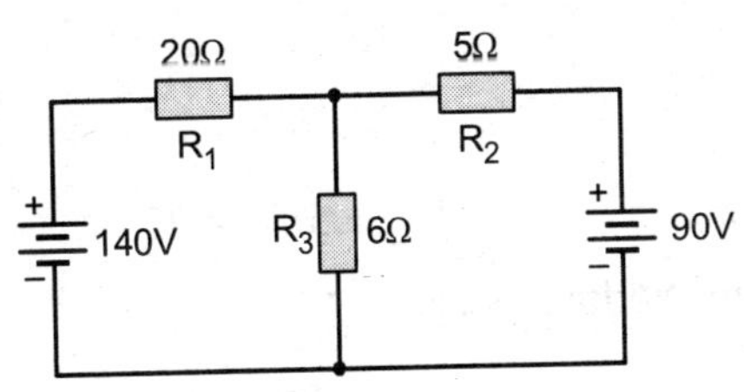

Fig.1.43: Given circuit

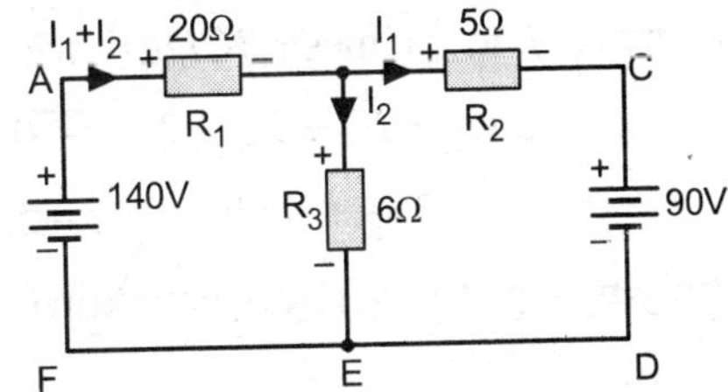

Fig.1.44: Circuit carrying current

Solution: The various junction points are marked in fig.1.44. According to Kirchhoff's first law, let the current flowing through various sections be as shown in fig.1.44.

Applying Kirchhoff's second law to mesh *ABEFA*, get,

$$-20(I_1 + I_2) - 6I_2 + 140 = 0 \quad \text{or} \quad 20I_1 + 26I_2 = 140$$

or $$10I_1 + 13I_2 = 70 \quad \text{.................}(i)$$

Applying Kirchhoff's second law to mesh *BCDEB*, we get,

$$-5I_1 - 90 + 6I_2 = 0 \quad \text{or} \quad -5I_1 + 6I_2 = 90$$

or $$-10I_1 + 12I_2 = 180 \quad \text{.............}(ii)$$

Adding equation (*i*), and (*ii*), we get,

$$25I_2 = 250$$

$$I_2 = \mathbf{10\ A}\ (Ans.)$$

Power in 6 ohm resistor = $I_2^2 R_3 = (10)^2 \times 6 =$ **600 W** (*Ans.*)

Example 1.8. *Fig.1.45 shows two batteries connected in parallel each represented by an emf along with its internal resistance. A load resistance of 5 Ω is connected across the ends of the batteries. Calculate the current through each battery and the load.*

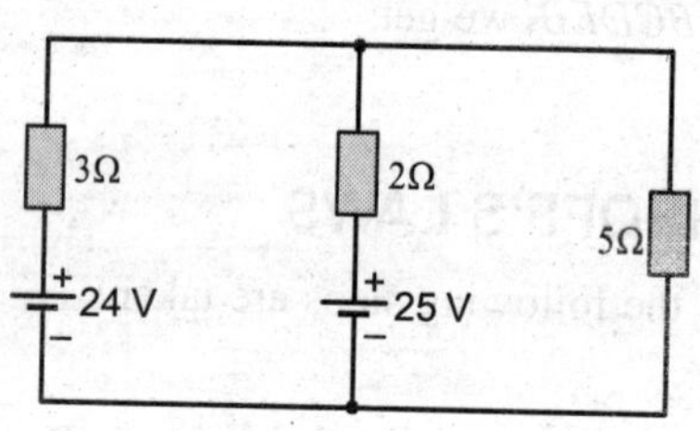

Fig.1.45: Given circuit.

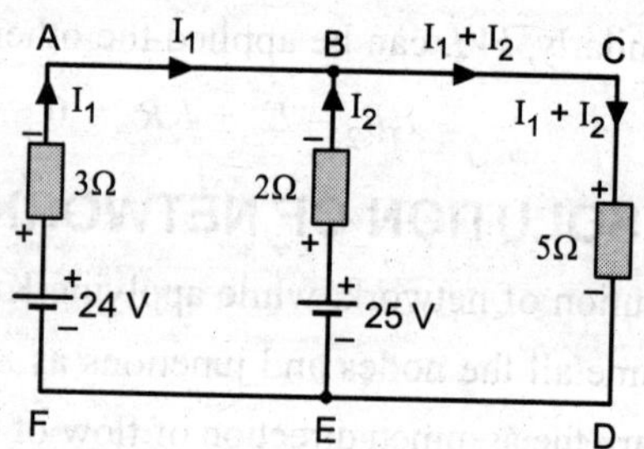

Fig.1.46: Circuit carrying current.

Solution: Assuming that the current flowing through various branches of the network is as marked in fig.1.46.

Applying *KVL* to mesh *ABEFA*, we get,

$$2I_2 - 25 + 24 - 3I_1 = 0$$

or $$2I_2 - 3I_1 = 25 - 24 \quad \text{or} \quad 2I_2 - 3I_1 = 1 \quad \text{.........}(i)$$

Applying *KVL* to the mesh *BCDEB*, we get,

$$-5(I_1 + I_2) + 25 - 2I_2 = 0$$

or $$2I_2 + 5(I_1 + I_2) = 25 \quad \text{or} \quad 7I_2 + 5I_1 = 25 \quad \text{........}(ii)$$

Multiplying equation (*i*) by 5, equation (*ii*) by 3 and adding, we get,

$$I_2 = \frac{80}{31} = \mathbf{2.58\ A}\ (Ans.)$$

Substituting the value of I_2 in equation (*i*),

$$I_1 = \frac{43}{31} = \mathbf{1.387\ A}\ (Ans.);$$

$$\text{Current through load, } I_L = I_1 + I_2 = \frac{43}{31} + \frac{80}{31} = \frac{123}{31}\text{ A} = \mathbf{3{\cdot}967\ A}\ (Ans.)$$

Example 1.9. *Solve the network shown in fig.1.47 for the current through 6 ohm resistor.*

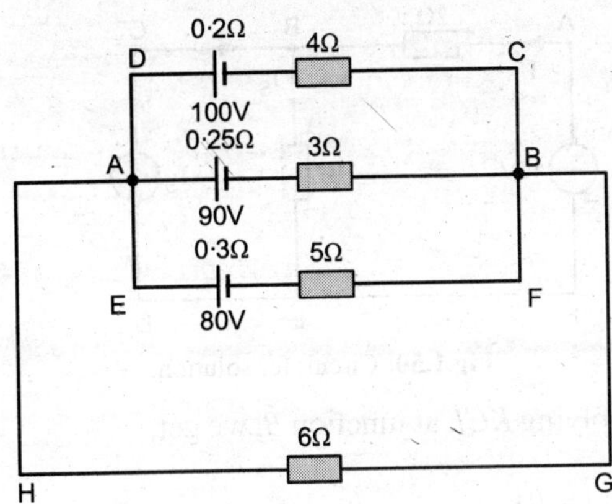

Fig.1.47: Given circuit

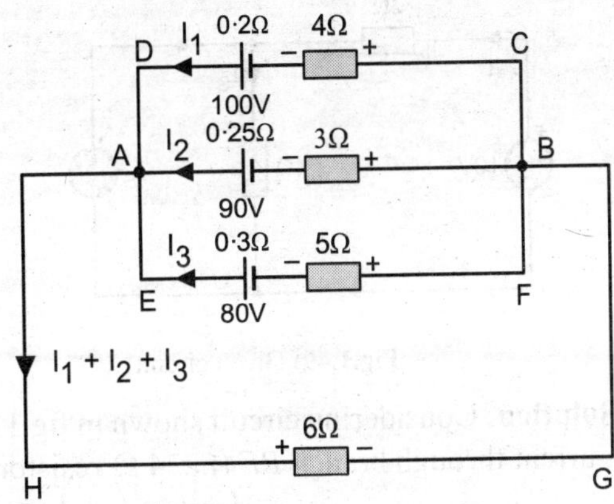

Fig.1.48: Circuit carrying current

Solution: Let the current flowing through various branches be as marked in fig.1.48.

Applying Kirchhoff's second law to the following closed circuits,

Circuit CDAHGBC

$$-4I_1 - 0{\cdot}2I_1 + 100 - 6(I_1 + I_2 + I_3) = 0$$

or $$10{\cdot}2I_1 + 6I_2 + 6I_3 = 100 \qquad \text{.............}(i)$$

Circuit BAHGB

$$-3I_2 - 0{\cdot}25I_2 + 90 - 6\,(I_1 + I_2 + I_3) = 0$$

or $$6I_1 + 9{\cdot}25I_2 + 6I_3 = 90 \qquad \text{............}(ii)$$

Circuit *FEAHGBF*

$$-5I_3 - 0{\cdot}3I_3 + 80 - 6\,(I_1 + I_2 + I_3) = 0$$

or $$6I_1 + 6I_2 + 11{\cdot}3\,I_3 = 80 \qquad \text{..........}(iii)$$

Subtracting (*ii*) from (*i*), we get,

$$4{\cdot}2I_1 - 3{\cdot}25I_2 = 10 \qquad \text{..........}(iv)$$

Solving (*ii*) and (*iii*), we get,

$$31{\cdot}8I_1 + 68{\cdot}525I_2 = 537 \qquad \text{...........}(v)$$

From equation (*iv*) and (*v*), we get,

$$391{\cdot}51I_2 = 137{\cdot}4$$

$$\therefore \qquad I_2 = \frac{1937 \cdot 4}{391 \cdot 15} = 4{\cdot}953\text{ A}$$

Solving equation (*v*), we get, $I_1 = 6{\cdot}21$ A

Solving equation (*iii*), we get, $I_3 = 1{\cdot}15$ A

Current in 6 ohm resistor, $I = I_1 + I_2 + I_3$

$= 6{\cdot}21 + 4{\cdot}953 + 1{\cdot}15$

$=$ **12·313 A** (*Ans.*)

Example 1.10. *In the circuit shown in fig.1.49 find the value of I_S for $I = 0$.*

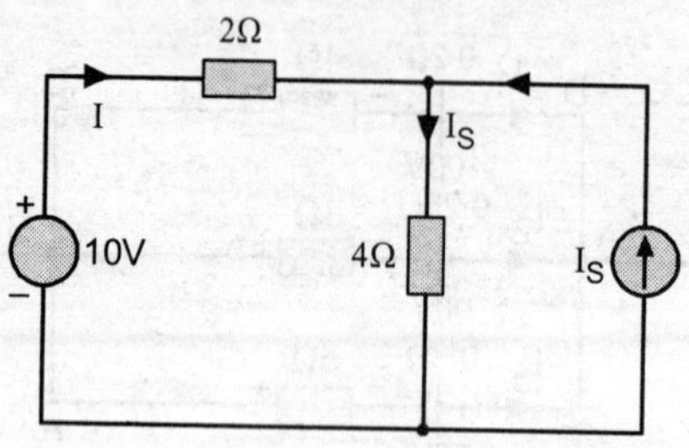

Fig.1.49: Given circuit

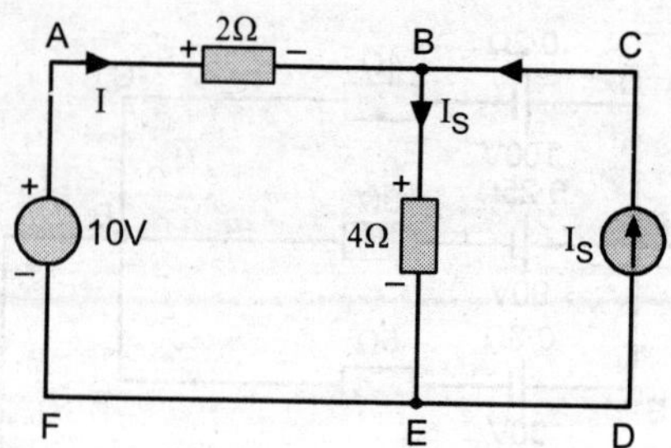

Fig.1.50: Circuit for solution.

Solution: Considering circuit shown in fig.1.50, applying *KCL* at junction *B*, we get,

Current through branch *BE* (*i.e.* 4 Ω resistor)

$$= I + I_S = 0 + I_S = I_S \quad (\because I = 0)$$

Appling *KVL* to mesh *ABEFA*, we get,

$$-2I - 4I_S + 10 = 0$$

or $$2I + 4I_S = 10$$

or $$I_S = \frac{10}{4} = \textbf{2.5 A} \ (\textit{Ans.}) \quad (\because I = 0)$$

Example 1.11. *Formulate the Kirchhoff's voltage law equation for the circuit shown in fig.1.51. Determine the values of currents I_1, I_2 and I_3.*

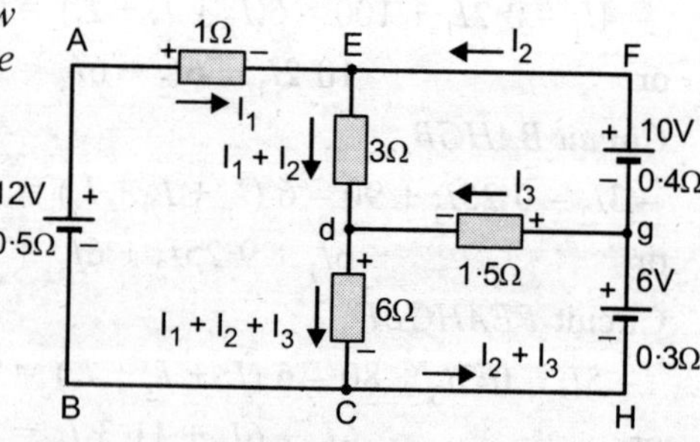

Fig.1.51: Given circuit.

Solution: Applying *KVL* to closed circuit *e f g d e*, we get,

$$-10 + 0{\cdot}4\,I_2 - 1{\cdot}5\,I_3 + 3\,(I_1 + I_2) = 0$$

$$3\,I_1 + 3{\cdot}4\,I_2 - 1{\cdot}5\,I_3 = 10 \quad \ldots\ldots\ldots(i)$$

Applying *KVL* to closed circuit *a e d c b a*, we get,

$$-I_1 - 3\,(I_1 + I_2) - 6\,(I_1 + I_2 + I_3) - 0{\cdot}5\,I_1 + 12 = 0$$

$$-10{\cdot}5\,I_1 - 9\,I_2 - 6\,I_3 + 12 = 0$$

$$10{\cdot}5\,I_1 + 9\,I_2 + 6\,I_3 = 12 \quad \ldots\ldots\ldots(ii)$$

Applying *K V L* to closed circuit *d g h c d*, we get,

$$1{\cdot}5 I_3 - 6 + 0{\cdot}3\,(I_2 + I_3) + 6\,(I_1 + I_2 + I_3) = 0$$

$$6\,I_1 + 6{\cdot}3\,I_2 + 7{\cdot}8\,I_3 = 6 \quad \ldots\ldots\ldots(iii)$$

Solving equation (*i*), (*ii*) and (*iii*), we get,

$$I_1 = -\,\textbf{1·242 A} \ (\textit{Ans.})$$

$$I_2 = 3{\cdot}537 \text{ A } (Ans.)$$
$$I_3 = -1{\cdot}1325 \text{ A } (Ans.)$$

The negative sign with current I_1 and I_3 shows that the actual flows of current is opposite to that of the direction marked in fig.1.51.

Example 1.12. *Find the magnitude and direction of the currents in all branches of the circuit shown in fig.1.52, using Kirchhoff's laws. All resistances are in ohm.*

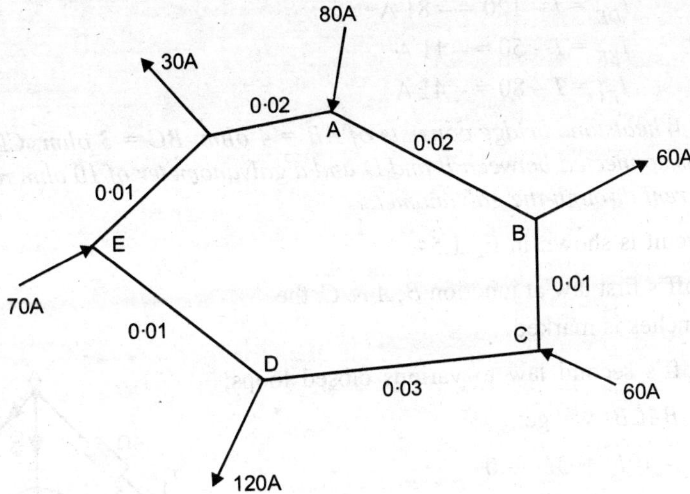

Fig.1.52: Given circuit

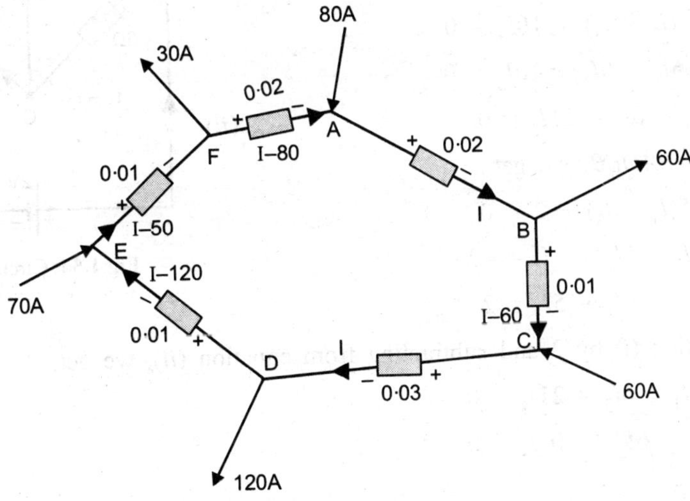

Fig.1.53: Circuit with assumed current.

Solution: Applying Kirchhoff's first law, let current flowing through various branches be as shown in fig.1.53.

Applying Kirchhoff's second law to a closed loop *ABCDEFA*, we get,

$-0{\cdot}02\,I - 0{\cdot}01\,(I - 60) - 0{\cdot}03I - 0{\cdot}01\,(I - 120) - 0{\cdot}01\,(I - 50) - 0{\cdot}02\,(I - 80) = 0$

or $0{\cdot}02\,I + 0{\cdot}01\,I + 0{\cdot}03\,I + 0{\cdot}01\,I + 0{\cdot}01\,I + 0{\cdot}02\,I = 0{\cdot}6 + 1{\cdot}2 + 0{\cdot}5 + 1{\cdot}6$

or $0{\cdot}1\,I = 3{\cdot}9$ or $I = 39$ A

Current in various branches is as under:

$I_{AB} = +39$ A (*i.e. A to B*)

$I_{BC} = I - 60 = -21$ A (*i.e. C to B*)

$I_{CD} = I = 39$ A (*i.e. C to D*)

$I_{DE} = I - 120 = -81$ A (*i.e. E to D*)

$I_{EF} = I - 50 = -11$ A (*i.e. F to E*)

$I_{FA} = I - 80 = -41$ A (*i.e. A to F*)

Example 1.13. *A Wheatstone bridge consists of AB = 4 ohm, BC = 3 ohm, CD = 6 ohm and DA = 5 ohm. A 2 volt cell is connected between B and D and a galvanometer of 10 ohm resistance between A and C. Find the current through the galvanometer.*

Solution: The circuit is shown in fig.1.54.

Applying Kirchhoff's first law at junction *B*, *A* & *C*, the current in various branches is marked.

Applying Kirchhoff's second law to various closed loops;

Considering loop *BACB*, we get,

$-4I_1 - 10I_3 + 3I_2 = 0$

or $4I_1 - 3I_2 + 10I_3 = 0$ ...(*i*)

Considering loop *ADCA*, we get,

$-5(I_1 - I_3) + 6\,(I_2 + I_3) + 10I_3 = 0$

or $-5I_1 + 5I_3 + 6I_2 + 6I_3 + 10I_3 = 0$

or $5I_1 - 6I_2 - 21I_3 = 0$ ....(*ii*)

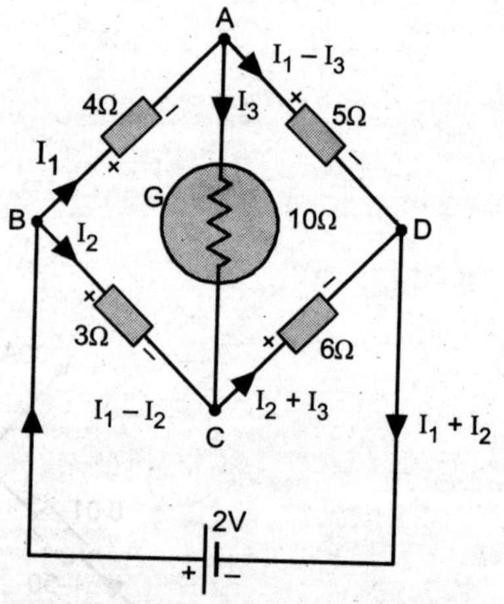

Considering loop *BADEB*, we get

$-4I_1 - 5(I_1 - I_3) + 2 = 0$

or $-4I_1 - 5I_1 + 5I_3 = -2$

or $9I_1 - 5I_3 = 2$ ....(*iii*)

Fig.1.54: Circuit for bridge.

Multiplying equation (*i*) by 2 and subtracting from equation (*ii*), we get,

$$5I_1 - 6I_2 - 21I_3 = 0$$

$$8I_1 - 6I_2 + 20\,I_3 = 0$$

$$- \quad + \quad -$$

$$-3I_1 \qquad - 41I_3 = 0$$

$$\therefore \quad I_1 = -\frac{41}{3}\,I_3$$

Substituting the value of I_1 in equation (*iii*), we get,

$$9\left(-\frac{41}{3}I_3\right) - 5I_3 = 2 \quad \text{or} \quad -123I_3 - 5I_3 = 2 \quad \text{or} \quad I_3 = \frac{-1}{64}\text{ A}$$

Current flowing through galvanometer is 1/64 ampere from C to A.

Example 1.14. *Determine the current I in 8 ohm resistance in the circuit shown in fig.1.55.*

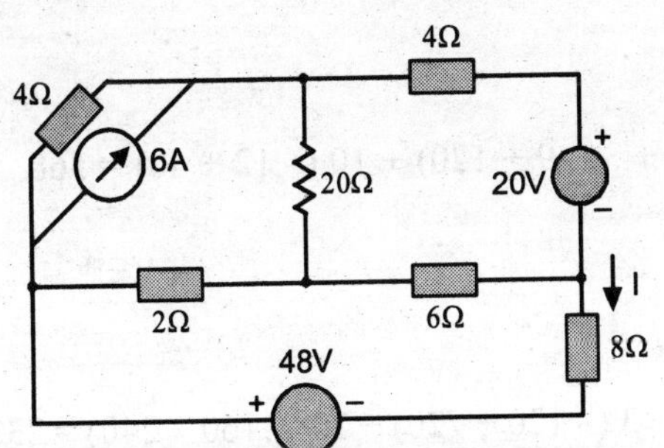

Fig.1.55: Given circuit.

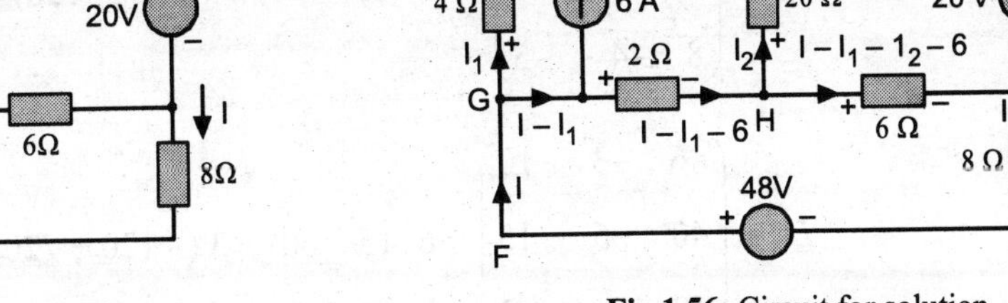

Fig.1.56: Circuit for solution.

Solution: A simplified circuit is shown in fig.1.56. Applying *KCL at various junctions*, different currents are marked in various sections.

Applying Kirchhoff's voltage law to

Loop *A B H G A*

$$20\,I_2 + 2\,(I - I_1 - 6) - 4I_1 = 0 \text{ or } 2I - 6I_1 + 20I_2 = 12$$

$$I - 3I_1 + 10\,I_2 = 6 \qquad \text{...........(i)}$$

Loop *B C D H B*

$$-4\,(I_1 + I_2 + 6) - 20 + 6\,(I - I_1 - I_2 - 6) - 20I_2 = 0$$

$$6I - 10\,I_1 - 30\,I_2 = 80 \quad \text{or} \quad 3I - 5I_1 - 15I_2 = 40 \qquad \text{...........(ii)}$$

Loop *G H D E F G*

$$-2\,(I - I_1 - 6) - 6\,(I - I_1 - I_2 - 6) - 8I + 48 = 0$$

$$16I - 8I_1 - 6I_2 = 96 \quad \text{or} \quad 8I - 4I_1 - 3I_2 = 48 \qquad \text{...........(iii)}$$

Eliminating, I_1 from equation (*i*) and (*ii*), we get,

$$4I - 95I_2 = 90 \qquad \text{...........(iv)}$$

Eliminating I_1 from equation (*ii*) and (*iii*), we get,

$$28I + 45I_2 = 80 \qquad \text{...........(v)}$$

Eliminating I_2 from equation (*iv*) and (*v*), we get,

$$568\,I = 2330 \text{ or } I = \frac{2330}{568} \; 4.1\text{ A}$$

∴ Current in 8Ω resistor, I = **4·1 A** (*Ans.*)

Alternatively

The three equation are:

$$I - 3I_1 + 10I_2 = 6$$

$$3I - 5I_1 - 15I_2 = 40$$

$$8I - 4I_1 - 3I_2 = 48$$

The three equation can be solved by the method of determinants *i.e.* by applying *Cramer's Rule*. The matrix from of the above equation is

$$\begin{pmatrix} 1 & -3 & 10 \\ 3 & -5 & -15 \\ 8 & -4 & -3 \end{pmatrix}\begin{pmatrix} I \\ I_1 \\ I_2 \end{pmatrix} = \begin{pmatrix} 6 \\ 40 \\ 48 \end{pmatrix}$$

$$D_0 = \begin{vmatrix} 1 & -3 & 10 \\ 3 & -5 & -15 \\ 8 & -4 & -3 \end{vmatrix} = 1\,(15 - 60) + 3\,(-9 + 120) + 10\,(-12 + 40) = 568$$

$$D = \begin{vmatrix} 6 & -3 & 10 \\ 40 & -5 & -15 \\ 48 & -4 & -3 \end{vmatrix} = 6\,(15 - 60) + 3\,(-120 + 720) + 10\,(-160 + 240) = 2330$$

$$I = \frac{D}{D_0} = \frac{2330}{568} = \mathbf{4{\cdot}1\ A}\ (Ans.)$$

PRACTICE EXERCISE

Q1. *Two batteries A and B are connected in parallel and a load of* 5 W *is connected across their terminals. A has an emf of* 6 *V and an internal resistance of* 1 W *and B has an emf of* 4 *V and an internal resistance of* 0.5 W. *Use Kirchhoff's laws to determine the magnitude of currents and also the directions in each of the batteries. Also determine the pd across external resistance.*

(Ans. *1.625A,–0.75A,0.875A,8.75V)*

Q2. *By using Kirchhoff's laws find the current in XY in the circuit shown in fig.1.57.* **(Ans.** *40A)*

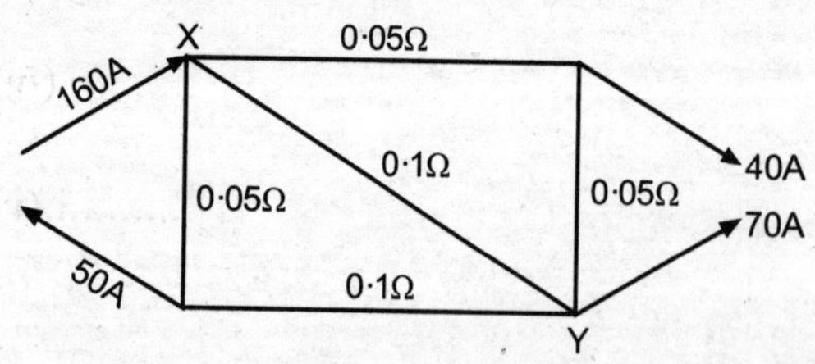

Fig.1.57: Given circuit

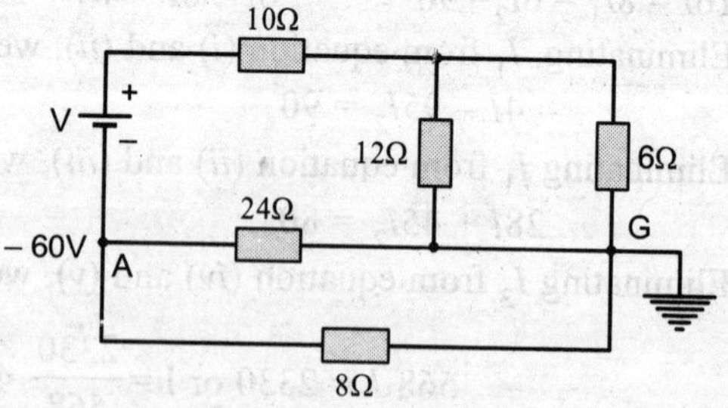

Fig.1.58: Given circuit

Q3. *In fig.1.58 the potential of point A =* – 60 *V. Using Kirchhoff's laws find (a) value of V and (b) power dissipated by* 8 W *resistance.* **(Ans.** *200V, 450W)*

Q4. *Find the value of R and the current through it in the network shown in fig.1.59, when the current is zero in branch OA.* *(Ans.6W, 0.5A)*

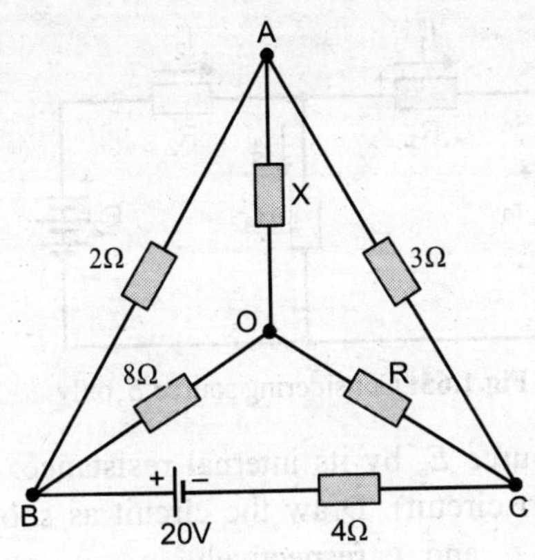

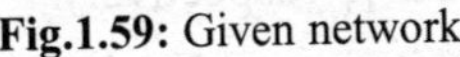

Fig.1.59: Given network

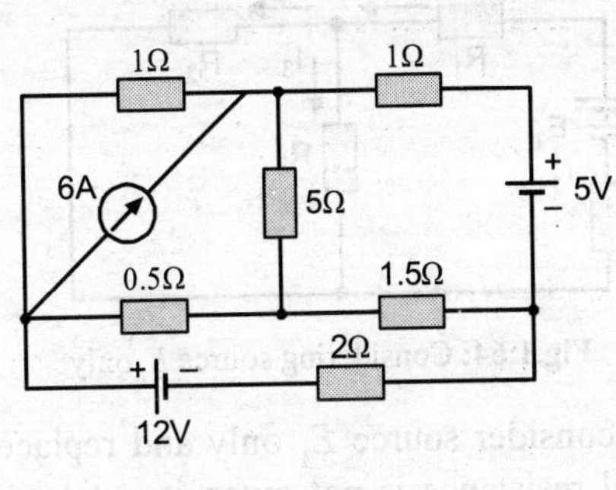

Fig.1.60: Given circuit

Q5. *Determine the current in the* 2-W resistance of circuit shown in fig.1.60. **(Ans.** *4.1A*)

Q6. *What is the difference of potential between X and Y in the network shown in fig.1.61.* **(Ans.** 9.4V)

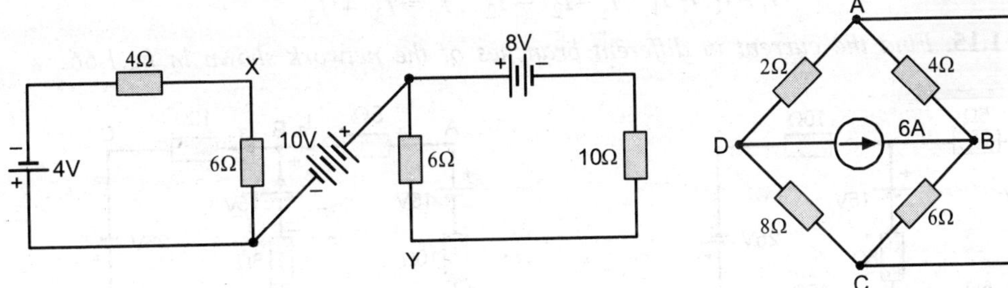

Fig.1.61: Given network

Fig.1.62: Given network

Q7. *Find the total power delivered to the circuit by two sources in fig.1.62.* **(Ans.** *164*W)

1.20 SUPERPOSITION THEOREM

This theorem is applicable to liner bilateral network having two or more than two sources.

According to this theorem, the current flowing through any section is the algebraic sum of all the currents which should flow in that section if each source of emf would be considered separately and other sources are replaced by their internal resistance (or impedance).

For instant, consider a circuit shown in fig.1.63. Let the current flowing through various branches be as marked in fig.1.63. According to superposition theorem.

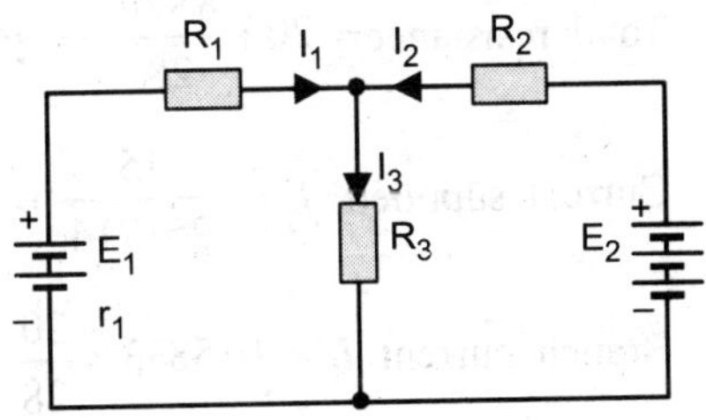

Fig.1.63: Given circuit

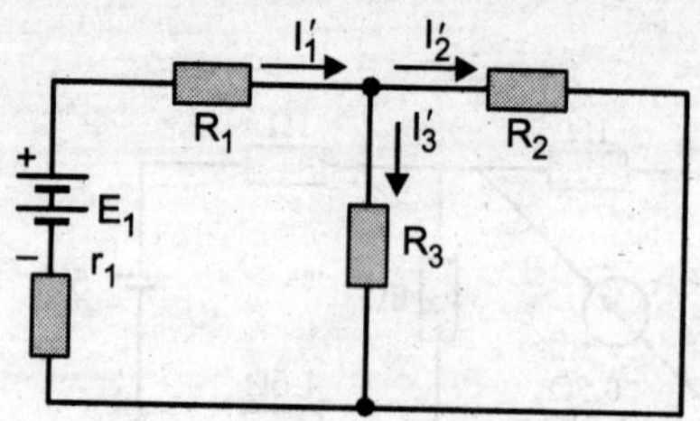

Fig.1.64: Considering source E_1 only

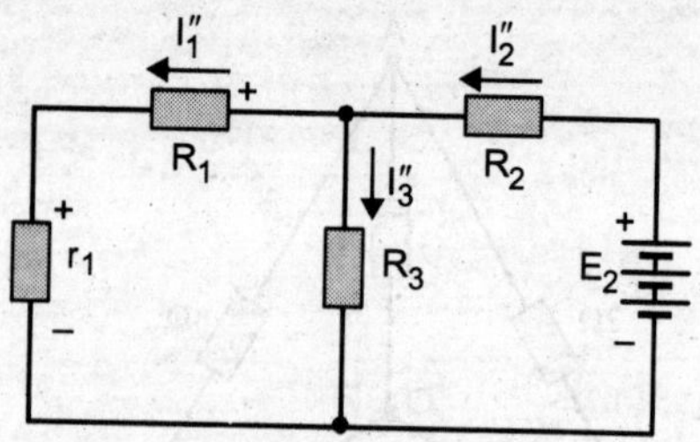

Fig.1.65: Considering source E_2 only

1. First, consider source E_1 only and replace the other source E_2 by its internal resistance. As its inernal resistance is not given it is taken as zero (short circuit). Draw the circuit as shown in fig.1.64 and determine currents in various section as I_1', I_2' and I_3' respectively.
2. Then consider the other source E_2 and replace the source E_1 by its internal resistance r_1 as shown in fig.1.65. Determine the currents in various sections as $I_1{}^2$, $I_2{}^2$ and $I_3{}^2$ respectively.

Actual flow of current in various sections;

$$I_1 = I_1' - I_1'' \; ; \; I_2 = I_2' - I_2'' \; ; \; I_3 = I_3' + I_3''$$

Example 1.15. *Find the current in different branches of the network shown in fig.1.66.*

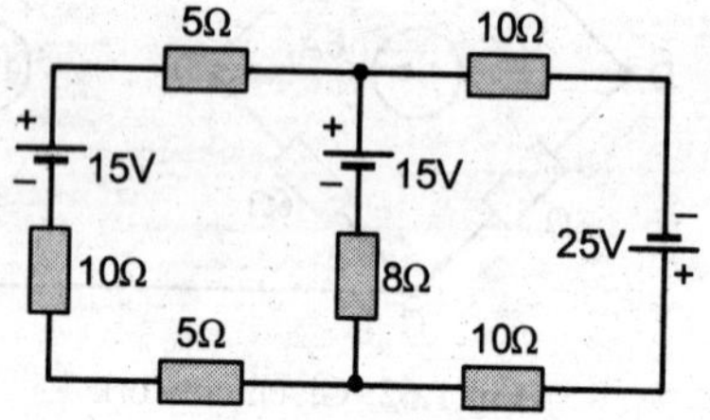

Fig.1.66: Given circuit.

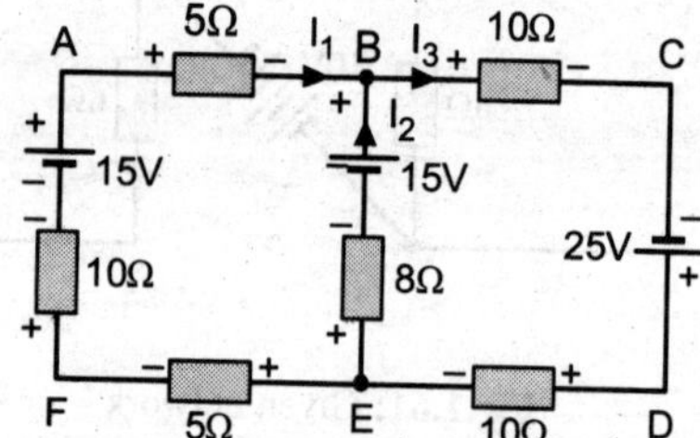

Fig.1.67: Circuit for solution.

Solution: Let the current flowing through different branches be as marked in fig.1.67. According to superposition theorem, reducing the circuit as shown in fig.1.68.

$$\text{Total resistance} = 20 + \frac{8 \times 20}{28} = 25 \cdot 714\Omega$$

$$\text{Current supplied, } I_1' = \frac{15}{25 \cdot 714} = 0 \cdot 5833 \text{ A}$$

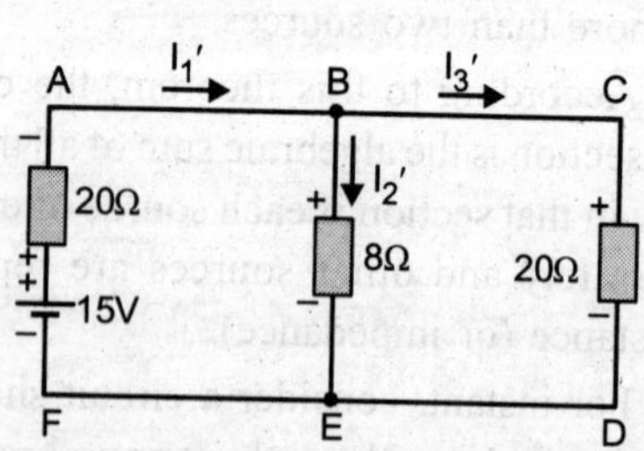

Fig.1.68: Considering one source only.

$$\text{Branch current } I_2' = 0 \cdot 5833 \times \frac{20}{28} = 0 \cdot 4167 \text{ A}$$

$$\text{Branch current } I_3' = 0 \cdot 5833 \times \frac{8}{28} = 1 \cdot 666 \text{ A}$$

Now, reducing the circuit as shown in fig.1.69.

$$\text{Total resistance} = 8 + \frac{20 \times 20}{40} = 18\Omega$$

Current supplied, $I_2'' = \frac{15}{18} = 0{\cdot}8333$ A

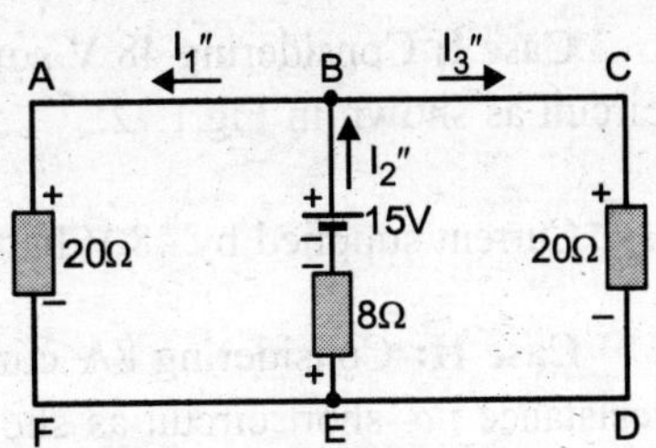

Fig.1.69: Considering second source only.

Branch current, $I_1'' = 0{\cdot}8333 \times \frac{20}{40} = 0{\cdot}4167$ A

Branch current, $I_3'' = 0{\cdot}8333 \times \frac{20}{40} = 0{\cdot}4167$ A

Again, reducing the circuit as shown in fig.1.70.

Total resistance $= 20 + \frac{20 \times 8}{28} = 25{\cdot}714\Omega$

Current supplied, $I_3''' = \frac{25}{25{\cdot}714} = 0{\cdot}9722$ A

Branch current, $I_1''' = 0{\cdot}9722 \times \frac{8}{28} = 0{\cdot}2778$ A

Branch current, $I_2''' = 0{\cdot}9722 \times \frac{20}{28} = 0{\cdot}6944$ A

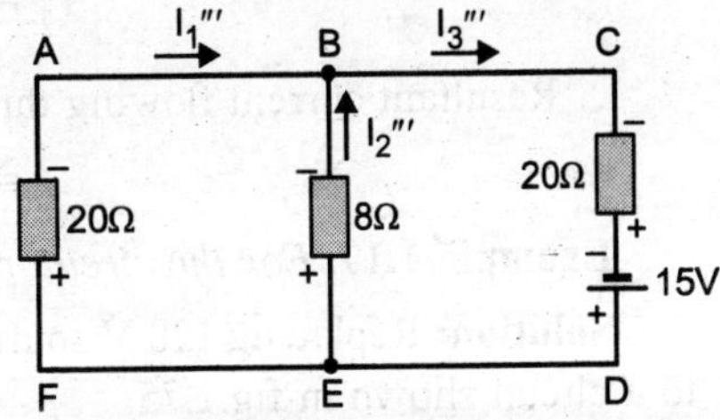

Fig.1.70: Considering third source only.

∴ Actual flow of current in different branches;

$I_1 = I_1' - I_1'' + I_1''' = 0{\cdot}5833 - 0{\cdot}4167 + 0{\cdot}2778 =$ **0·4444 A** *(Ans.)*

$I_2 = -I_2' + I_2'' + I_2''' = -0{\cdot}4167 + 0{\cdot}8333 + 0{\cdot}6944 =$ **1·1111 A** *(Ans.)*

$I_3 = I_3' + I_3'' + I_3''' = 0{\cdot}1666 + 0{\cdot}4167 + 0{\cdot}9722 =$ **1·5555 A** *(Ans.)*

Example 1.16. *Find the current in the circuit given in fig.1.71.* **(UPTU)**

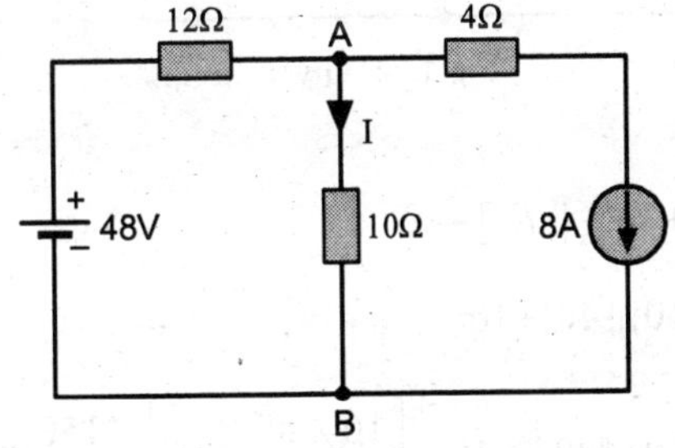

Fig.1.71: Given circuit

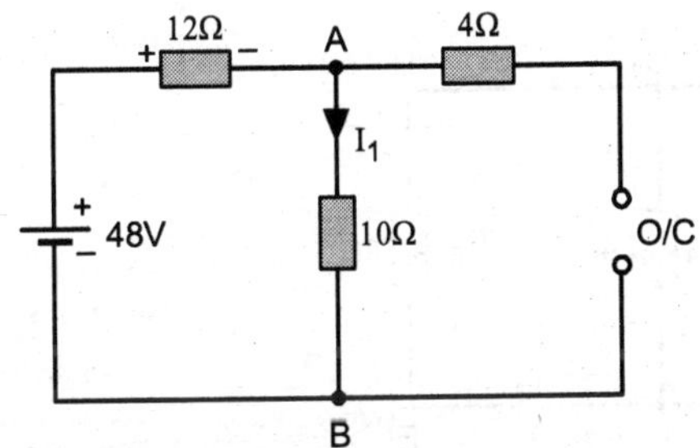

Fig.1.72: Considering voltage source only.

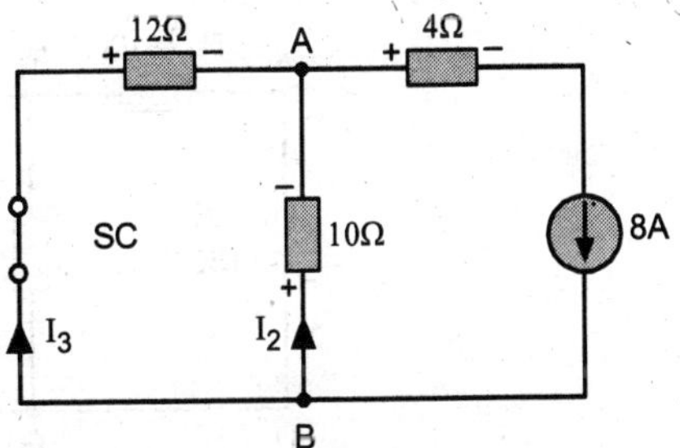

Fig.1.73: Considering current source only.

Solution: Applying superposition theorem

Case I: Considering 48 V source only and denoting 8 A source by its internal resistance *i.e.* open circuit as shown in Fig.1.72.

$$\text{Current supplied by 48 V battery, } I_1 = \frac{48}{12+10} = 2.18\text{A} \ (\textit{from A to B})$$

Case II: Considering 8A current source only and replacing 48 V voltage battery by its internal resistance *i.e.* short circuit as shown in fig.1.73.

Current is divided in two paths; current in 10 W resistor,

$$I_2 = \frac{12}{10+12} \times 8 = 4.36\text{A} \ (\textit{from B to A})$$

∴ Resultant current flowing through 10 W resistor

i.e. $I = 2{\cdot}18 - 4{\cdot}36 = \mathbf{-2.18\ A}$ (*Ans.*) (from B to A)

Example 1.17. *For the circuit of fig.*1.74. *find I using superposition theorem.*

Solution: Replacing 128 V source by its internal resistance (*i.e.* short circuit), the circuit is reduced to a circuit shown in fig.1.75.

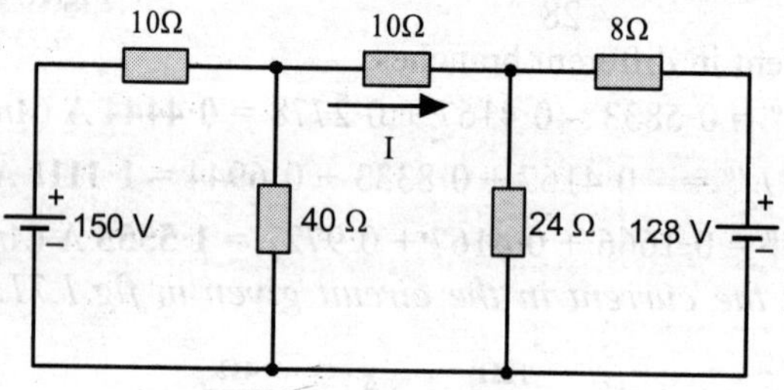

Fig.1.74: Given circuit.

Equivalent resistance,

$$R'_{eq} = [\{(R_4 \| R_5) + R_2\} \| R_3] + R_1$$

$$= [\{(8 \| 24) + 10\} \| 40] + 10$$

$$= \left[\left\{\frac{8\times 24}{8+24} + 10\right\} \| 40\right] + 10 = \left[\frac{16\times 40}{16+40} + 10\right] = \frac{150}{7}\ \Omega$$

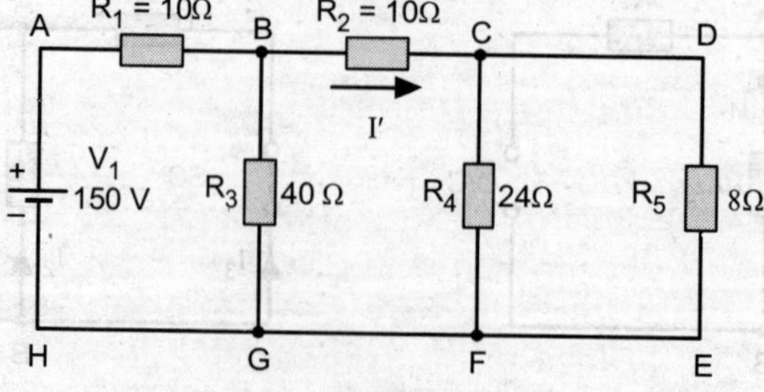

Fig.175:

$$\text{Current through branch } AB = \frac{V_1}{R'_{eq}} = \frac{150}{150/7} = 7\text{A}$$

$$\text{Current through branch } BC,\ I' = I_{AB} \times \frac{R_{BG}}{R_{BG}+R_{BF}} = 7 \times \frac{40}{40+16} = 5\text{A (from } B \text{ to } C)$$

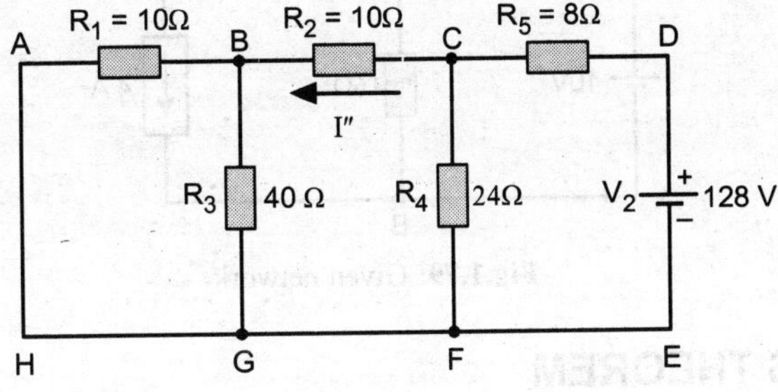

Fig.1.76:

Replacing 150 V source by its internal resistance (*i.e.* short circuit), the circuit is reduced to a circuit shown in fig.1.76.

$$\text{Equivalent resistance, } R''_{eq} = [\{(R_1 \| R_3) + R_2\} \| R_4] + R_5 = [\{(10 \| 40) + 10\} \| 24] + 8$$

$$= \left[\left\{\frac{10 \times 40}{10+40} + 10\right\} \| 24\right] + 8 = \left[\frac{18 \times 24}{18+24} + 8\right] = \frac{128}{7}\ \Omega$$

$$\text{Current through branch } DC = \frac{V_2}{R''_{eq}} = \frac{128}{128/7} = 7\text{A}$$

$$\text{Current through branch } CB,\ I'' = I_{DC} \times \frac{R_{CF}}{R_{CG}+R_{CF}} = 7 \times \frac{24}{18+24} = 4A \text{ (from } C \text{ to } B)$$

$$\text{Current, } I = I' - I'' = 5 - 4 = \mathbf{1\ A}\ (Ans.)$$

PRACTICE EXERCISE

Q1. *By using superposition theorem, find the current in resistance R shown in fig.1.77. $R_1 = 0{\cdot}15$ W, $R_2 = 0{\cdot}12$ W, $R = 3$ W, $E_1 = 6.15$ V, $E_2 = 6{\cdot}45$ V. Internal resistance of cells are negligible.*

(Ans.2.06A**)**

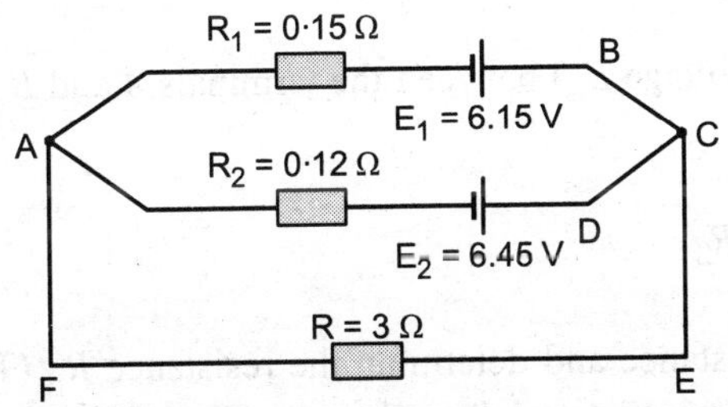

Fig.1.77: Given network.

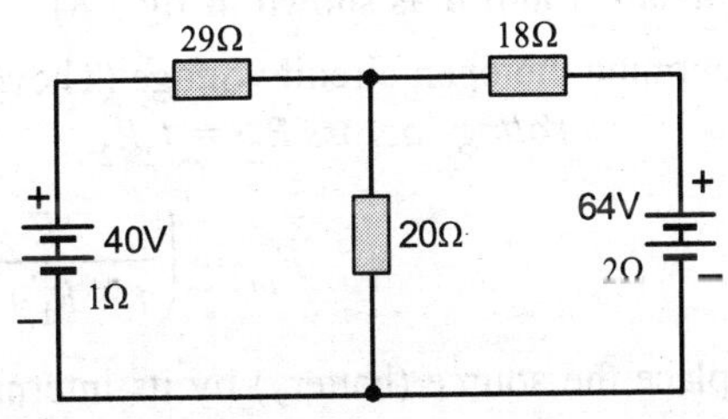

Fig.1.78: Given network

Q2. *Determine the branch currents by superposition theorem in the network shown in fig.1.78.*

(Ans.*0.2*A,*1.5*A, *1.7*A**)**

Q3. *Determine current through* 16 W *resistor in the following network (fig.*1.79*) using superposition theorem.*

(Ans.*1.2*A from *B* to *A***)**

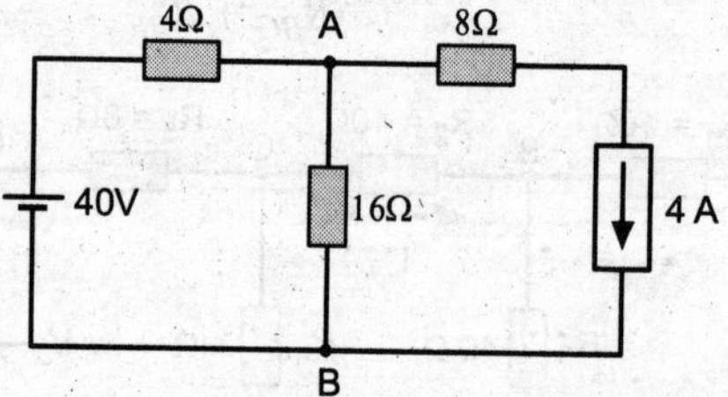

Fig.1.79: Given network

1.21 THEVENIN'S THEOREM

Thevenin's theorem states that:

The current flowing through a resistor connected across any two terminals of a network can be determined by converting the remaining network into a voltage source of emf, E_{th} and internal resistance, R_{th}.

Where, E_{th} = The open circuit voltage across the two terminals of the given branch, called Thevenin voltage.

R_{th} = The equivalent resistance of the network across the given terminals with all other sources replaced by their internal resistances, called Thevenin resistance.

For illustration, consider a circuit shown in fig.1.80. To determine the current through load resistance R_L, proceed with the following steps:

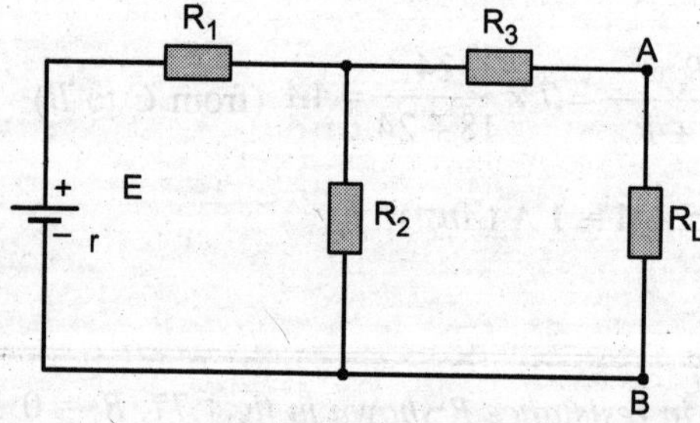

Fig.1.80: Given circuit.

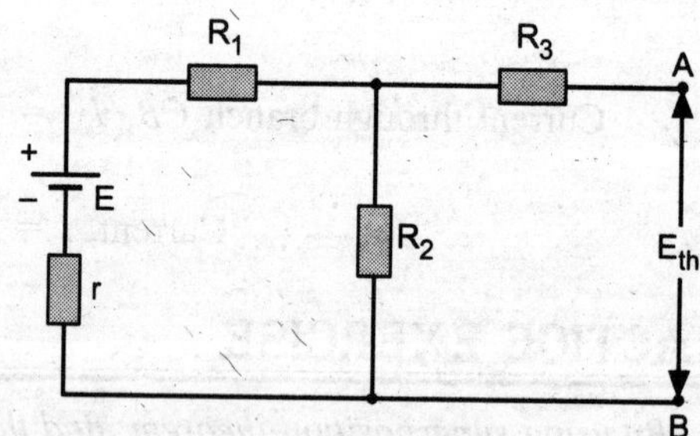

Fig.1.81: Circuit to determine E_{th}

- Remove the resistance R_L in which current is to be determined, thus creating an open circuit between terminals *A* and *B* as shown in fig.1.81.
- Determine the open circuit voltage (Thevenin voltage E_{th}) between the terminals *A* and *B i.e.*,

$$\text{Voltage across } R_2 = I_2R_2$$

$$E_{th} = \left(\frac{E}{r + R_1 + R_2}\right) R_2$$

- Replace the source (battery) by its internal resistance and determine the resistance R_{th} (Thevenin resistance) of the network as seen from the terminals *A* and *B* as shown in fig.1.82.

$$R_{th} = \frac{(r + R_1)\, R_2}{(r + R_1) + R_2} + R_3$$

- Replace the entire network by a single Thevenin voltage source having an e.m.f. E_{th} and internal resistance R_{th} as shown in fig.1.83.

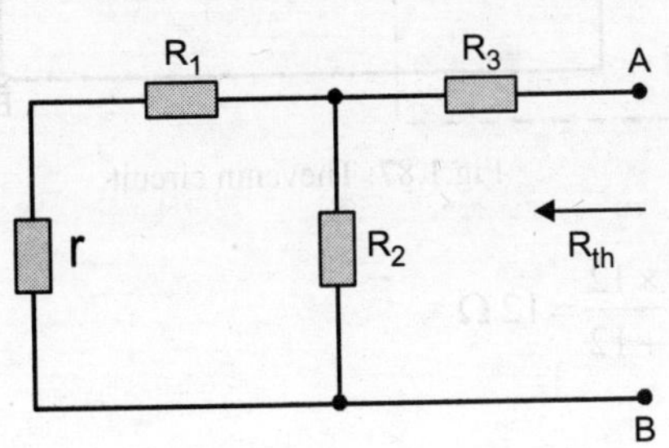

Fig.1.82: Circuit to determine R_{th}

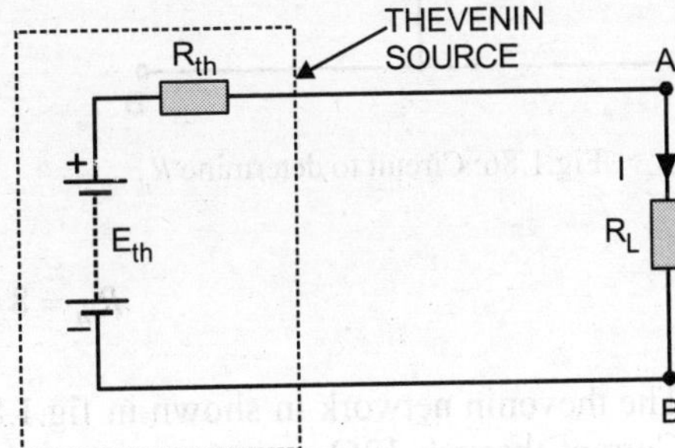

Fig.1.83: Thevenin circuit.

- Connect the load resistance R_L back to its terminals A and B from where it was removed.
- Determine the current flowing through the load resistance R_L by applying ohm's law *i.e.*

$$I = \frac{E_{th}}{R_{th} + R_L}$$

Example 1.18. *Determine current through 12Ω resistor connected across terminals A-B in the electric circuit shown in fig.1.84 using Thevenin's Theorem.*

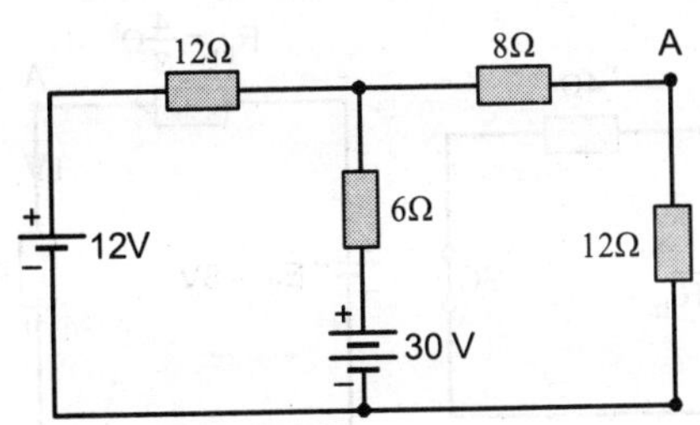

Fig.1.84: Given circuit.

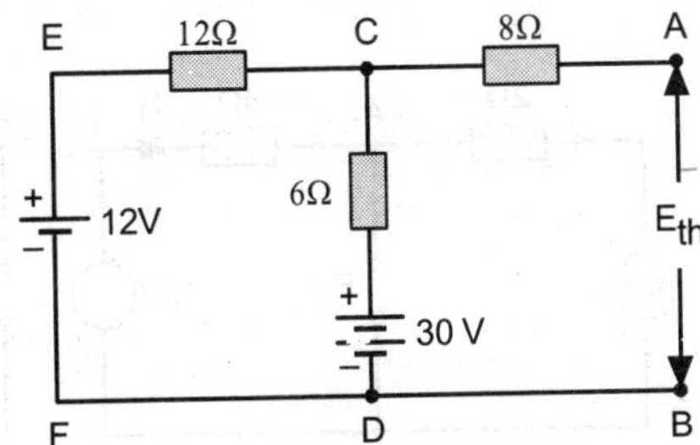

Fig.1.85: Circuit to determine E_{th}

Solution: To determine E_{th}, open the terminals A and B as shown in fig.1.85.

Current in the circuit $DCEFD$,

$$I = \frac{30 - 12}{12 + 6} = \frac{18}{18} = 1\text{A}$$

Open circuit voltage across the terminals AB, *i.e.*

$$E_{th} = 30 - 1 \times 6 = 24 \text{ V}$$

Thevenin resistance R_{th} of the network when viewed through the terminals A and B with all voltage sources are being short circuited as shown in fig.1.86.

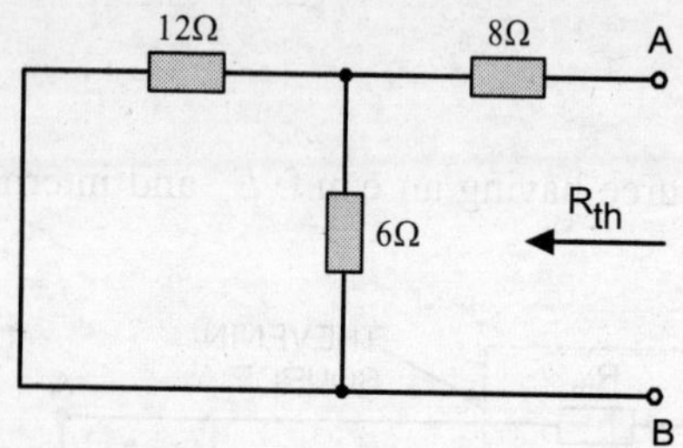

Fig.1.86: Circuit to determine R_{th}

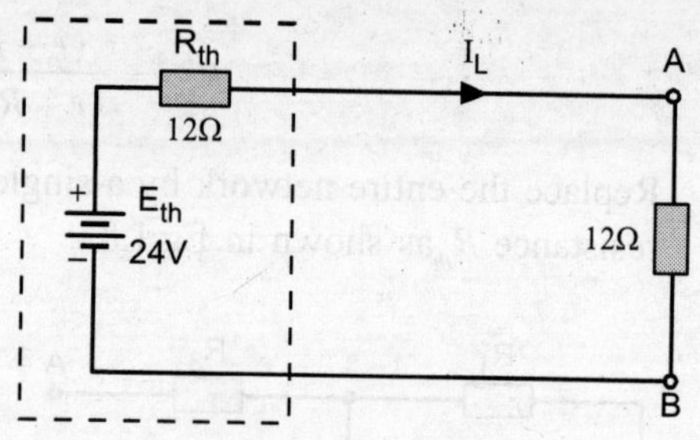

Fig.1.87: Thevenin circuit.

$$R_{th} = 8 + (6 \| 12) = 8 + \frac{6 \times 12}{6+12} = 12\,\Omega$$

The thevenin network in shown in fig.1.87.

Current through 12Ω resistor connected across terminal *AB*.

$$I_L = \frac{E_{th}}{R_{th} + R_L} = \frac{24}{12+12} = \mathbf{1\ A}\ (Ans.)$$

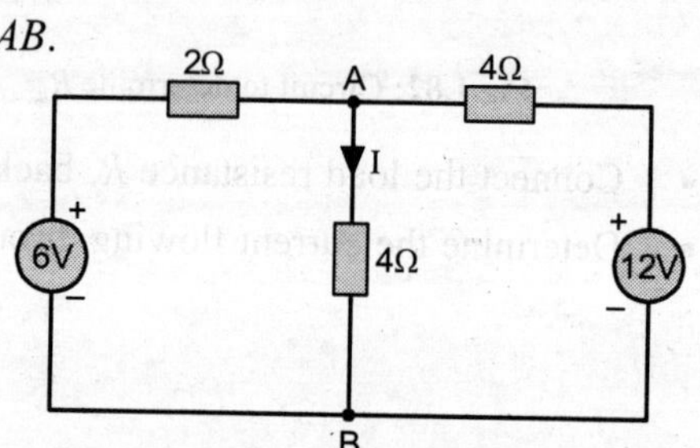

Fig.1.88: Given circuit.

Example 1.19. *Find current I using thevenin's theorem in the circuit shown in fig.*1.88.

Solution: To determining thevenin voltage or open circuit voltage across *AB*, remove 4Ω resistor connected across *AB* as shown in fig.1.89.

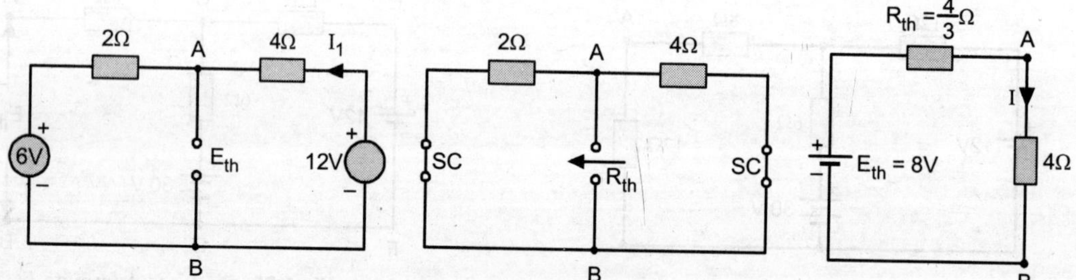

Fig.1.89: Circuit to determine E_{th} **Fig.1.90:** Circuit to determine R_{th} **Fig.1.91:** Thevenin circuit

Current in the circuit shown in fig.1.89.

$$I_1 = \frac{12-6}{2+4} = \frac{6}{6} = 1\text{ A}$$

Open circuit voltage across *AB i.e.* thevenin voltage,

$$E_{th} = 12 - 4 \times 1 = 8\text{ V}$$

To determine thevenin resistance looking into terminal *AB*, replace the voltage source by their internal resistance *i.e.* short circuit as shown in fig.1.90.

$$R_{th} = 2 \| 4 = \frac{2 \times 4}{2+4} = \frac{4}{3}\,\Omega$$

The thevenin equivalent circuit is shown in fig.1.91.

Current in 4Ω resistor connected across *AB*,

$$I = \frac{E_{th}}{R_{th} + R_L} = \frac{8}{\frac{4}{3}+4} = \frac{8}{\frac{16}{3}} = \frac{8}{16} \times 3 = \mathbf{1 \cdot 5\ A}\ (Ans.)$$

Example 1.20. *Find Thevenin's equivalent circuit at terminals BC of fig 1.92. Hence determine current through the resistor* $R = 1W$.

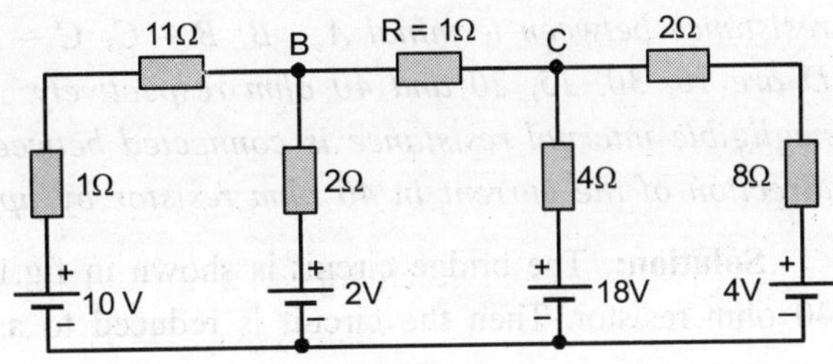

Fig.1.92: Given circuit

Solution: To determine thevenin equivalent resistance of given circuit with reference to terminals *BC*, replace voltage sources by this internal resistance as shown in fig.1.93.

$$R_{th} = [(11 + 1) \parallel 2] \parallel [(2 + 8) \parallel 4]$$

$$= \left[\frac{2 \times 12}{2+12}\right] \parallel \left[\frac{10 \times 4}{10+4}\right] = \left(\frac{12}{7}\right) \parallel \left(\frac{20}{7}\right) = \frac{15}{14} = 1 \cdot 0714\ \Omega$$

To determine open circuit voltage across the terminals *BC* (*i.e.* thevenin voltage), redraw the circuit as shown in fig.1.94.

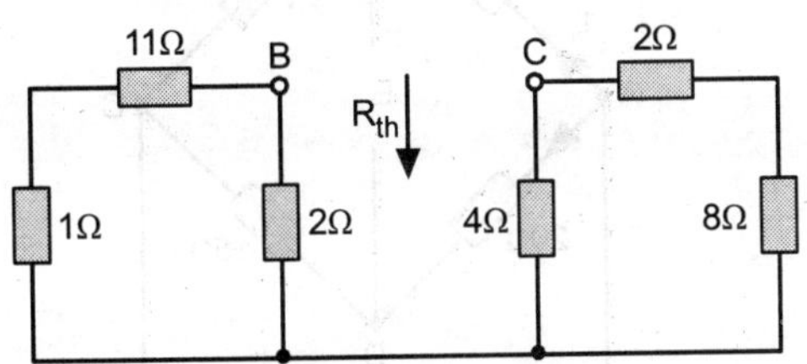

Fig.1.93: Circuit to determine R_{th}

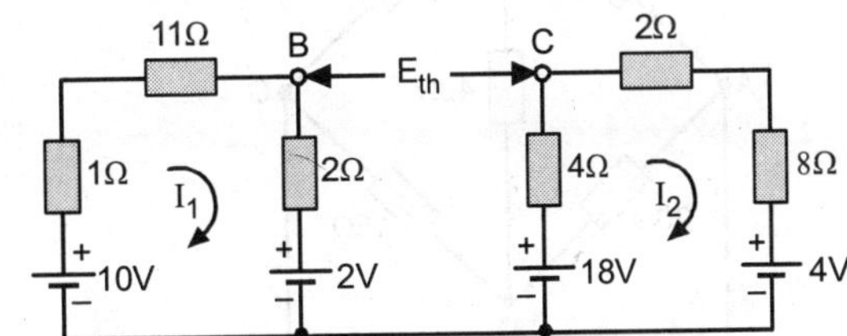

Fig.1.94: Circuit to determine E_{th}

$$\text{Current in the loop of terminal } B,\ I_1 = \frac{10-2}{11+1+2} = \frac{8}{14} = \frac{4}{7}\text{ A}$$

$$\text{Voltage at point } B,\ V_B = 10 - \frac{4}{7}(1+11) = \frac{22}{7}\text{ V}$$

$$\text{Current in the loop of terminal } C,\ I_2 = \frac{18-4}{4+2+8} = \frac{14}{14} = 1\text{ A}$$

$$\text{Voltage at point } C,\ V_C = 18 - 4 \times 1 = 14\text{ V}$$

$$\text{Thevenin voltage, } E_{th} = V_{CB} = V_C - V_B = \left(14 - \frac{22}{7}\right) = \frac{76}{7} = 10 \cdot 857\text{ V}$$

The Thevenin's equivalent circuit is shown in fig.1.95.

Current through resistor R of 1 Ω, $I = \dfrac{E_{th}}{R_{th}+R} = \dfrac{10\cdot857}{1\cdot0714+1}$

$= \mathbf{5\cdot24\ A}$ (*Ans.*)

Fig.1.95: Thevenin circuit.

Example 1.21. *A bridge network ABCD is arranged as follows: resistance between terminal A – B; B – C; C – D; D – A and B – D are 10, 30, 15, 20 and 40 ohm respectively. A 2 volt battery of negligible internal resistance is connected between terminals A and C. Determine the value and direction of the current in 40 ohm resistor by applying Thevenin theorem.*

Solution: The bridge circuit is shown in fig.1.96. To determine Thevenin voltage E_{th}, remove 40 ohm resistor. Then the circuit is reduced to as shown in fig.1.97.

Current in branch ABC, $I_1 = \dfrac{2}{10+30} = 0\cdot05\text{A}$

Current in branch ADC, $I_2 = \dfrac{2}{20+15} = 0\cdot05714\ \text{A}$

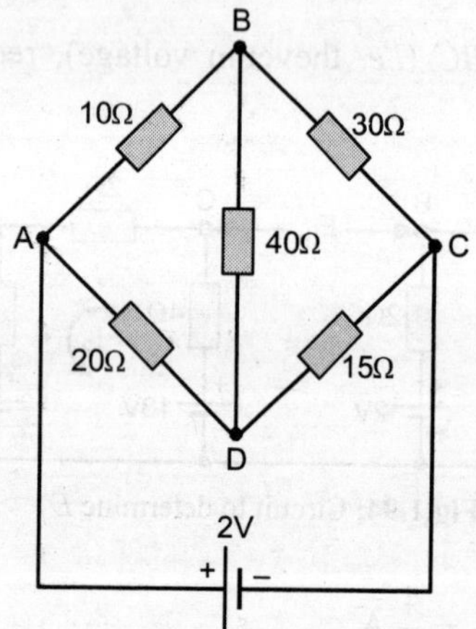

Fig.1.96: Given circuit as per data.

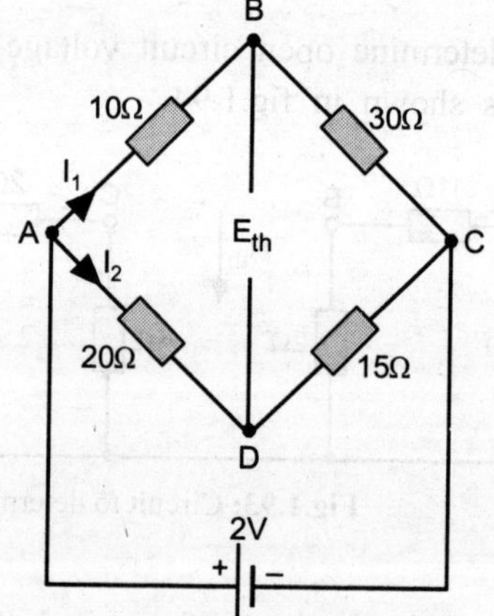

Fig.1.97: Circuit dermine E_{th}

Voltage at point B, $V_B = 2 - 0\cdot05 \times 10 = 1\cdot5$ V

Voltage at point D, $V_D = 2 - 0\cdot05714 \times 20 = 0\cdot857$ V

Potential difference across terminal B and D *i.e.*

Thevenin voltage, $E_{th} = 1\cdot5 - 0\cdot857 = 0\cdot643\text{V}$

To determine resistance R_{th} looking between terminals B and D, the source is replaced by its internal resistance (zero in this case) as shown in fig.1.98(a). The circuit is further simplified as shown in fig.1.98(b).

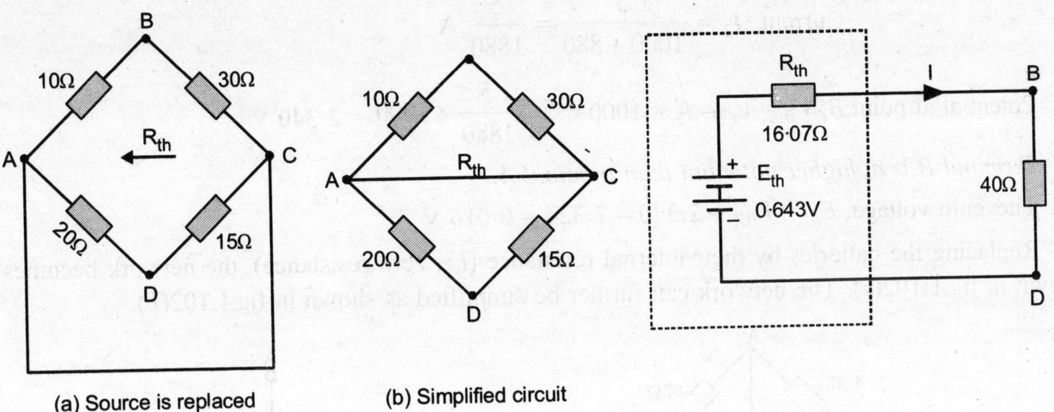

Fig.1.98: Circuit to determine R_{th}

Fig.1.99: Thevenin circuit.

A little thought will reveal that the resistance across terminal B and D *i.e.* Thevenin resistance,

$$R_{th} = \frac{10 \times 30}{10 + 30} + \frac{20 \times 15}{20 + 15} = 7 \cdot 5 + 8 \cdot 57 = 16 \cdot 07 \ \Omega$$

Thevenised circuit is shown in fig.1.99.

$\therefore$ Current in 40 ohm resistor. $I = \dfrac{0 \cdot 643}{16 \cdot 07 + 40} = 11 \cdot 467$ mA (*Ans.*)

Example 1.22. *State Thevenin's theorem and calculate current in a* 1000 Ω *resistor connected between terminals A and B, as shown in fig.*1.100.

Solution: To determine open circuit voltage across AB *i.e.* E_{th}, consider fig.1.101.

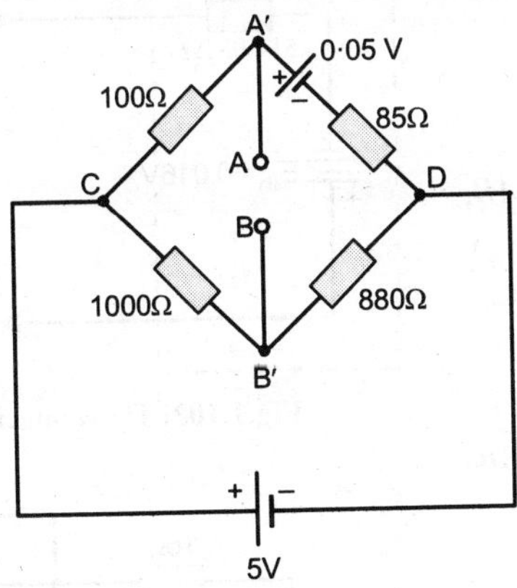

Fig.1.100: Given circuit.

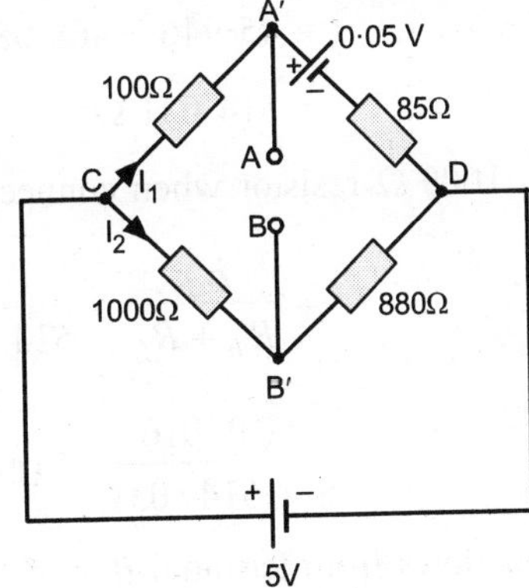

Fig.1.101: Circuit to determine E_{th}

$$\text{Current, } I_1 = \frac{5 - 0 \cdot 05}{100 + 85} = \frac{4 \cdot 95}{185} \text{ A}$$

Potential at point A, $V_A = V_C - I_1 \times 100 = 5 - \dfrac{4 \cdot 95}{185} \times 100 = 2 \cdot 324$ V

$$\text{Current, } I_2 = \frac{5}{1000+880} = \frac{5}{1880} \text{ A}$$

$$\text{Potential at point } B,\ V_B = V_C - I_2 \times 1000 = 5 - \frac{5}{1880} \times 1000 = 2{\cdot}340 \text{ V}$$

Terminal B is at higher potential than terminal A.

Thevenin voltage, $E_{th} = V_{BA} = 2{\cdot}340 - 2{\cdot}324 = 0{\cdot}016$ V

Replacing the batteries by their internal resistance (*i.e.* zero resistance), the network becomes as shown in fig.1.102(a). The network can further be simplified as shown in fig.1.102(*b*).

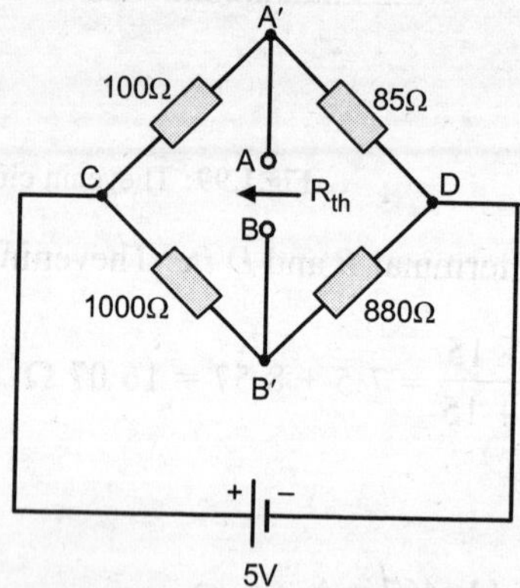

Fig.1.102(a): Circuit to determine R_{th}

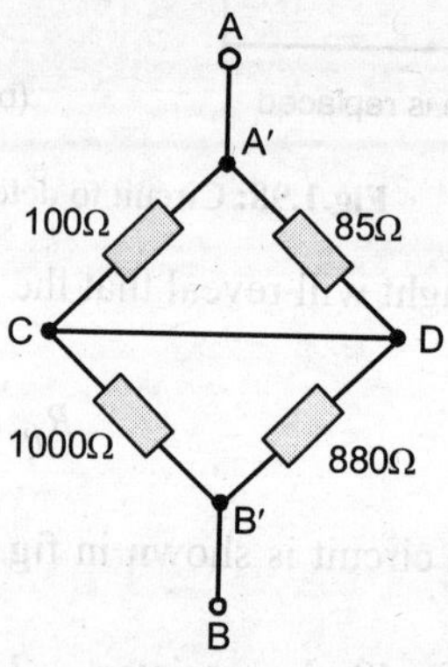

Fig.1.102(b): Simplified circuit.

$$R_{th} = R_{AB} = 100 \parallel 85 + 1000 \parallel 880$$

$$= \frac{100 \times 85}{100+85} + \frac{1000 \times 880}{1000+880}$$

$$= 45{\cdot}946 + 468{\cdot}085$$

$$= 514{\cdot}031\ \Omega$$

Current in 1000 Ω resistor when connected across *AB*,

$$I_L = \frac{E_{th}}{R_{th}+R_L} = \frac{0{\cdot}016}{514{\cdot}031+1000}$$

$$= \frac{0{\cdot}016}{1514{\cdot}031} = \mathbf{10{\cdot}6344\ \mu A}\ (\textit{Ans.})$$

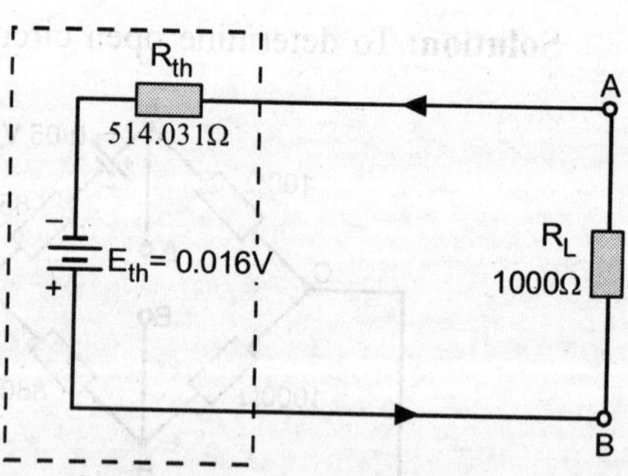

Fig.1.103: Thevenin circuit.

The current flows from terminal *B* to *A*.

Example 1.23. *Find the thevenin's equivalent circuit for terminal pair AB of the network shown in fig.1.104.*

Solution: For determining the Thevenin's equivalent resistance R_{th} of the circuit with reference to terminals *AB* the voltage source is short-circuited and current source is open-circuited, as shown in fig.1.105.

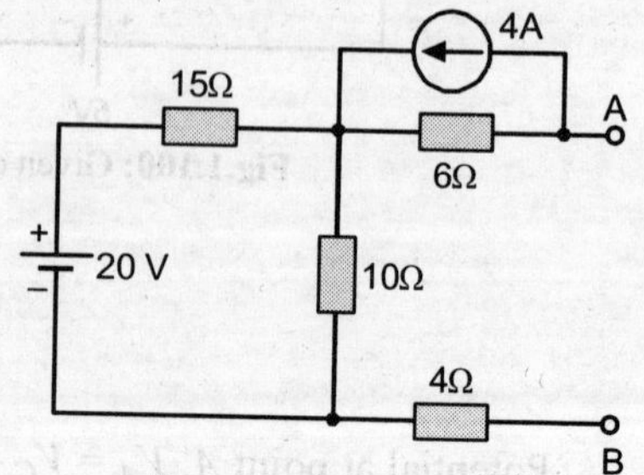

Fig.1.104: Given circuit.

$$R_{th} = 6 + (15 \parallel 10) + 4 = 6 + \frac{15 \times 10}{15+10} + 4 = 16\ \Omega$$

Converting current source of 4A connected across 6W resistor source into the equivalent voltage source, the circuit becomes as shown in fig.1.106.

Considering terminal *A* and *B* open circuited there is no current through resistor of 4 W and 6 W (hence no voltage drop occurs across them),

$$\text{Voltage across } 10\ \Omega \text{ resistor} = \frac{20}{15+10} \times 10 = 8\ \text{V}$$

Thevenin voltage across *AB*, $E_{th} = 8 - 24 = -16$ V

Thevenin's equivalent circuit is shown in fig.1.107.

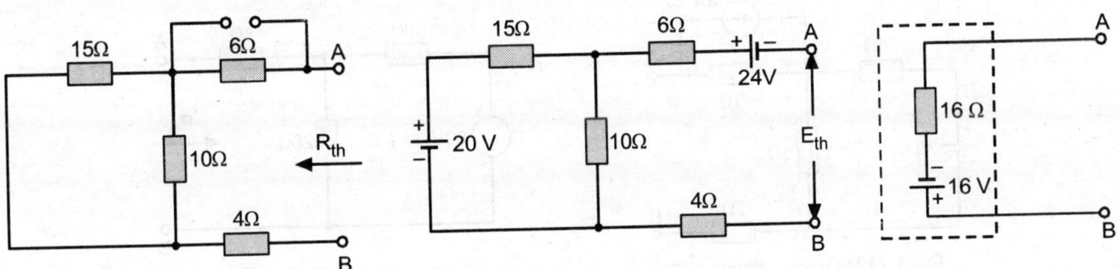

Fig.1.105: Circuit to determine R_{th} **Fig.1.106:** Circuit to determine E_{th} **Fig.1.107:** Thevenin equivalent

PRACTICE EXERCISE

Q.1 *Find current I using thevenin's theorem in the circuit shown in fig.*1.108. (*Ans.*2.4A)

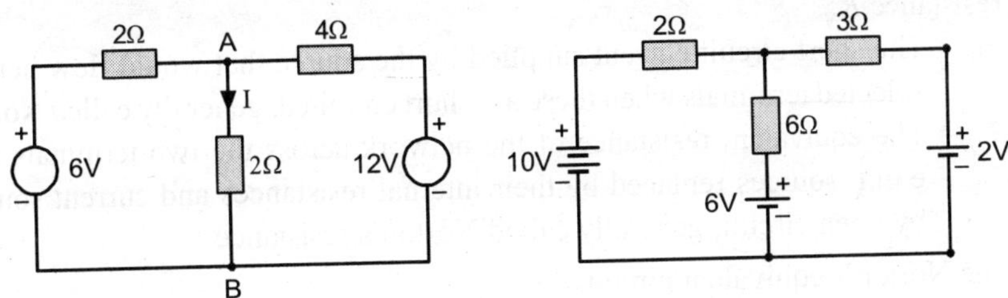

Fig.1.108: Given circuit **Fig.1.109:** Given circuit

Q.2 *Using thevenin theorem find the current in 3W resistor in Fig.1.109.* (*Ans. 1.556A*)

Q.3 *Using Mesh equation method,* find current in the resistance R_1 *of the network shown in Fig.*1.110. (*Ans.* 0.625A)

Q.4 *By applying thevenin theorem, find the current in* 15W *resistor of the network shown in fig.1.111.* (*Ans.* 3.5A)

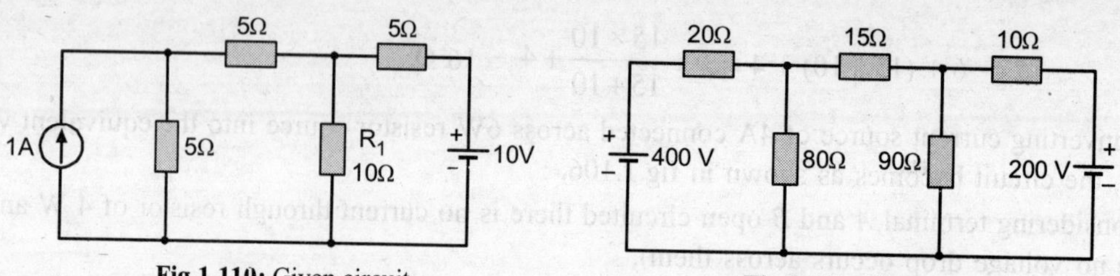

Fig.1.110: Given circuit

Fig.1.111: Given circuit

Q.5 *Find the thevenin's equivalent circuit for terminal pair AB of the network shown in fig.*1.112.

(*Ans.* E_{th}= –6V, R_{th}=9W)

Q.6. *Draw the thevenins equivalent circuit across AB, and determine current flowing through 12W resistor for the network shown in fig.1.113.* (*Ans.* 4.533 A)

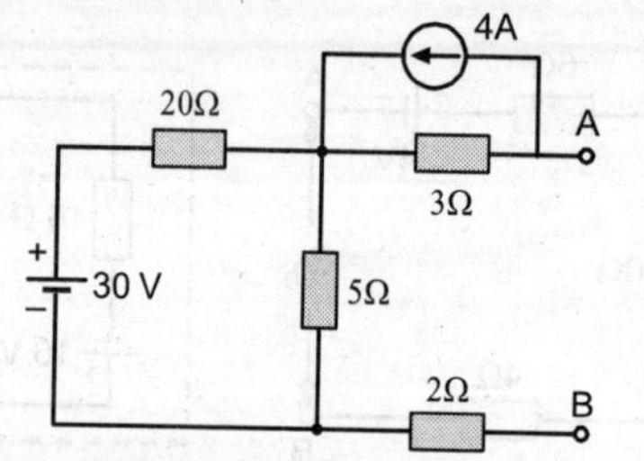

Fig.1.112: Given circuit

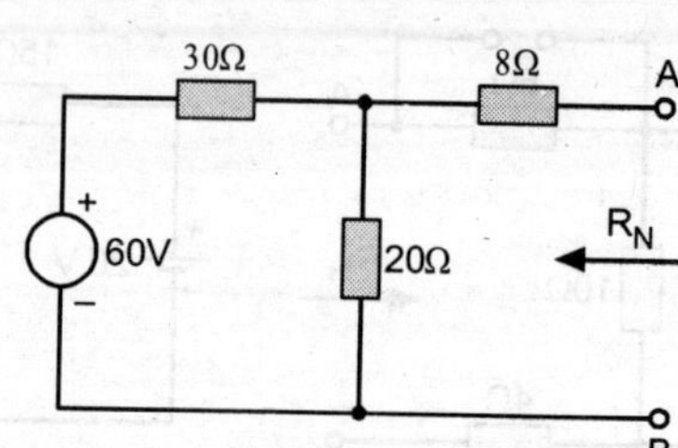

Fig.1.113: Given circuit

1.22 NORTON'S THEOREM

Norton's theorem states that:

The current flowing through a resistor connected across any two terminals of a network can be determined by converting the whole network into a current source having a current output of I_N in parallel with a resistance R_N.

Where, I_N = The short circuit current supplied by the source that would flow between the two selected terminals when these are short circuited, generally called Norton's current.

R_N = The equivalent resistance of the network across the two terminals with all other e.m.f. sources replaced by their internal resistances and current sources replaced by open circuit, generally called Norton's resistance.

Steps to determine Norton's equivalent circuit.

- Short circuit the terminals across which the load resistor in connected and calculate the current which would flow between them. This is the Norton current I_N.
- Redraw the network replacing each voltage source by a short circuit in series with its internal resistance if any and each current source by an open circuit in parallel with its internal resistance.
- Determine the resistance R_N of the network as seen from the network terminals. (Its value is the same as that of R_{th}).

1.23 CONVERSION OF THEVENIN'S EQUIVALENT INTO NORTON'S EQUIVALENT AND VICE-VERSA

A Thevenin's equivalent can be converted into its Norton's equivalent and vice-versa. According to statement, Norton's current source equals the current I_{SC} or I_N which flows across the terminals *A* and *B* when they are short circuited.

Hence $$I_{SC} = \frac{E_{th}}{R_{th}} \qquad \text{...(i)}$$

Likewise a Norton's circuit can be converted into its Thevenin's equivalent. The Thevenin's equivalent source V_{OC} or E_{th} is the voltage on open circuit and is given as

$$V_{OC} \text{ or } E_{th} = I_{SC}\, R_{th} \qquad \text{...(ii)}$$

Each theorem is dual of the other.

30Ω 8Ω A 60V 20Ω R_N B

Fig.1.114: Given circuit.

Example 1.24. *For the circuit shown in fig.1.114, obtain Norton current and equivalent resistance seen from 'AB'.*

Solution: Equivalent resistance of the network with reference to terminals *A* and *B*, as shown in fig.1.115.

$$R_N = (30 \parallel 20) + 8 = \frac{30 \times 20}{30+20} + 8 = 12 + 8 = \mathbf{20\ \Omega}\ (Ans.)$$

The current supplied by 60 V source, when terminals *A* and *B* are short circuited as shown in fig.1.116.

$$I = \frac{60}{30 + (8 \parallel 20)} = \frac{60}{30 + \frac{8 \times 20}{8+20}} = \frac{42}{25}\ \text{A}$$

Norton's current, I_N *i.e.*, the current from terminal *A* to *B* when these are short circuited as shown in fig.1.117.

$$I_N = \frac{20}{20+8} \times I = \frac{20}{28} \times \frac{42}{25} = \mathbf{1{\cdot}2\ A}\ (Ans.)$$

The Norton's equivalent circuit is shown in fig.1.117.

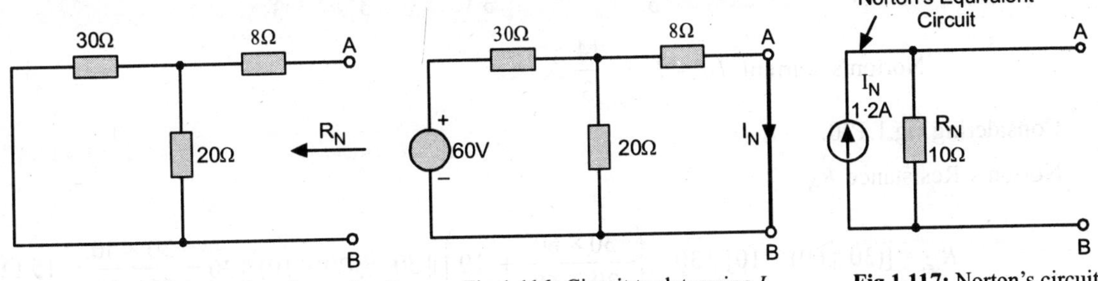

Fig.1.115: Circuit to determine R_N **Fig.1.116:** Circuit to determine I_N **Fig.1.117:** Norton's circuit.

Example 1.25. *Find the Norton's equivalent circuit, as seen by R_L in the circuit shown in fig.1.118.*

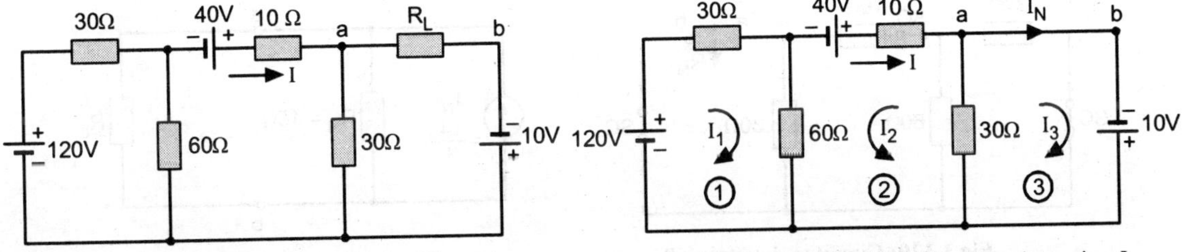

Fig.1.118: Given circuit. **Fig.1.119:** Circuit to determine I_N

Solution: Let the Norton's current be I_N, *as shown in fig.1.119.*

Applying *KVL* in Loop (1), we get,

$$30I_1 + 60(I_1 - I_2) = 120$$

or $$90I_1 - 60I_2 = 120$$

$$3I_1 - 2I_2 = 4 \quad \text{..........}(i)$$

Applying *KVL* in Loop (2), we get,

$$60(I_2 - I_1) + 10I_2 + 30(I_2 - I_3) = 40$$

or $$-60I_1 + 100I_2 - 30I_3 = 40 \quad \text{..........}(ii)$$

or $$-6I_1 + 10I_2 - 3I_3 = 4$$

Applying *KVL* in Loop (3), we get,

$$30(I_3 - I_2) = 10$$

or $$-30I_2 + 30I_3 = 10 \quad \text{or} \quad -3I_2 + 3I_3 = 1 \quad \text{..........}(iii)$$

Solving eqn., (*ii*) and (*iii*), we get,

$$-6I_1 + 10I_2 - 1 - 3I_2 = 4 \text{ (since } 3I_3 = 1 + 3I_2)$$

$$6I_1 + 7I_2 = 5 \quad \text{............(iv)}$$

Multiplying eqn. (*i*) by 2 and adding eqn. (*iv*), we get

$$I_2 = \frac{13}{3}\text{ A}$$

from eqn. (*iii*), $$I_3 = \frac{1}{3}(1+3I_2) = \frac{1}{3}\left(1+3\times\frac{13}{3}\right) = \frac{14}{3}\text{ A}$$

$\therefore$ Norton's current $I_N = I_3 = \dfrac{14}{3}$ A

Considering fig.1.120.

Norton's Resistance R_N

$$R_N = [(30 \parallel 60) + 10] \parallel 30 = \left(\frac{30\times 60}{30+60} + 10\right) \parallel 30 = (20+10) \parallel 30 = \frac{30\times 30}{30+30} = 15\ \Omega$$

Norton's equivalent circuit is shown in fig.1.121.

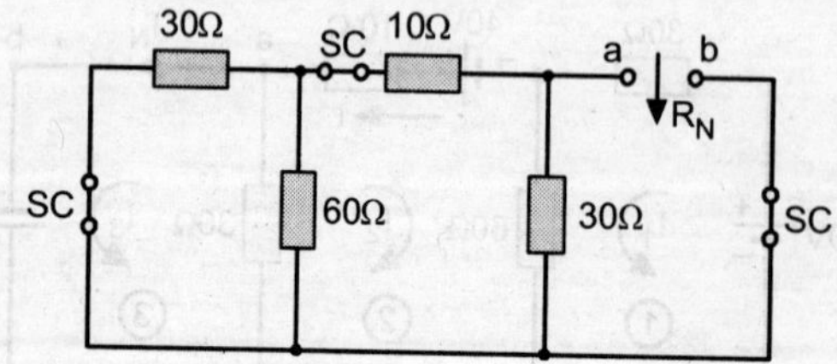

Fig.1.120: Circuit to determine R_N

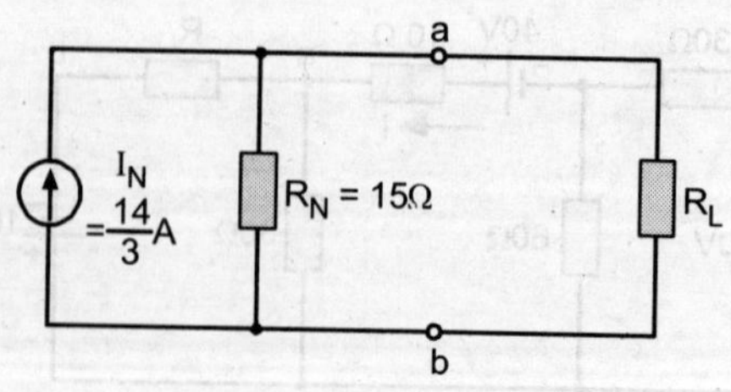

Fig.1.121: Norton's circuit

Example 1.26. *For the network shown in fig.1.122 draw a Norton's equivalent circuit and determine the current flowing through 15Ω resistor.*

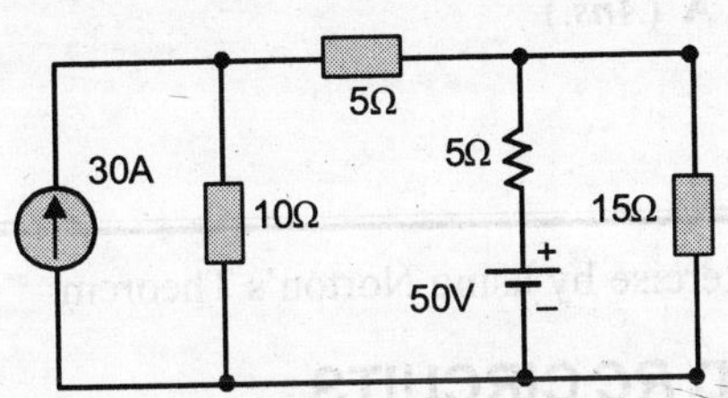

Fig.1.122: Given circuit

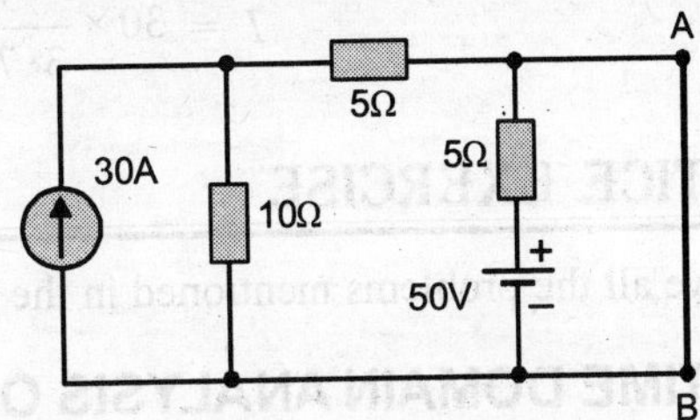

Fig.1.123: Circuit to determine I_N

Solution: To determine I_N, the 15Ω resistor is removed and terminals *A* and *B* are short circuited as shown in fig.1.123. Since, there are two sources due to which current will flow between terminal *A* and *B*, therefore applying superposition theorem

$$I_N = I_1 + I_2$$

(*i*) Considering only current source, the circuit will be reduced as shown in fig.1.124.

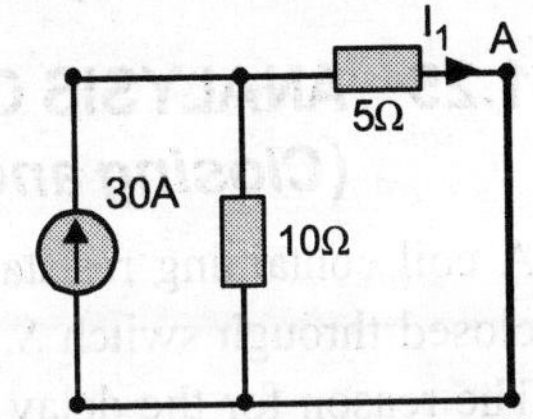

Fig.1.124: Considering only current source

$$\text{Current } I_1 = 30 \times \frac{10}{10+5} = 20\text{A}$$

(*ii*) Considering only voltage source, The circuit will be as shown in fig.1.125.

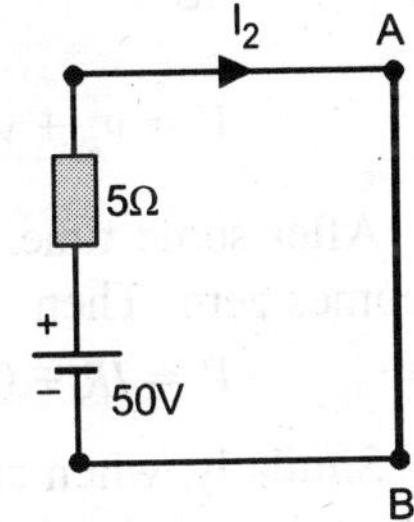

Fig.1.125: Considering only voltage source.

$$I_2 = \frac{50}{5} = 10\text{A}$$

According to superposition theorem:

Norton current, $I_N = I_1 + I_2 = 20 + 10 = \mathbf{30}$ **A** (*Ans.*)

Now replacing current source with open circuit and voltage source with short circuit, the circuit will reduce to as shown in fig.1.126. Looking between the terminals *A* and *B*, the Norton resistance,

$$R_N = \frac{5 \times (5+10)}{5+(5+10)} = \mathbf{3{\cdot}75\Omega} \ (\textit{Ans.})$$

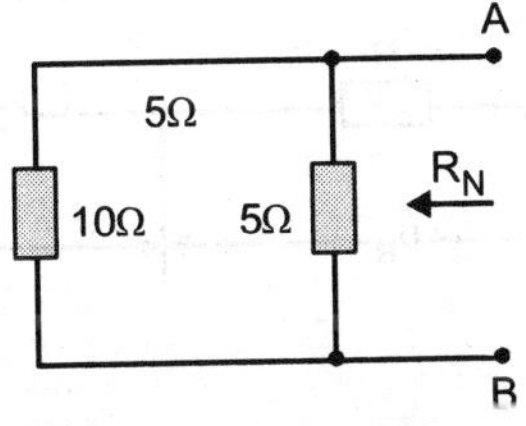

Fig.1.126: **Circuit to determine R_N**

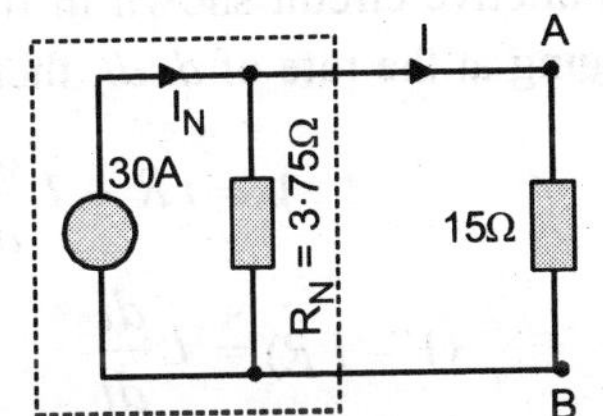

Fig.1.127: **Norton's circuit.**

The Norton's equivalent circuit is shown in fig.1.127.

Current flowing through 15Ω resistor,

$$I = 30 \times \frac{3\cdot 75}{3\cdot 75+15} = \mathbf{6\ A}\ (Ans.)$$

PRACTICE EXERCISE

Q.1 Solve all the problems mentioned in the previous exercise by using Norton's Theorem.

1.24 TIME DOMAIN ANALYSIS OF *RL* AND *RC* CIRCUITS

To analyse *RL* and *RC* circuits under time domain, the *RL* and *RC* circuits are opened and closed and their effects are analysed.

1.25 ANALYSIS OF RL CIRCUITS (*Closing and opening of RL circuit*)

A coil containing resistance *R* W and inductance *L* henry is shown in fig.1.128. When the circuit is closed through switch *S*, the current increases gradually and takes some time to attain its final value. The reason for the delay is that when current increases to attain its final value, an e.m.f. (self-induced e.m.f.) is induced in the inductance of the coil which opposes the increase in current.

Let at any instant, the current in the circuit be *i* ampere and is increasing at the rate of *di*/*dt*. Then applied voltage.

$$V = v_R + v_L = i\,R + L\frac{di}{dt}\left(v_L = e_L = L\frac{di}{dt}\right)$$

After some time, the current (*i*) obtains its final value (*I*) and rate of change of current (*di*/*dt*) becomes zero. Then

$V = IR + 0$ or $I = V/R$.

Similarly, when an inductive circuit (or coil) is opened, the current does not attain its zero value instantaneously but it falls gradually.

Thus, the role of an inductance, in a d.c. circuit to delay the change (rise or fall) in current. Its behaviour is just analogous to inertia in mechanics. The delay in change of current, in either case, depends upon the value of *R* and *L* as explained in the next articles.

1.26 RISE OF CURRENT IN AN INDUCTIVE CIRCUIT

Consider an inductive circuit shown in fig.1.128. At any instant, let the value of current be *i* ampere which is changing at the rate of *di*/*dt*, then

$$V = i\,R + L\frac{di}{dt}$$

or
$$(V - i\,R) = L\frac{di}{dt}$$

or
$$\frac{di}{(V - i\,R)} = \frac{dt}{L}$$

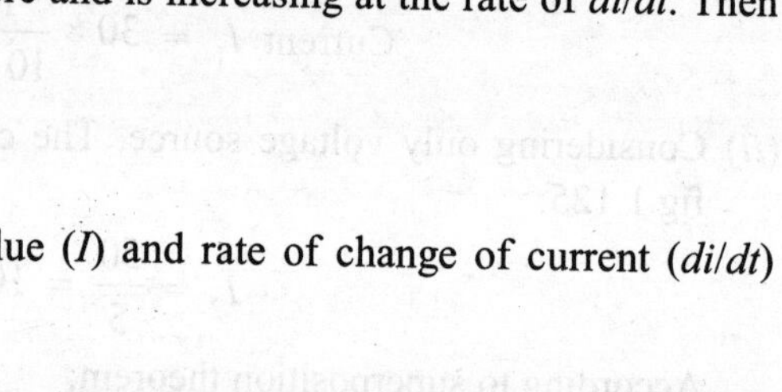

R
L
v_R
v_L
i
S
+
–
V

Fig.1.128: Closing of RL circuit

Multiplying both sides by – R and then integrating, we get,

$$\int \frac{-R\,di}{(V - i\,R)} = -\frac{R}{L}\int dt$$

or $$\log_e (V - i\,R) = -\frac{R}{L}t + K \qquad ...(i)$$

where K is a constant of integration whose value can be determined from the initial known conditions, *i.e.* at $t = 0$; $i = 0$. Substituting these values in the exp. (*i*), we get,

$$\log_e V = K$$

∴ Expression (*i*) becomes;

$$\log_e (V - i\,R) = -\frac{R}{L}t + \log_e V$$

or $$\log_e \frac{V - i\,R}{V} = -\frac{R}{L}t$$

or $$\frac{V - i\,R}{V} = e^{-Rt/L} \quad \text{or} \quad V - i\,R = V e^{-Rt/L}$$

or $$i = \frac{V}{R}\left(1 - e^{-Rt/L}\right)$$

Now, V/R represents the final value of the steady current I attained by the circuit.

∴ $$i = I\left(1 - e^{-Rt/L}\right) \qquad ...(ii)$$

or $$i = I\left(1 - e^{-t/\lambda}\right)$$

where λ is called time constant of the circuit.

Time constant

Expression (*ii*) shows the rise of current w.r.t. time t. The exponent of e is Rt/L. The quantity L/R must have the dimensions of time so that exponent of e (*i.e.* Rt/L) is a number.

The quantity L/R is called the *time constant* of the circuit and effects the rise of current in the circuit. It is represented by λ.

∴ Time constant, $\lambda = L/R$ second

If time interval, $t = \lambda$ (or L/R), then

$$i = I\,(1 - e^{-1}) = 0{\cdot}632\ I \ (Rt/L = R \times L/L \times R = 1)$$

Hence, the time required for the current in an R – L circuit, to reach 0·632 times of its steady value (I) while rising is called the **time constant** *of the circuit.*

1.27 DECAY OF CURRENT IN AN INDUCTIVE CIRCUIT

An inductive circuit containing resistance R and inductance L in series is shown in fig.1.129. When switch is thrown to position 1, the current in the circuit starts rising and attains its final value I (= V/R) after some time. If now, the switch is thrown to position 2, the current in the circuit does not cease at once but gradually reduces to zero.

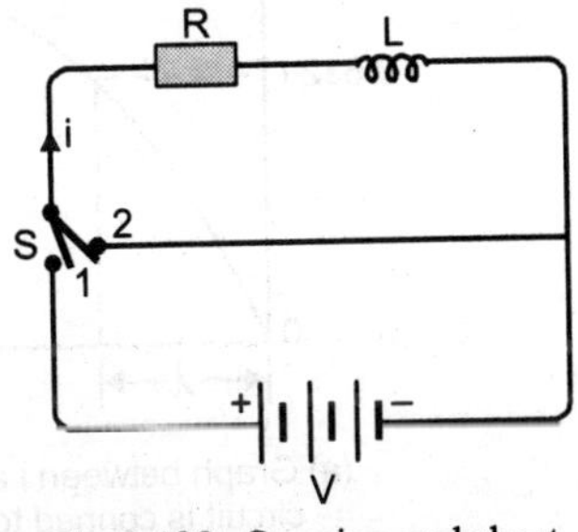

Fig.1.129: Opening and short circuiting of RL circuit

Let at any instant (during decay), the current in the circuit be i ampere and is decreasing at the rate of di/dt, then

$$0 = iR + L\frac{di}{dt}$$

or $$\frac{di}{i} = -\frac{R}{L}dt$$

Integrating both sides, we get,

$$\log_e i = -\frac{R}{L}t + K \quad ...(i)$$

where K is constant of integration whose value can be determined from the initial conditions, *i.e.* at $t = 0$; $i = I$ (= V/R). Substituting these values in exp. (*i*), we get,

$$\log_e I = 0 + K \text{ or } K = \log_e I$$

∴ Expression (*i*) becomes;

$$\log_e i = -\frac{R}{L}t + \log_e I$$

or $$\log_e \frac{i}{I} = -\frac{R}{L}t \text{ or } \frac{i}{I} = e^{-Rt/L}$$

or $$i = I\,e^{-Rt/L}$$

$$i = I\,e^{-t/\lambda} \quad ...(ii)$$

where λ (= L/R) is called time constant of the circuit.

Time constant

The quantity L/R is called the *time constant* of the circuit and affects the fall of current in the circuit. It is represented by l.

∴ Time constant, $\lambda = L/R$ second

$$i = I\,e^{-1} = 0{\cdot}368\,I$$

Hence, the time required for the current, in an R – L circuit, to reach 0·368 times of its steady value (I) while decaying is called the **time constant** *of the circuit.*

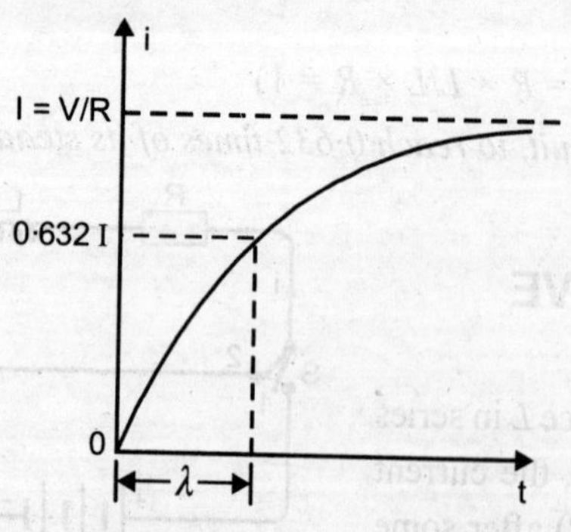

(a) Graph between i and t when RL circuit is conned to dc source

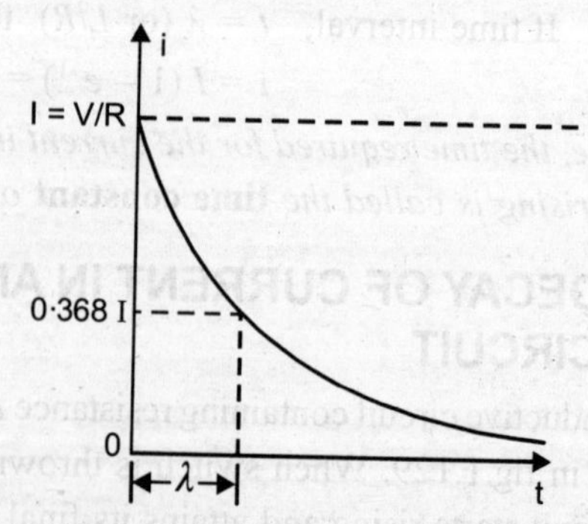

(b) Graph between i and t when RL circuit is disconnected from source and short circuited

Fig.1.130: Graph between current and time

Fig.1.130 (*a*) shows a graph between current and time in an $R - L$ circuit, when it is connected across a d.c. source. The current rises in the circuit following exponential law. Whereas, fig.1.130 (*b*) shows a graph between current and time in an $R - L$ circuit, when it is short circuited after attaining its final steady value of current. The current decays in the circuit following exponential law.

Example 1.27. *The resistance and inductance of a coil are 4 Ω and 0·15 m H respectively. What potential difference exists at the terminals of the solenoid at the instant when the current is 1.5 A, but increasing at the rate of 10000 A per second ?*

Solution: Voltage across the solenoid, $E = IR + L\dfrac{di}{dt}$

where, $I = 1.5\text{A}$; $R = 4\ \Omega$; $L = 0{\cdot}5 \times 10^{-3}$ H; $di/dt = 10000$ A/s;

$\therefore\ E = 1.5 \times 4 + 0{\cdot}15 \times 10^{-3} \times 10000 =$ **7.5 V** (*Ans.*)

Example 1.28. *A coil having a resistance of 10 Ω and inductance of 6 H is connected to a constant supply voltage of 100 V. Find (i) the rate of change of current at the instant of closing the switch; (ii) final steady value; (iii) time constant of the circuit and (iv) the time taken for the current to reach a value of 4 A.*

Solution: We know that $V = i\,R + L\dfrac{di}{dt}$

(*i*) At the time of closing the switch, $i = 0$

$$\therefore\ 100 = 0 + 6\frac{di}{dt} \quad \text{or} \quad \frac{di}{dt} - \frac{100}{6} = \mathbf{16.67\ A/s}\ (\textit{Ans.})$$

(*ii*) Final steady value, $I = V/R = 100/10 =$ **10 A** (*Ans.*)

(*iii*) Time constant, $\lambda = L/R = 12/20 =$ **0·6** (*Ans.*)

(*iv*) Now, $\quad i = I\left(1 - e^{-t/\lambda}\right)$

$$\text{or} \quad 4 = 10\left(1 - e^{-t/0.6}\right) \quad \text{or} \quad 0.4 = 1 - e^{-t/0{\cdot}6}$$

$$\text{or} \quad e^{-t/0{\cdot}6} = 0{\cdot}6 \quad \text{or} \quad e^{t/0{\cdot}6} = 1/0{\cdot}6 = 1.67$$

$$\text{or} \quad (t/0{\cdot}6)\log_e e = \log_e 1.67 \quad \text{or} \quad t/0{\cdot}6 = \log_e 1.67$$

$$\text{or} \quad t = 0{\cdot}6 \times 0{\cdot}5128 = \mathbf{0{\cdot}3077\ s}\ (\textit{Ans.})$$

Example 1.29. *Consider a series R – L circuit connected to a battery source of 5 volt. If R = 5 ohm and L = 5 henry, find the current i(t) through the inductance. Find the voltage across resistance and inductance separately at the time of switching the supply and when sufficient time has elapsed after switching the source.*

Solution: The current flowing through the circuit is given by the expression,

$$i\,(t) = \frac{V}{R}\left(1 - e^{-Rt/L}\right)$$

where, $\quad V = 5$ V; $R = 5\ \Omega$ and $L = 5$ H.

$$\therefore \quad i\,(t) = \frac{5}{5}\left(1 - e^{-t}\right) = 1 - e^{-t}$$

(*i*) At the time of switching, $t = 0$

$$\therefore \quad i\,(t) = 1 - e^{-0} = 0$$

Now, $$V = iR + L\frac{di}{dt}$$

$\therefore$ Voltage across resistor, $i\,R = \mathbf{0\ V}$ (*Ans.*)

Voltage across inductor, $L\frac{di}{dt} = \mathbf{5\ V}$ (*Ans.*)

(*ii*) When the sufficient time is elapsed *i.e.* $t \to \infty$

$$i\,(t) = 1 - e^{-\infty} = 1$$

$\therefore$ Voltage across resistor, $i\,R = \mathbf{5\ V}$ (*Ans.*)

Voltage across inductor, $L\frac{di}{dt} = \mathbf{0\ V}$ (*Ans.*)

Example 1.30. *Consider the series R – L circuit connected to a battery and a switch as shown in fig.1.131. Switch S_1 closes at t = 2 second; switch S_2 closes at t = 5 second. (i) Find the current i (t) in the resistance R; (ii) Plot the current i (t) for 0 £ t £ 10 second to the scale. Assume R/L equal to 100.*

Solution: When switch S_1 is closed and switch S_2 remains open the circuit becomes as shown in fig.1.132.

The switch is closed after 2 seconds, therefore upto 2 seconds no current flows (*OA*) through the circuit.

After 2 second, when switch S_1 is closed current starts rising exponentially (*AB*) given by the equation;

$$i\,(t) = \frac{V}{R}\left(1 - e^{-Rt/L}\right) = \frac{V}{R}\left(1 - e^{-100t}\right)$$

After $t = 5$ seconds, the switch S_2 is closed and the circuit becomes as shown in fig.1.133. The current starts decaying following the exponential law (*BC*) and is given by the equation;

$$i\,(t) = \frac{V}{R}\left(e^{-Rt/L}\right) = \frac{V}{R}\left(e^{-100t}\right)$$

A graph between current and time is shown in fig.1.134.

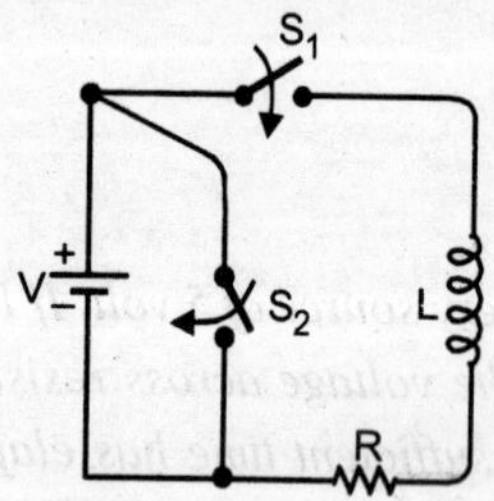

Fig.1.131: Given Circuit

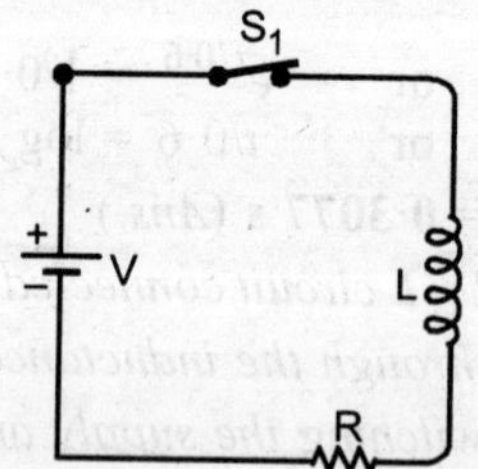

Fig.1.132: When S_1 is closed and S_2 is opened

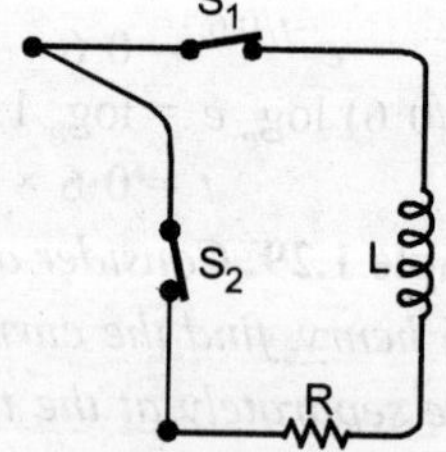

Fig.1.133: When S_2 is closed and S_1 is opened

Example 1.31. *A coil having L = 5 H and R = 10 Ω is connected to a constant 180 V supply source. How long does it take the voltage across the resistor to reach 90 V ?*

Solution: $$i = \frac{V}{R}\left(1 - e^{-t/\lambda}\right)$$

or $$i\,R = V\left(1 - e^{-t/\lambda}\right)$$

where, $i R = 90$ V; $V = 180$ V; $\lambda = L/R = 5/10 = 0{\cdot}5$

$\therefore \quad 90 = 180\left(1-e^{-t/0.5}\right)$

or $\quad 0{\cdot}5 = 1 - e^{-t/0{\cdot}5}$ or $\quad e^{-t/0{\cdot}5} = 0{\cdot}5$

or $\quad e^{t/0{\cdot}5} = 1/0{\cdot}5$ or $\quad (t/0{\cdot}5)\log_e e = \log_e 2$

or $\quad t = 0{\cdot}5 \times 0{\cdot}6931 = \mathbf{0{\cdot}3466\ s}$ *(Ans.)*

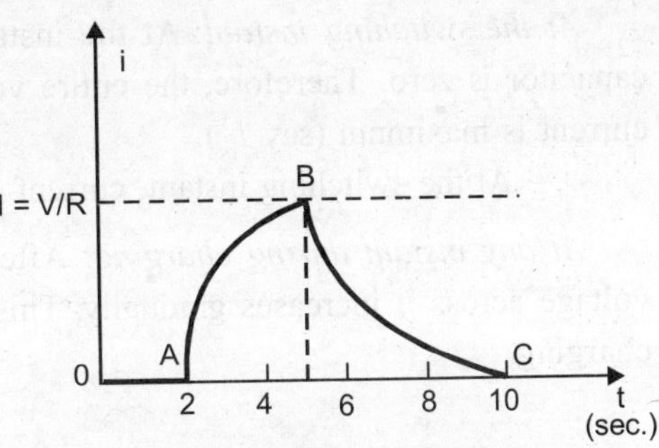

Fig.1.134: Graph between i & t

PRACTICE EXERCISE

Q.1 *The resistance and inductance of a coil are 3 W and 0·1 m H respectively. What potential difference exists at the terminals of the solenoid at the instant when the current is 1 A, but increasing at the rate of 10000 A per second ?* (*Ans.* 4 V)

Q.2 *A coil having a resistance of 20 W and inductance of 12 H is connected to a constant supply voltage of 100 V. Find (i) the rate of change of current at the instant of closing the switch; (ii) final steady value; (iii) time constant of the circuit and (iv) the time taken for the current to reach a value of 4 A.* *(Ans. 8·33 A/s ,5 A , 0·6, 0·9656 s)*

Q.3 *Consider a series R – L circuit connected to a battery source of 1 volt. If R = 1 ohm and L = 1 henry, find the current i(t) through the inductance. Find the voltage across resistance and inductance separately at the time of switching the supply and when sufficient time has elapsed after switching the source.* *(Ans. 0V, 1V, 1V, 0V)*

Q.4 *Consider the series R – L circuit connected to a battery and a switch as shown in fig.1.131. Switch S_1 closes at t = 1 second; switch S_2 closes at t = 2.5 second. (i) Find the current i (t) in the resistance R; (ii) Plot the current i (t) for 0 £ t £ 5 second to the scale. Assume R/L equal to 100.*

Q.5 *A coil having L = 6·4 H and R = 8 W is connected to a constant 200 V supply source. How long does it take the voltage across the resistor to reach 100 V ?* *(Ans. 0.5545 s)*

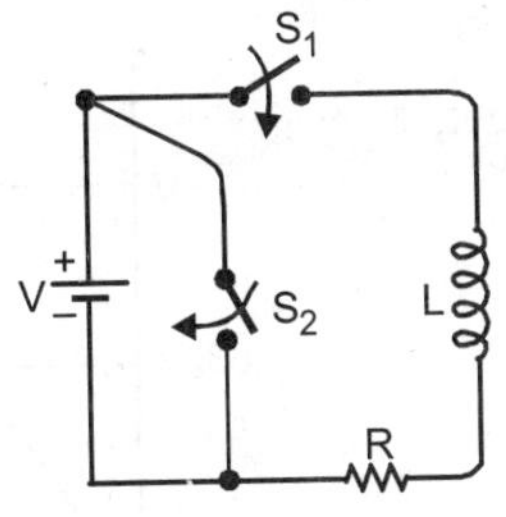

Fig.1.135:

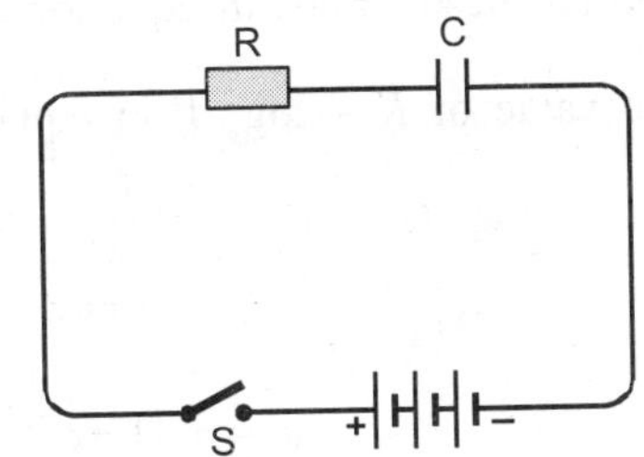

Fig.1.136: Closing of RC circuit.

1.28 ANALYSIS OF RC CIRCUITS *(Closing of RC circuit)*

In an RC circuit, when the circuit is closed, the capacitor is charged from a fixed value of d.c. supply. The rate of charging is controlled by the resistor connected in series with the capacitor.

Consider a capacitor of capacitance C farad connected in series with a resistor of R ohms, to be charged from a d.c. source of V volt. When the switch S is open the voltage across the capacitor is zero as shown in fig.32. However, when the switch is closed, the charge on the capacitor starts rising and so does the potential difference across it.

At the switching instant: At the instant when switch is closed, the voltage built up across the capacitor is zero. Therefore, the entire voltage V is dropped across the resistor R and the charging current is maximum (say I_m).

$\therefore$ At the switching instant, current $I_m = \dfrac{V}{R}$ ...(*i*)

At any instant during charging: After closing the switch, the capacitor starts charging and the voltage across it increases gradually. This decreases the charging current. Let at any instant during charging,

v = p.d. across the capacitor

i = charging current

q = charge on the capacitor = $C\,v$

According to Kirchhoff's second law,

Applied voltage = voltage across capacitor + voltage across resistor

$$V = v + i\,R \qquad ...(ii)$$

or $$V = v + C\frac{dv}{dt}R \qquad \left(\because i = \frac{dq}{dt} = \frac{d}{dt}(Cv) = C\frac{dv}{dt}\right)$$

or $$V = v + C\,R\,\frac{dv}{dt}$$

or $$-\frac{dv}{V-v} = -\frac{dt}{RC}$$

Integrating both sides w.r.t. 't';

$$\int -\frac{dv}{V-v} = \int -\frac{dt}{RC}$$

$$\log_e (V - v) = -\frac{t}{RC} + K \qquad ...(iii)$$

where K is a constant of integration whose value can be determined from the initial conditions. At the instant of closing the switch S, $t = 0$; $v = 0$

$\therefore$ Substituting these values in equation (*iii*), we get, $\log_e V = K$

Putting the value of $K = \log_e V$ in equation (*iii*), we get,

$$\log_e (V - v) = -\frac{t}{RC} + \log_e V$$

or $$\log_e \frac{V-v}{V} = e^{-t/RC}$$

or $$v = V\left(1-e^{-t/RC}\right)$$

The above expression shows the variation of voltage (v) developed across the capacitor w.r.t. time (t). It is represented graphically in fig.1.137.

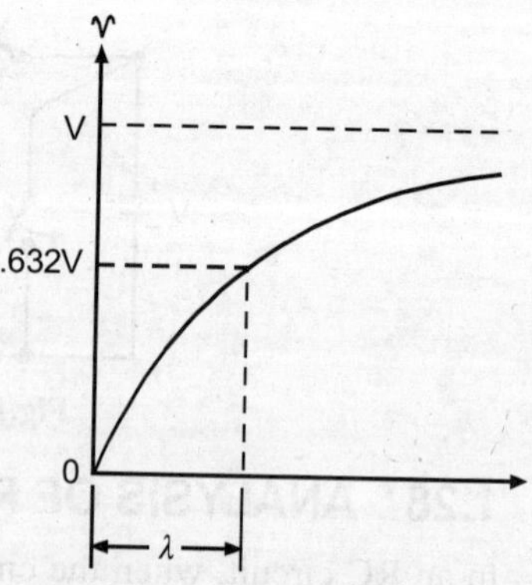

Fig.1.137: Graph between v and i when circuit is closed.

The voltage across the capacitor is increasing following an* exponential law. The other quantities vary as under:

* A quantity is said to vary following exponential law, if it changes to half of its value in say t seconds and in the next t seconds the value changes to half of the new value and so on.

(*i*) *Variation of charge stored in the capacitor with time*

We know that $v = q/C$ and $V = Q/C$;

$$\therefore \quad \frac{q}{C} = \frac{Q}{C}\left(1-e^{-t/RC}\right)$$

$$\text{or} \quad q = Q\left(1-e^{-t/RC}\right) \quad ..(v)$$

where q is the charge at any instant 't' and Q is the full charge on the capacitor.

(*ii*) *Variation of charging with time*

From equation (*ii*), we have, $V - v = iR$

From equation (*iv*), we have, $V - v = V\,e^{-t/RC}$

$$\therefore \quad i\,R = V\,e^{-t/RC}$$

$$\text{or} \quad i = \frac{V}{R}e^{-t/RC}$$

From equation (*i*) above, $I_m = \dfrac{V}{R}$

$$\therefore \quad i = I_m\;e^{-t/RC} \quad ...(iv)$$

where I_m (= V/R) is the initial charging current.

This charging current decreases following the exponential law and is represented graphically in fig.1.138.

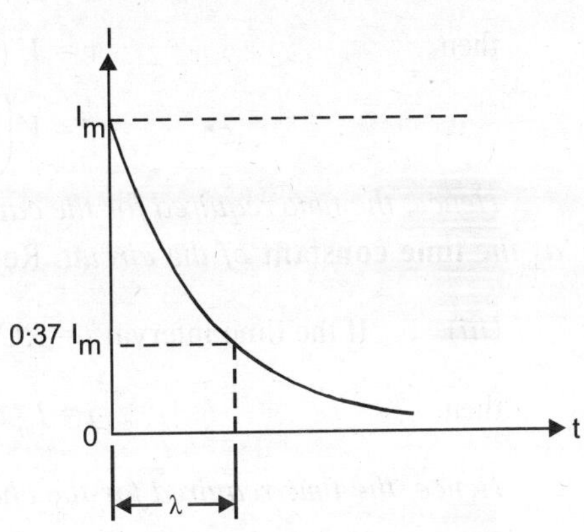

Fig.1.138: Graph between i and t when circuit is opened.

1.29 TIME CONSTANT

The rise of p.d. across the capacitor is given by the equation;

$$v = V\,(1 - e^{-t/RC})$$

In this equation, the exponent of e is t/RC. The exponent of e must be a number, therefore, the quantity RC should have the ** dimensions of time. Hence, the quantity RC is called the *time constant* of the circuit and affects the charging (or discharging) time. It is generally represented by λ (or T or τ).

$\therefore$ Time constant, $\lambda = RC$ second.

The timc constant will be in seconds if R and C are in ohm and farad respectively. The time constant λ affects the time period of the charging process of capacitor. The larger the time constant, the longer the charging (or discharging) period and *vice-versa*.

In fact, the time constant indicates the rate at which a capacitor is charged. It can be defined in the following ways:

(*i*) We have seen the expression in the previous article

$$V = v + CR\,\frac{dv}{dt}$$

** Units of RC = $\dfrac{\text{volt}}{\text{ampere}} \times \dfrac{\text{coulomb}}{\text{volt}} = \dfrac{\text{volt}}{\text{coulomb / sec}} \times \dfrac{\text{coulomb}}{\text{volt}}$ = second.

At the instant of closing the switch S, $v = 0$

$$\therefore \qquad V = CR\frac{dv}{dt} \quad \text{or} \quad \frac{dv}{dt} = \frac{V}{CR}$$

i.e. Rate of rise of voltage $= \dfrac{V}{CR}$

If this rate of rise of voltage could continues the capacitor voltage will reach to final value V in time $= V \div V/RC = RC$ second *i.e.* time constant λ

Hence, the time required for the capacitor voltage to rise to its final steady value (V volt) if it continued rising at its initial rate (i.e. V/RC) is called the **time constant** *of the circuit.*

(*ii*) If the time internal $t = RC$ (*i.e.* λ),

then

$$v = V(1 - e^{-t/t}) = V(1 - e^{-1})$$

$$= V\left(1 - \frac{1}{e}\right) = V\left(1 - \frac{1}{2 \cdot 718}\right) = 0 \cdot 632\, V$$

Hence, the time required for the capacitor voltage to rise to 0·632 of its final steady value is known as the **time constant** *of the circuit.* Refer fig.1.137.

(*iii*) If the time interval $t = RC$ (*i.e.* λ))

then

$$i = I_m\, e^{-t/t} = I_m\, e^{-1} = I_m \times \frac{1}{e} = \frac{I_m}{2 \cdot 718} = 0 \cdot 37\, I_m$$

Hence, the time required for the charging current to fall to 0·37 of its initial maximum value Im (= V/R) is called the taken constant of the circuit. Refer fig.1.138.

1.30 DISCHARGING OF A CAPACITOR

Consider a capacitor of C farad connected in series with a resistor of R ohm and a switch S. When the switch is open, the capacitor is charged to V volt and the current in the circuit is zero as shown in fig.1.139.

When the switch S is closed, the charge on the capacitor starts decreasing and so does the voltage across it.

Let at any instant during discharging;

v = p.d. across the capacitor

i = discharging current

q = charge on the capacitor $= Cv$

According to Kirchhoff's second law, $0 = v + iR$

$$\text{or} \qquad 0 = v + CR\frac{dv}{dt}$$

$$\text{or} \qquad v = -CR\frac{dv}{dt}$$

$$\text{or} \qquad \frac{dv}{v} = -\frac{dt}{RC}$$

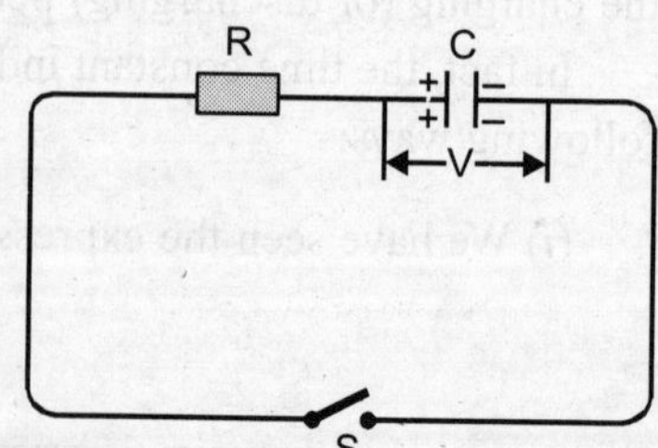

Fig.1.139: Charged capacitor with open switch without supply

Integrating both sides, we get

$$\int \frac{dv}{v} = -\frac{1}{RC}\int dt \;\therefore\; \log_e v = -\frac{t}{RC} + K \quad ...(i)$$

where K is a constant of integration whose value can be determined from the initial conditions. At the instant of closing the switch S, $t = 0$; $v = V$.

$\therefore$ Substituting these values in equation (*i*), we get,

$$\log_e V = K$$

Putting the value of $K = \log_e V$ in equation (*i*), we get,

$$\log_e v = -\frac{t}{RC} + \log_e V$$

or $$\log_e \frac{v}{V} = -\frac{t}{RC}$$

or $$\frac{v}{V} = e^{-t/RC}$$

$\therefore$ $$v = V e^{-t/RC} \quad ...(ii)$$

The above expression shows the variation of voltage (v) across the capacitor w.r.t. time (t) while discharging and is represented graphically in fig.1.140.

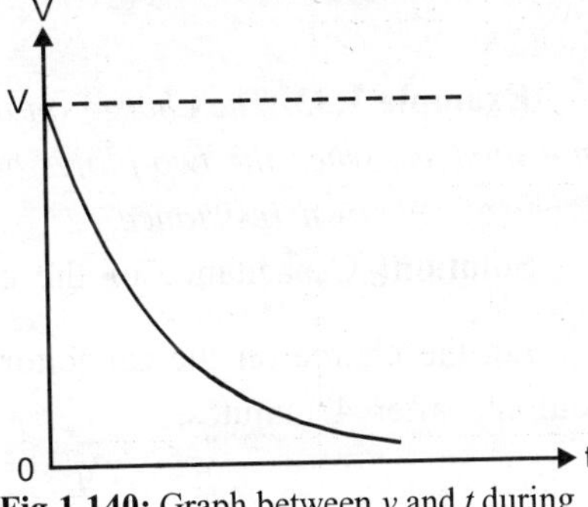

Fig.1.140: Graph between v and t during discharging of capacitor

The other quantities vary as under:

(*i*) *Variations of charge on the capacitor with time*

Since $v = \frac{q}{C}$ and $V = \frac{Q}{C}$

Substituting these values in equation (*ii*), we get,

$$\frac{q}{C} = \frac{Q}{C} e^{-t/RC}$$

or $$q = Q e^{-t/RC}$$

(*ii*) *Variation of discharging current with time*

We know that $0 = v + iR$

or $iR = -v$

or $i = -\frac{V}{R} e^{-t/RC} = -I_m e^{-t/RC}$

where $I_m = \frac{V}{R}$ = initial current

The variation of discharging current in shown graphically in fig.1.141.

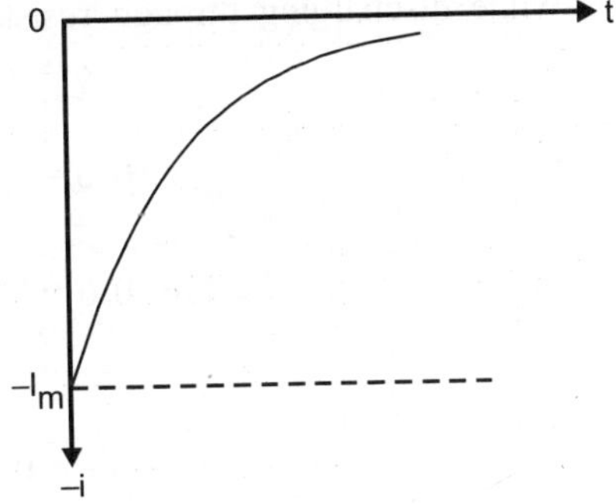

Fig.1.141: Graph between t and i during discharging of capacitor.

Example 1.32. *A capacitor is charged by a d.c. source through a resistor of 1.2 mega ohm. If in one second the p.d. across the capacitor reaches 75% of the final value, calculate the capacity of the capacitor.*

Solution: Series resistance, $R = 1.2\ \text{M}\ \Omega = 1.2 \times 10^6\ \Omega$; Time, $t = 1$ second

P.d. across capacitor, $v = \dfrac{75}{100} V = 0{\cdot}75\ V$

While charging, the p.d. across capacitor is given by the equation

$$v = V\left(1 - e^{-t/RC}\right)$$

$$0{\cdot}75\ V = V\left(1 - e^{-t/RC}\right) \quad \text{or} \quad 0{\cdot}75 = \left(1 - e^{-t/RC}\right)$$

or
$$\log_e 0{\cdot}25 = \frac{-t}{RC} \log_e e$$

or
$$-1{\cdot}3863 = -\frac{1}{1.2 \times 10^6 C} \times 1 \quad \text{or} \quad C = \frac{1}{1.3863 \times 1.2 \times 10^{-6}} = \mathbf{0{\cdot}6011\ \mu F}\ (\textit{Ans.})$$

Example 1.33. *The charge on a perfect capacitor of capacitance 2 μ F falls to 60% of its value in 4 minutes, when the two plates of the capacitor are joined by an unknown resistance. What is the value of unknown resistance.*

Solution: Capacitance of the capacitor, $C = 2\mu\text{F} = 2 \times 10^{-6}$ F

Let the charge on the capacitor be Q coulomb when it is fully charged. This charge falls to q coulomb after 4 minutes.

$$\frac{q}{Q} = 60\% = 0{\cdot}6$$

and
$$t = 4 \times 60 = 240 \text{ second}$$

while discharging through resistance of R ohm,

$$q = Q\ e^{-t/RC}$$

or
$$0{\cdot}6 = e^{-240/R \times 2 \times 10^{-6}}$$

or
$$\log_e 0{\cdot}6 = \frac{240}{R \times 2 \times 10^{-6}}$$

or
$$R = \frac{240}{0.5108 \times 10^{-6}} = 234{\cdot}9 \times 10^6 \text{ ohm} = \mathbf{234{\cdot}9\ M\ \Omega}\ (\textit{Ans.})$$

Example 1.34 *An 16 μ F capacitor is connected in series with a 1 M Ω resistor across a 240 V d.c. supply. Calculate (i) the time constant, (ii) the initial charging current, (iii) the time taken for the p.d. across the capacitor to grow to 168 V and (iv) the current and the p.d. across the capacitor in 4 second after it is connected to the supply.*

Solution: Time constant, $\lambda = RC = 1 \times 10^6 \times 16 \times 10^{-6}$ = **16 second** (*Ans.*)

Initial charging current, $I_m = V/R = 240/1 \times 10^6$ = **240 μA** (*Ans.*).

$$v = V(1 - e^{-t/RC})$$

$$168 = 240\,(1 - e^{-t/16})$$

or $e^{-t/16} = 0{\cdot}3$ or $e^{t/16} = 1/0{\cdot}3$ or $e^{t/16} = 3.33$

or $(t/16)\log_e e = \log_e 3.33$

or $t = 16 \log_e 3.33 = 16 \times 1{\cdot}203$ = **19.247 second** (*Ans.*)

$i = I_m\, e^{-t/RC} = 240 \times 10^{-6}\, e^{-4/16}$ = **308·17 µA** (*Ans.*)

Now, $v = 240\,(1 - e^{-4/16})$ = **53.1 V** (*Ans.*)

PRACTICE EXERCISE

Q.1 *A capacitor is charged by a d.c. source through a resistor of one mega ohm. If in one second the p.d. across the capacitor reaches 80% of the final value, calculate the capacity of the capacitor.* (*Ans,0·6213 mF*)

Q.2. *The charge on a perfect capacitor of capacitance 1 m F falls to 50% of its value in 5 minutes, when the two plates of the capacitor are joined by an unknown resistance. What is the value of unknown resistance.* (*Ans.432·9 M W*)

Q.3. *An 8 m F capacitor is connected in series with a 0·5 M W resistor across a 200 V d.c. supply. Calculate (i) the time constant, (ii) the initial charging current, (iii) the time taken for the p.d. across the capacitor to grow to 160 V and (iv) the current and the p.d. across the capacitor in 4 second after it is connected to the supply.*

(*Ans. 4 second ,400 mA, 6·4376 second, 147·15 mA,126·424 V*)

SUMMARY

1. *Circuit Elements:* In general, the circuit elements can be categorised as:
   (i) Active and passive elements. (ii) Unilateral and bilateral elements.
   (iii) Linear and non-linear elements. (iv) Lumped and distributed elements.
2. *Resistors:* Resistor is a component, used to limit the amount of current or divide the voltage in an electronic circuit. The ability of a resistor to oppose the current is called resistance.
3. *Classification of resistors:* The resistors may be classified as fixed and variable resistors.
4. *Inductors:* A component that opposes the change of current in circuit is called an inductor. The ability of an inductor or coil due to which it opposes the change of current flowing through it is called its inductance.
5. *Classification Inductors:* Inductors can be classified broadly as fixed and variable inductors.
6. *Capacitors:* Two conducting plates separated by an insulating material (or dielectric) forms a capacitor. Capacitor has the ability to store electric charge. The capacity of a capacitor to store charge per unit potential difference is called its capacitance.

7. *Classification of capacitors*: The capacitors can also be classified broadly as fixed and variable capacitors.

8. *Voltage and current sources:* To deliver electrical energy to electrical/electronic circuits, a source is required. The source may be either a d.c. (direct current) source or an a.c. (alternating current) source.

9. *Constant voltage source:* An electrical source is called a constant voltage source if it supplies power at almost constant voltage to a load irrespective to its value.

10. *Constant Current source:* An electrical source is called a constant current source if it supplies power at almost constant current to a load irrespective to its value.

11. *Source transformation (Conversion of Voltage Source to Current Source and Vice-Versa):* It has been seen that, the same electrical source behaves as a constant voltage source and as a constant current source.

12. *Ohm's law:* This law states that the current flowing between any two points of a conductor (or circuit) is directly proportional to the potential difference across them, provided physical conditions (i.e. temperature etc.) remain the same.

13. *D.C circuit:* The closed path for the flow of direct current is called d.c. circuit.

14. *Effective value of the resistance (R)*

(*i*) in series circuit; $R = R_1 + R_2 + \ldots\ldots + R_n$

(*ii*) in parallel circuit; $\frac{1}{R} = \frac{1}{R_1} + \frac{1}{R_2} + \ldots\ldots + \frac{1}{R_n}$

15. *Kirchhoff's laws*: KCL states that $\Sigma I = 0$ at a junction. KVL states that $\Sigma E + \Sigma IR = 0$ in a loop. While applying KCL, incoming currents are taken as +ve and outgoing currents as – ve.

16. *Superposition theorem*: In this method, If there are two of more than two sources, current is determined in the required branch by considering each source separately and then resultant value is determined by superimposing them.

17. *Thevenin's theorem*: According to this theorem, to determine current in a resistor, the resistor is removed and an open circuit voltage across the two terminals is determined called thevenin voltage E_{th}. Then, resistance of the whole network is determined across the terminals called Thevenin resistance R_{th} replacing all the voltage sources by their internal resistances, $I = E_{th}/(R_{th} + R_L)$.

18. *Norton's theorem*: It is similar to Thevenin's theorem but in this case, the network is reduced to a current source having a current output of I_N when the given terminals are short circuited in parallel with a resistance R_N similar to R_{th}. Then, $I = I_N \; R_N/(R_N + R_L)$.

19. Closing and opening of an inductive circuit: $V = iR + L\frac{di}{dt}$

(*i*) *Rise of current when circuit is closed*: at any instant.

$i = I\,(1 - e^{-Rt/L})$ where $I = V/R$

(*ii*) *Decay of current when circuit is opened*: at any instant

$$i = I\, e^{-Rt/L} \text{ where } I = V/R$$

20. *Time constant*: $\lambda = L/R$ second

21. Charging of a capacitor

(*i*) Current at the switching instant, $I_m = \dfrac{V}{R}$

(*ii*) Voltage across capacitor at any instant, $v = V\,(1 - e^{-t/RC})$

(*iii*) Charge stored in the capacitor at any instant, $q = Q\,(1 - e^{-t/RC})$

(*iv*) Charging current at any instant, $i = I_m\,(1 - e^{-t/RC})$

22. Discharging of a capacitor

(*i*) Voltage across capacitor at any instant, $v = V\,e^{-t/RC}$.

(*ii*) Charge on the capacitor at any instant, $q = Q\,e^{-t/RC}$.

(*iii*) Discharge current at any instant, $i = -\,I_m\,e^{-t/RC}$.

SHORT ANSWER QUESTIONS

1. What do you mean by elements of an electric circuit?
2. How will you differentiate between active and passive elements?
3. How will you differentiate between unilateral and bilateral elements?
4. How will you differentiate between linear and non-linear elements?
5. How will you differentiate between lumped and distributed elements?
6. What are resistors? How will you classify them?
7. What are inductors? How will you classify them?
8. What are capacitors? How will you classify them?
9. What do you mean by constant voltage source?
10. How will you differentiate between ideal and real constant voltage source?
11. What do you mean by constant current source?
12. How will you differentiate between ideal and real constant current source?
13. State Ohm's law and give its limitations.
14. In series circuits, show that effective resistance is sum of the individual resistances.
15. In parallel circuits show that $\dfrac{1}{R_{eff}} = \dfrac{1}{R_1} + \dfrac{1}{R_2} + \text{.........} + \dfrac{1}{R_n}$
16. How will you differentiate between node and funcation in an electric network?
17. How will you differentiate between loop and mesh in an electric network?

18. State Kirchhoff's current and voltage laws.

19. State superposition theorem.

20. State Thevenin's theorem.

21. State Norton's theorem.

22. How will you justify that the unit of *L/R* is second?

23. How will you justify that the unit of *RC* is second?

NUMERICAL FOR PRACTICE

1. There are two wires *A* and *B* of the same material. *A* is 10 times longer than *B* and has twice the cross-section as that of *B*. If the resistance of *A* is one ohm, what is the resistance of *B* ? **(Ans.** *1/5 ohm*)

2. Calculate the resistance between the points *X* and *Y* in network shown in fig.1.142. All the resistors are given in ohms. **(Ans.** *4 ohm*)

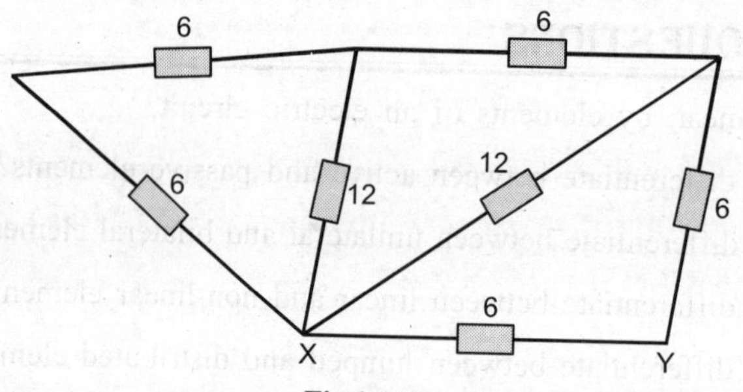

Fig.1.142

3. A resistance of 15 Ω is connected in series with two resistances each of 30 Ω arranged in parallel. What resistance must be connected across the parallel combination that the total current taken shall be 3 A with 75 V applied. **(Ans.** *30 ohm*)

4. A letter *A* is constructed of a uniform wire of resistance 3 ohm per cm. The sides of the letter are 20 cm long and the cross piece is 10 cm long while the apex angle is 60°. Find the resistance of the letter between the two ends of the legs. **(Ans.** *80 Ω*

5. Two batteries *A* and *B* connected in parallel and load of 5Ω is connected across their terminals. A has an e.m.f. of 6 V and an internal resistance of 1Ω; *B* has an e.m.f. of 4V and an internal resistance of 0.5 Ω. Using Kirchhoff's laws determine the value and direction of flow of current in each battery and in the external resistance. Also determine the potential difference across the external resistance.

(Ans. *1·625 A charging; 0·75 A discharging; 0·875 A; 4·375 V*)

6. A wheatstone bridge *ABCD* is arranged as follows:

Resistance between *A – B*, *B – C*, *C – D*, *D – A* and *B – D* are 20, 4, 16, 8 and 10 ohm respectively. A 200 V supply is connected between terminals *A* and *C*. Determine the currents in branches *AB* and *AD* of the circuit and total current taken from the supply. **(Ans.** *7·44 A; 11·905 A; 19·345 A*)

7. In the circuit shown in fig.1.143, determine the current supplied by the d.c. generator of 280V. **(Ans.** *4 A*)

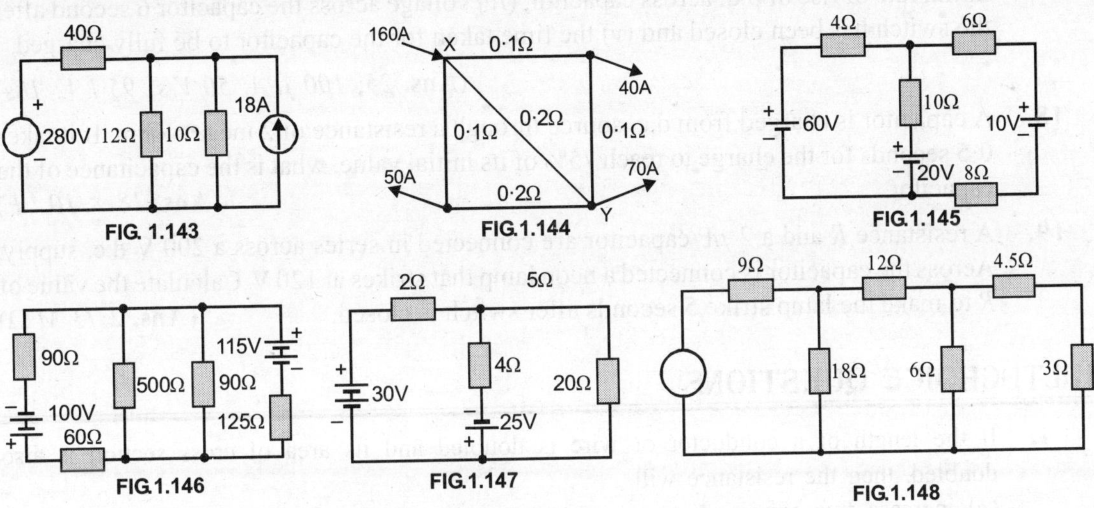

FIG. 1.143 FIG.1.144 FIG.1.145 FIG.1.146 FIG.1.147 FIG.1.148

8. By using Kirchhoff's laws, determine current in *XY* in the circuit shown in fig.1.144. **(Ans.** *40 A*)

9. Using superposition theorem, find the current in all the branches of the circuit shown in fig.1.145. **(Ans.** *4·4915 A, 2·2881 A, 2·2084 A*)

10. Using superposition theorem, determine potential difference across 60Ω resistor in the network shown in fig.1.146. **(Ans.** *87·3 V*)

11. Using Thevenin's theorem, determine the current is 20 ohm resistor of the network shown in fig.1.147. **(Ans.** *0·433 A*)

12. Using Thevenin's theorem, determine the current in 3 ohm resistor in the network shown in fig.1.148. What will be the value of current if this resistance is replaced by 15 ohm resistor? **(Ans.** *2 A, 1 A*)

13. Using Norton's theorem, determine current in 60Ω resistor of the network shown in fig.1.146. **(Ans.** *0·433 A*)

14. Using Norton's theorem, determine current in 3Ω resistor of the network shown in fig.1.148. **(Ans.** *2 A*)

15. The resistance and inductance of a coil are 5 Ω and 0·15 m H respectively. What potential difference exists at the terminals of the solenoid at the instant when the current is 3 A but increasing at the rate of 10^5 A/s. **(Ans.** *30 V*)

16. The resistance and inductance of an inductive coil are 5Ω and 20 H respectively. At the instant of closing the supply, the current increases at the rate of 4 A/s. Calculate (*i*) the applied voltage (*ii*) the rate of growth of current when the current is 5 A (*iii*) the time taken to reach 10 A after closing the circuit and (*iv*) the time taken to reach 10 A while decaying after reaching the maximum value. **(Ans.** *80 V; 2·75 A/s; 3·923 s; 1·88 s*)

17. A 2μF capacitor is connected, by closing switch, to a supply of 100 V through a 1 M ohm series resistance. Calculate (*i*) the time constants (*ii*) initial charging current, (*iii*) the initial rate of rise of p.d. across capacitor, (*iv*) voltage across the capacitor 6 second after the switch has been closed and (*v*) the time taken for the capacitor to be fully charged.

(**Ans.** *25; 100 μ A; 50 V/s; 95·1 V; 10s*)

18. A capacitor is charged from d.c. source through a resistance of 2 mega – ohm. If it takes 0·5 seconds for the charge to reach 75% of its initial value, what is the capacitance of the capacitor.

(**Ans.** *18 × 10^{-4} F*)

19. A resistance *R* and a 2 μF capacitor are connected in series across a 200 V d.c. supply. Across the capacitor is connected a neon lamp that strikes at 120 V. Calculate the value of *R* to make the lamp strike 5 seconds after switch is closed. (**Ans.** 2·73 *M* Ω)

MULTI-CHOICE QUESTIONS

1. If the length of a conductor or wire is doubled and its area of cross section is also doubled, then the resistance will

(*a*) increase four times
(*b*) remains uncharged
(*c*) decrease to four times
(*d*) change at random

2. A certain piece of aluminium is to be shaped into a conductor of minimum resistance, its length and cross-sectional area shall be respectively.

(*a*) *L and A* (*b*) *2L and A/2* (*c*) *L/2 and 2A* (*d*) *L/2 and A/2*

3. A man has three resistances, each of value 1/3 ohm. What is the minimum resistance he can obain by combining them ?

(*a*) 1 ohm (*b*) 3 ohm (*c*) 1/6 ohm (*d*) 1/9 ohm

4. If n identical resistors are connected in series, their total value is K ohm. What will be their effective value if they are now connected in parallel.

(*a*) K/n^2 (*b*) K^2/n (*c*) K/n (*d*) n^2/K

5. A parallel arrangement of 3 ohm and 6 ohm resistors is placed in series with an 8 ohm resistor. If p.d. of 60 V is connected across the whole circuit, the current in the three ohm resistor is

(*a*) 6 A (*b*) 4 A (*c*) 2 A (*d*) 10 A

6. Kirchhoff's first law states that at a junction in an electric network

(*a*) Algebraic sum of all the current's is zero.
(*b*) Algebraic sum of all the *emfs* is zero.
(*c*) Algebraic sum of all the voltage drops is zero.
(*d*) both (a) and (b).

7. An ideal voltage sources should have

(*a*) zero internal resistance
(*b*) infinite internal reistance
(*c*) large value of e.m.f.
(*d*) low value of current

8. The polarity of the voltage drop across a resistor is determine by

(*a*) value of resistor
(*b*) value of current
(*c*) direction of current
(*d*) all *a*, *b*, and *c*

9. An ideal current source should have
   (*a*) zero internal resistance (*b*) infinite internal resistance
   (*c*) large value of e.m.f. (*d*) none of above

10. While determining R^{th} of a network
    (*a*) The voltage and current source should be short circuited.
    (*b*) All current and voltage sources must be open circuited.
    (*c*) The voltage sources should be open circuit and current sources should be short circuited.
    (*d*) All sources should be replaced by their source resistance.

11. A network which does not have either voltage sources or current sources is called
    (*a*) Active network (*b*) Passive network (*c*) Resistive network (*d*) Dummy network

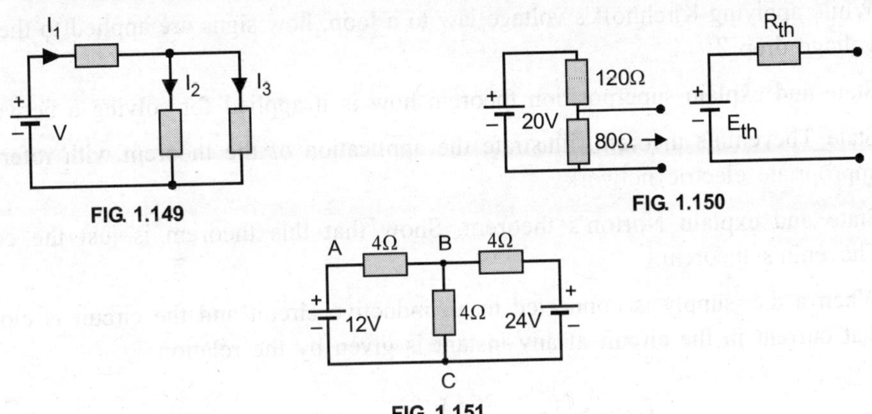

FIG. 1.149

FIG. 1.150

FIG. 1.151

12. In the circuit shown in fig.1.149, according to Kirchhoff's first law which of the following relation is correct
    (*a*) $I_1 + I_2 = I_3$ (*b*) $I_1 + I_3 = I_2$ (*c*) $I_1 - I_2 = I_3$ (*d*) all of above

13. A voltage divider and its Thevenin's equivalent circuit is shown in fig.1.150. What will be the value of E_{th} and R_{th}
    (*a*) 8V, 48 Ω (*b*) 4V, 48 Ω (*c*) 4V, 24 Ω (*d*) 8V, 24 Ω

14. *In the circuit shown in fig.1.151, the power consumed in 2 ohm resistor connected across A—B is*
    (*a*) 16 W (*b*) 0 W (*c*) 36 W (*d*) 64 W

15. *In an inductive coil, the delay in change of current depends upon the value of*
    (*a*) resistance (*b*) inductance (*c*) both *a* and *b* (*d*) none of above

TEST QUESTIONS

1. What are the various electric circuit elements? How will you distinguish between.
   (*i*) lincar and non-linear elements (*ii*) Active and passive elements.

2. Giving examples, explain what do you mean by
   (i) Fixed and variable resistors. (ii) Fixed and variable capacitors

3. Explain, what do you mean by real voltage and current sources.
4. State and explain Ohm's law. Give its limitations.
5. Show that:

   (*i*) in series circuit, the effective value of resistance; $R = R_1 + R_2 + \text{..........} + R_n$.

   (*ii*) in parallel circuit, the reciprocal of the effective value of resistance, is equal to the sum of the reciprocal of individual resistances.
6. Distinguish between

   (*i*) Node and junction; (*ii*) loop and mesh
7. State and explain Kirchhoff's current and voltage law.
8. While applying Kirchhoff's voltage law to a loop, how signs are applied to the e.m.f. and voltage drop ?
9. State and explain superposition theorem how is it applied for solving a network ?
10. State Thevenin's theorem, illustrate the application of the theorem with reference to an appropriate electric network.
11. State and explain Norton's theorem. Show that this theorem is just the converse of Thevenin's theorem.
12. When a d.c. supply is connected to an inductive circuit and the circuit is closed, prove that current in the circuit at any instant is given by the relation

$$i = \frac{V}{R}\left(1 - e^{-Rt/L}\right)$$

13. A d.c. voltage V is applied across a circuit consisting of resistance R ohms in series with a capacitance C farads. Derive expression for variation of voltage across C with time.
14. A condenser and a resistor are connected in series with a d.c. source. Find the charge on the condenser at a time 't' after switching the circuit and hence determine the rate of growth of charge ?
15. A condenser and a resistor are connected in series with a d.c. source of V volt. Derive an expression for the voltage across the capacitor after 't' second when the capacitor is discharged.

ANSWER TO MULTI-CHOICE QUESTIONS

Multi-Choice questions:

1. (*b*)	2. (*c*)	3. (*d*)	4. (*a*)	5. (*c*)
6. (*a*)	7. (*a*)	8. (*c*)	9. (*b*)	10. (*d*)
11. (*b*)	12. (*c*)	13. (*d*)	14. (*b*)	15. (*c*)

AC CIRCUITS

LEARNING OBJECTIVES

- What is ac (alternating current)? How it is different to dc.
- Why sinusoidal alternating voltage and current is preferred?
- How sinusoidal alternating voltage and current is produced?
- What are the terns such as frequency, tune period, cycle, alternation, wave form used in ac system?
- What is the meaning of instantaneous, average, *rms* and peak values of alternating quantities?
- Why alternating quantities (voltage and current) are the vector quantities?
- What is phase and phase difference?
- How alternating quantities are added and subtracted?
- How a pure resistor, inductor or capacitor behaves when connected across ac supply?
- How much power is consumed by these elements?
- How an ac circuit behaves when it carries resistance and inductance, resistance and capacitance or resistance, inductance and capacitance is series?
- How much power is consumed by such circuits?
- What is power factor and how it affects the power system?
- What is a series resonant circuit and its applications,
- How to analyse ac parallel circuits?
- What is a parallel resonant circuit and how its behavior is different to a series resonant circuit?
- How the power factor of an ac circuit is improved?
- Three-phase system and its advantages.
- How 3-phase power is generated and how the three-phases are abbreviated?
- What are star and delta connections and how line and phase voltages are related?

INTRODUCTION

The use of d.c. is limited to a few applications *e.g.* electroplating, charging of batteries, electric traction, electronic circuits etc. For large scale power generation, transmission, distribution and utilisation, a.c. system is invariably adopted. In a.c. system, voltage acting in the circuit changes polarity and magnitude at regular interval of time and hence the current.

Alternating supply is invariably used for domestic and industrial applications. *The path for the flow of alternating current is called an* **a.c. circuit**. In d.c. circuits, the opposition to the flow of current is only resistance of the circuit. Whereas, in a.c. circuits, the opposition to the flow of current is due to resistance (R), inductive reactance ($X_L = 2p\ fL$) and capacitive reactance ($X_C = 1/2\ p\ fC$) of the circuit.

In a.c. circuits frequency plays an important role.

Although single-phase system is employed for the operation of almost all the domestic and commercial appliances *e.g.* lamps, fans, electric irons, refrigerators, TV sets, washing machines, exhaust fans, computers etc. But it has its own limitations in the field of generation, transmission, distribution and industrial applications. Due to this it has been replaced by poly-phase system.

In this chapter, we shall confine our attention to ac supply system, ac circuits and their practical utility in the field of engineering.

2.1 ALTERNATING VOLTAGE AND CURRENT

A voltage that changes its polarity and magnitude at regular intervals of time is called an *alternating voltage*.

When an alternating voltage source is connected across a load resistor R, as shown in fig.2.1, the current flows through it in one direction and then in opposite direction when the polarity is reversed.

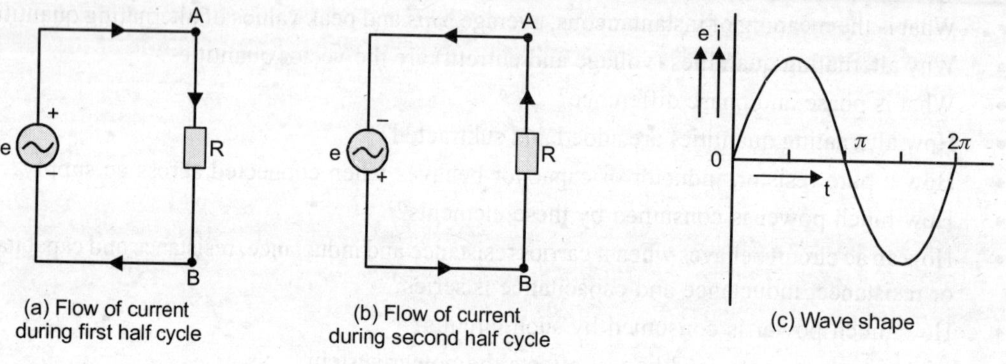

Fig.2.1: Alternating voltage and current.

Fig.2.1 (*c*) shows the wave shape of the source voltage (representing the variation of voltage w.r.t. time) and current flowing through the circuit (*i.e.* load resistor R).

Wave form: The graph representing the manner in which an alternating voltage or current changes w.r.t. time is known as *wave-form* or *wave-shape*. While plotting a graph, usually the instantaneous values of the alternating quantities are taken along *y-axis* and time along *x-axis*. The alternating voltage or current may vary in different manner, as shown in fig.2.2, accordingly their wave shapes are named in different ways such as irregular wave; triangular wave; square wave; periodic wave; saw-tooth wave; sine wave etc.

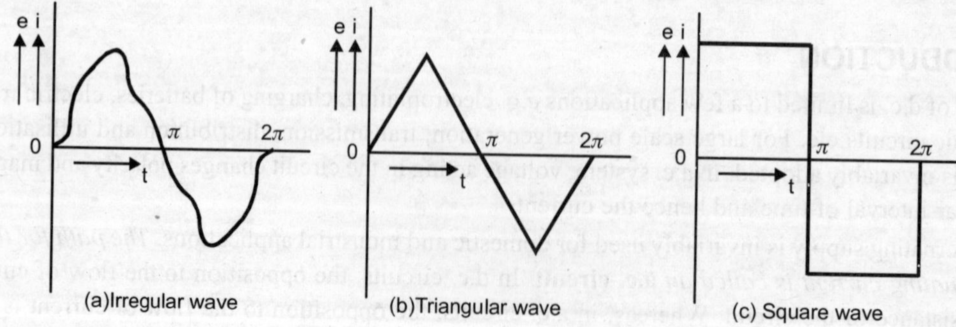

Fig.2.2: Wave shape of ac quantities (i)

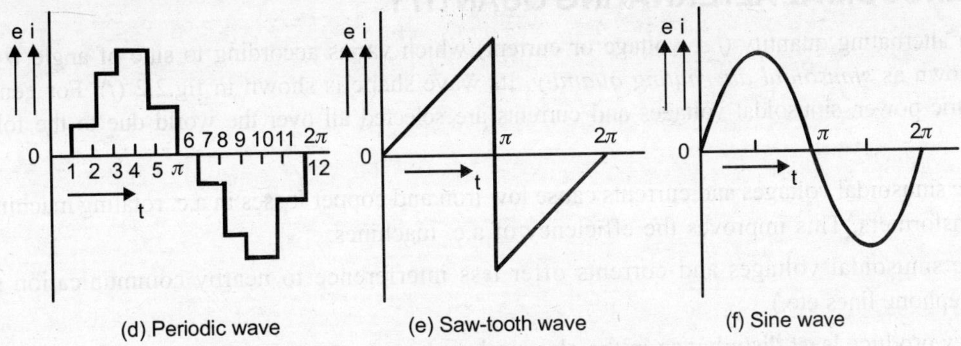

(d) Periodic wave (e) Saw-tooth wave (f) Sine wave

Fig.2.2: Wave shape of ac quantities

2.2 ADVANTEGES OF AC SYSTEM OVER DC SYSTEM

(*i*) The alternating voltage can be stepped up and stepped down efficiently by means of transformer. To transmit huge power over a long distance, the voltages are stepped up (upto 400 kV) for economical reasons at the generating stations. Whereas, they are stepped down to a very low level (400/230 V) for utilisation of electrical energy from safety point of view.

(*ii*) The *a.c.* motors (*i.e.* induction motors) are cheaper in cost, simple in construction, more efficient and robust as compared to d.c. motors.

(*iii*) The switchgear (*e.g.* switches, circuit breakers etc.) for *a.c.* system is simpler than *d.c.* system.

Thus, *a.c.* system is universally adopted for generation, transmission, distribution and utilisation of electrical energy.

2.3 DIFFERENCE BETWEEN AC AND DC

AC current	*DC* current
1. Alternating current reverses periodically and its magnitude changes.	1. Direct current flows only in one direction and remains unaltered.
2. Amplitude and polarities are varying continuously.	2. Amplitude and polarities are fixed.
3. It has a particular frequency.	3. It is independent of frequency.
4. *AC* can be generated at higher voltages.	4. *DC* cannot be generated at high voltages because of commutation difficulties.
5. In case of *ac*, the cost of generation is less.	5. In case of *dc*, the cost of generation is more.
6. Alternating voltage can be increased (stepped up) or decreased (stepped down) easily with the help of a transformer.	6. Direction voltage cannot be increased or decreased easily.
7. *AC* motors are of less cost, more robust and durable.	7. *DC* motors are costly and less durable.
8. The maintenance cost of ac equipment and appliances is less.	8. The maintenance cost of *dc* equipment and appliances is more.
9. *AC* cannot be used directly for electroplating	9. Only *dc* can be used directly for electroplating.
10. The speed of *ac* motors cannot be controlled easily.	10. The speed control of *dc* motor is very easy and economical.

2.4 SINUSOIDAL ALTERNATING QUANTITY

An alternating quantity (*i.e.* voltage or current) which varies according to sine of angle θ ($\theta = \omega t$) is known as *sinusoidal alternating quantity*. Its wave shape is shown in fig.2.2 (*f*). For generation of electric power, sinusoidal voltages and currents are selected all over the world due to the following reasons:

(*i*) The sinusoidal voltages and currents cause low iron and copper losses in a.c. rotating machines and transformers. This improves the efficiency of a.c. machines.

(*ii*) The sinusoidal voltages and currents offer less interference to nearby communication system (telephone lines etc.)

(*iii*) They produce least disturbance in the electrical circuits.

Whenever the word '*alternating voltage or current*' is used in this text, it means sinusoidal alternating voltage or current unless stated otherwise.

2.5 GENERATION OF ALTERNATING VOLTAGE AND CURRENT

An alternating voltage can be generated either

(*i*) by rotating a coil in a uniform magnetic field at constant speed as shown in fig.2.3 or

(*ii*) by rotating a uniform magnetic field within a stationary coil at a constant speed as shown in fig.2.4.

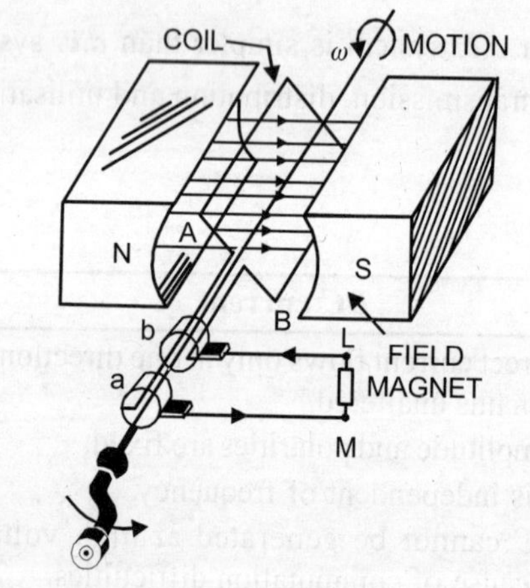

Fig.2.3: Coil rotating in the stationary magnetic field.

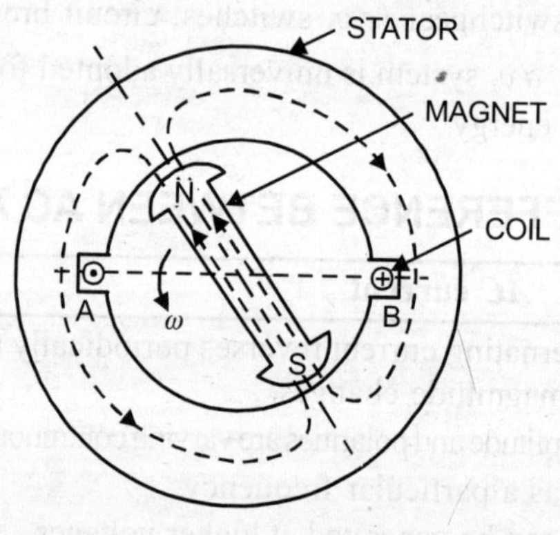

Fig.2.4: Field rotating in the stationary coil.

The first method is generally applied in small a.c. generators, whereas, second method is applied in large a.c. generators due to economical considerations. In both the cases, magnetic field is cut by the conductors (or coil sides) and an e.m.f. is induced in them. The direction and magnitude of the induced e.m.f. in the conductors depend upon the position of the conductors as explained below:

For simplicity, consider a coil placed in a uniform magnetic field to which a load (*LM*) is connected through brushes and slip rings as shown in fig.2.3. When it is rotated in anticlockwise direction at a constant angular velocity of *w* radians per second, an e.m.f. is induced in the coil sides. The cross-sectional view of the coil and its different positions at different instants are shown in fig.2.5.

The magnitude of induced e.m.f. depends upon the rate at which the flux is cut by the conductors. At (*i*), (*iii*) and (*v*) instants, induced e.m.f. in the conductor *A* and *B* is zero as they are moving parallel to the magnetic lines of force and the rate of flux cut is zero. Whereas, the magnitude of e.m.f. induced in the conductor *A* and *B* is maximum at instant (*ii*) and (*iv*) as the conductors are moving perpendicular to the magnetic lines of force and the rate of flux cut is maximum.

The direction of e.m.f. induced in the conductors is determined by applying *Fleming's right hand rule*. At instant (*ii*), the direction of e.m.f. induced in conductor *A* is outward whereas, at instant (*iv*), the direction of induced e.m.f. in the conductor *A* is inward (*i.e.* the direction of induced e.m.f. at this instant is opposite to that of the direction of induced e.m.f. at instant (*ii*).

The wave shape of the e.m.f. induced in the coil is also shown in fig.2.5.

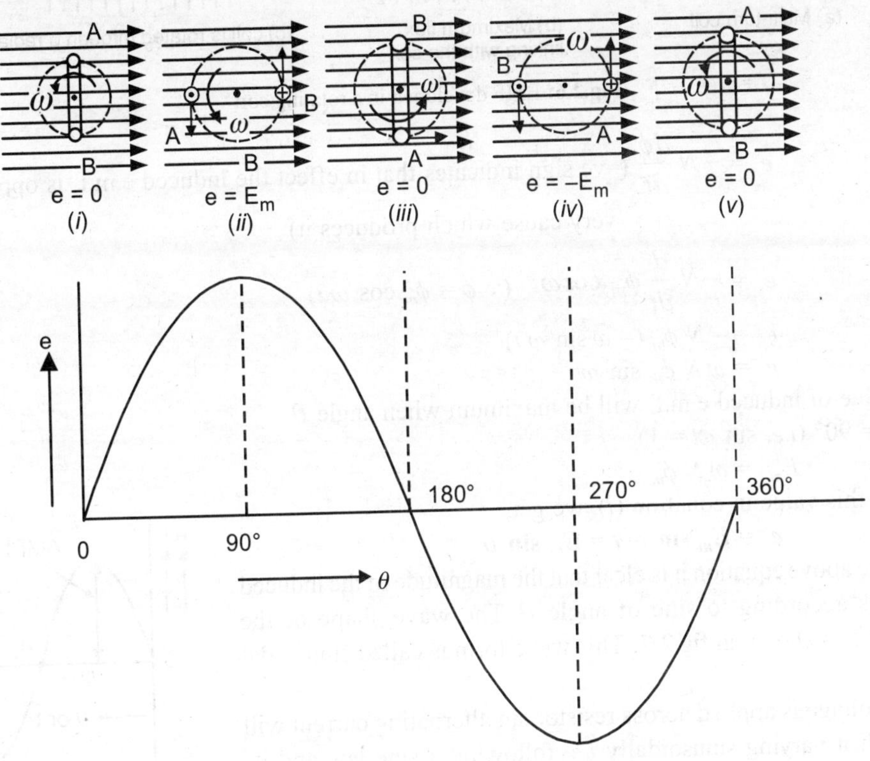

Fig.2.5: EMF induced in a coil at various instants

2.6 EQUATION OF ALTERNATING E.M.F AND CURRENT

Consider a coil having *N* turns rotating in a uniform magnetic field of density *B* Wb/m² in the counter-clockwise direction at an angular velocity of *w* radians per second as shown in fig.2.6.

At the instant, as shown in fig.2.6 (*b*) maximum flux ϕ_m is linking with the coil. After *t* seconds, the coil is rotated through an angle $q = wt$ radians. The component of flux linking with he coil at this instant is $\phi_m \cos wt$. Whereas, the other component $\phi_m \sin wt$ is parallel to the plane of the coil.

According to Faraday's laws of electromagnetic induction, the magnitude of e.m.f. induced in the coil at this instant *i.e.*

Instantaneous value of e.m.f. induced in the coil,

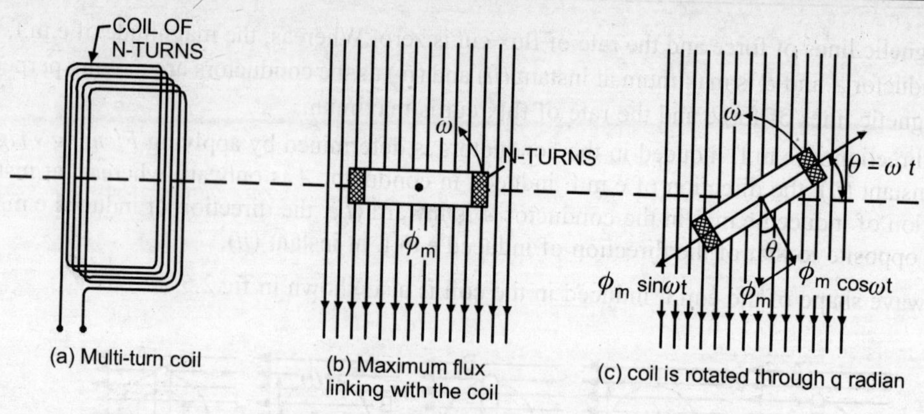

Fig.2.6: EMF developed in a rotating coil

$$e = -N\frac{d\phi}{dt}$$ (–ve sign indicates that in effect the induced e.m.f. is opposite to the very cause which produces it)

or $$e = -N\frac{d}{dt}\phi_m \cos\omega t \quad (\because \phi = \phi_m \cos \omega t)$$

or $$e = -N\,\phi_m\,(-\,\omega \sin \omega t)$$

or $$e = \omega N\,\phi_m \sin \omega t \qquad \text{...(i)}$$

The value of induced e.m.f. will be maximum when angle θ or $\omega t = 90°$ (*i.e.* $\sin \omega t = 1$)

$\therefore$ $$E_m = \omega N\,\phi_m \qquad \text{...(ii)}$$

Putting this value in equation (*i*), we get,

$$\boldsymbol{e = E_m \sin \omega t = E_m \sin \theta}$$

From the above equation it is clear that the magnitude of the induced e.m.f. varies according to sine of angle θ. The wave shape of the induced e.m.f. is shown in fig.2.7. This wave form is called sinusoidal wave.

If this voltage is applied across resistor, an alternating current will flow through it varying sinusoidally *i.e.* following a sine law and its wave shape will be same as shown in fig.2.7.

This alternating current is given by the equation:

$$i = I_m \sin \omega t = I_m \sin \theta$$

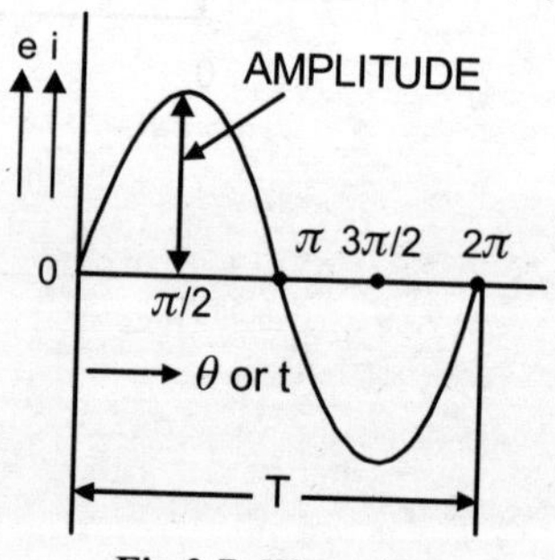

Fig.2.7: Wave shape of induced emf

2.7 IMPORTANT TERMS

An alternating voltage or current changes its magnitude and direction at regular intervals of time. A sinusoidal voltage or current varies as a sine function of time t or angle q (= w t). The following important terms are generally used in alternating quantities:

(*i*) **Wave form:** The shape of the curve obtained by plotting the instantaneous values of alternating quantity (voltage or current) along *y*-axis and time or angle ($\theta = \omega t$) along *x*-axis is called *wave form* or *wave shape*. Fig.2.7 shows the waveform of an alternating quantity varying sinusoidally. The graphical representation of an alternating quantity is called its *wave form.*

(ii) **Instantaneous value:** The value of an alternating quantity *i.e.* voltage or current at any instant is called its *instantaneous value* and is represented by *e* or *i* respectively.

(iii) **Cycle:** When an alternating quantity goes through a complete set of +ve and –ve values or goes through 360 electrical degrees, it is said to have completed one *cycle*.

(iv) **Alternation:** One half cycle is called *alternation*. An alternation spans 180 electrical degrees.

(v) **Time period:** The time taken in seconds to complete one cycle by an alternating quantity is called *time period*. It is generally denoted by *T*.

(vi) **Frequency:** The number of cycles made per second by an alternating quantity is called *frequency*. It is measured in cycles per second (c/s) or hertz (Hz) and is denoted by *f*.

(vii) **Amplitude:** The maximum value (positive or negative) attained by an alternating quantity in one cycle is called its *amplitude* or *peak value* or *maximum value*. The maximum value of voltage and current is generally denoted by E_m (or V_m) and I_m respectively.

2.8 IMPORTANT RELATIONS

Some of the terms used in a.c. terminology have definite relations among themselves as given below:

(i) **Relation between frequency and time period:** Consider an alternating quantity having a frequency of *f* c/s. Then

Time taken to complete *f* cycle = 1 s

Time taken to complete 1 cycle = 1/*f* second

Hence time period, $T = 1/f$ second or $f = 1/T$ c/s

(ii) **Relation between frequency and angular velocity:** Consider an alternating quantity having a frequency of *f* c/s.

Angular distance covered in one cycle $= 2\pi$ radian

∴ Angular distance covered per second in *f* cycles $= 2\pi$ radian

Hence, $\omega = 2\pi f$ radian/s

2.9 VALUES OF ALTERNATING VOLTAGE AND CURRENT

The voltage and current in d.c. system are constant so that there is no problem of specifying their magnitudes. Whereas, in a.c. system, the alternating voltage and current varry from instant to instant. So the question arises how to express the magnitude of alternating voltage and current. The following three ways are adopted to express the magnitude of these quantities:

(i) Peak Value.

(ii) Average value or Mean value.

(iii) Effective value or R.M.S. value.

The r.m.s. value of an alternating quantity (voltage or current) represents the real magnitude. Whereas, the peak and average values are important in some of the engineering applications.

2.10 PEAK VALUE

The maximum value attained by an alternating quantity during one cycle is called is *peak value*. This is also called *maximum value* or *crest value* or *amplitude*. A sinusoidal alternating quantity obtains its maximum value at 90° as shown in fig.2.7. The peak of an alternating voltage and current is represented by E_m and I_m. The knowledge of peak value is important in case of testing dielectric strength of insulating materials.

2.11 AVERAGE VALUE

The arithmetic average of all the instantaneous values considered of an alternating quantity (current or voltage) over one cycle is called **average value.**

In case of symmetrical waves (like sinusoidal current or voltage wave) the +ve half is exactly equal to the –ve half, therefore, the average value over a complete cycle is zero. Since work is being done by the current in the +ve as well as in –ve half cycle, therefore, average value is determined regardless of signs. Hence, to determine average value of alternating quantities having symmetrical waves, only (+ve half) cycle is considered.

Divide the +ve half cycle into n number of equal parts as shown in fig.2.8. Let i_1, i_2, i_3 .... i_n be the mid-ordinates.

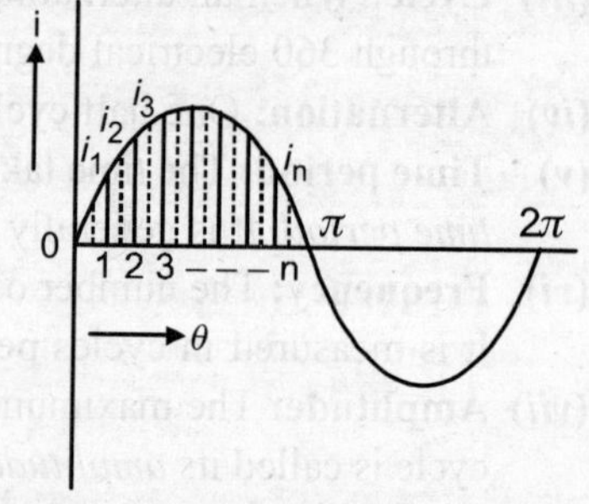

Fig.2.8: Considering half cycle of an alternating quantity for a symmetrical wave

Average value of current, I_{av} = mean of mid-ordinates.

$$= \frac{i_1 + i_2 + i_3 + \ldots + i_n}{n}$$

$$= \frac{\text{Area of alternation}}{\text{base}}$$

2.12 AVERAGE VALUE OF SINUSOIDAL CURRENT

The alternating current varying sinusoidally, as shown in fig.2.9, is given by the equation:

$$i = I_m \sin \theta$$

Consider an elementary strip of thickness $d\theta$ in the +ve half cycle, i be its mid-ordinate. Then

$$\text{Area of strip} = i\, d\,\theta$$

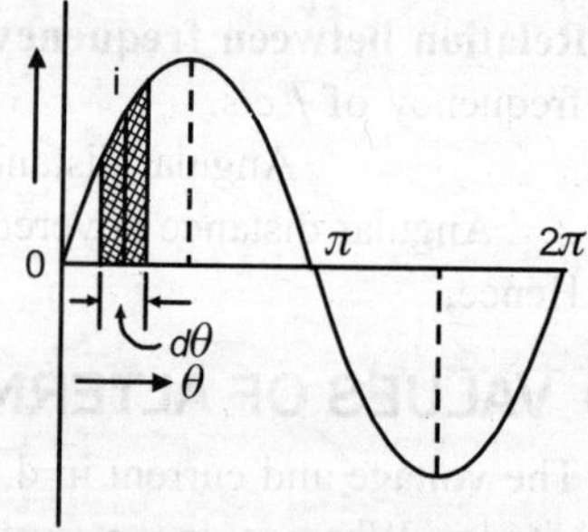

Fig.2.9: Sinusoidal alternating current

$$\text{Area of half cycle} = \int_0^{\pi} i\, d\,\theta = \int_0^{\pi} I_m \sin \theta d\,\theta$$

$$= I_m \left(-\cos \theta\right)_0^{\pi} = I_m \left(-(\cos \pi - \cos 0)\right) = I_m\left[-(-1-1)\right] = 2\,I_m$$

$$\text{Base} = 0 \text{ to } \pi = \pi - 0 = \pi$$

$$\therefore \quad \text{Average value, } I_{av} = \frac{\text{Area of alternation}}{\text{base}} = \frac{2\,I_m}{\pi} = \frac{I_m}{\pi/2} = 0 \cdot 637\; I_m$$

2.13 EFFECTIVE OR R.M.S VALUE

That steady current which when flows through a resistor of known resistance for a given time produces the same amount of heat as produced by the alternating current when flows through the same resistor for the same time is called **effective or r.m.s. value** *of the alternating current.*

Let i be the alternating current flowing through a resistor of resistance R for time t seconds which produces the same amount of heat energy as produced by I_{eff} (a direct current). The base of one alternation is divided into n equal parts, as shown in fig.2.10, so that interval is of $\frac{t}{n}$ second. Let i_1, i_2, i_3, ... i_n be the mid-ordinate.

Then heat energy produced in the

First interval = $i_1^2\ Rt/n$ joule

Second interval = $i_2^2\ Rt/n$ joule

Third interval = $i_3^2\ Rt/n$ joule

nth interval = $i_n^2\ Rt/n$ joule

Fig.2.10: Wave shape of an alternating current

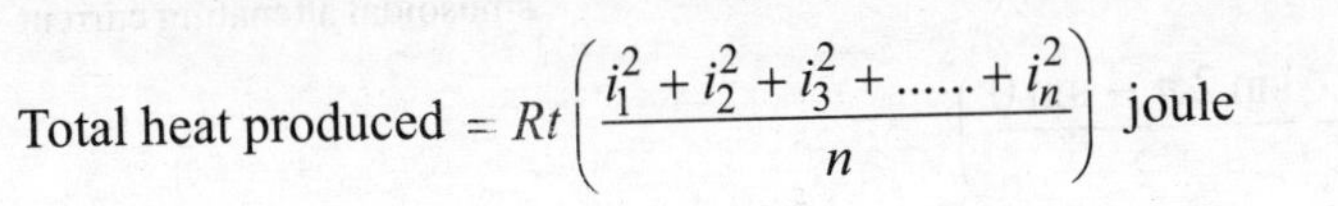

Total heat produced $= Rt\left(\frac{i_1^2 + i_2^2 + i_3^2 + ... + i_n^2}{n}\right)$ joule ...(i)

Since I_{eff} is considered as the effective value of this current.

Then total heat energy produced by this current = $I_{eff}^2\ Rt$ joule ...(ii)

Equating equation (i) and (ii), we get,

$$I_{eff}^2\ Rt = Rt\left(\frac{i_1^2 + i_2^2 + i_3^2 + ...i_n^2}{n}\right)$$

or $$I_{eff} = \sqrt{\frac{i_1^2 + i_2^2 + i_3^2 + ... + i_n^2}{n}} = \sqrt{\text{mean of squares of instantaneous values}}$$

or I_{eff} = Square root of mean of squares of instantaneous values
= root-mean-square (*r.m.s.*) value

It is the actual value of an alternating quantity which tells us the energy transfer capability of an a.c. source. For example, if we say that 5A alternating current is flowing through a circuit, it means the *r.m.s.* value of alternating current which flows through the circuit is 5A. It transfers the same amount of energy as is transferred by 5A d.c.

The ammeters and voltmeters record the *r.m.s.* values of alternating currents and voltages respectively. The domestic single-phase a.c. supply is 230 V, 50 Hz. Where 230 V is the *r.m.s.* value of alternating voltage.

2.14 R.M.S VALUE OF SINUSOIDAL CURRENT

An alternating current varying sinusoidally is given by the equation;

$$i = I_m \sin\theta$$

To determine the r.m.s. value, the squared wave of the alternating current is drawn as shown in fig.2.11. Considering an elementary strip of thickness $d\theta$ in the first half cycle of the squared wave. Let i^2 be its mid ordinate. Then

$$\text{Area of strip} = i^2\, d\theta$$

Area of first half cycle of squared wave

$$= \int_0^{\pi} i^2 \, d\theta = \int_0^{\pi} (I_m \sin\theta)^2 \, d\theta$$

$$= I_m^2 \int_0^{\pi} \sin^2\theta \, d\theta = I_m^2 \int_0^{\pi} \frac{1-\cos 2\theta}{2} \, d\theta$$

$$= \frac{I_m^2}{2} \int_0^{\pi} (1-\cos 2\theta) \, d\theta = \frac{I_m^2}{2}\left(\theta - \frac{\sin 2\theta}{2}\right)_0^{\pi}$$

$$= \frac{I_m^2}{2}\left((\pi - 0) - \frac{\sin 2\pi - \sin 0}{2}\right)$$

$$= \frac{I_m^2}{2}[(\pi-0)-(0-0)] = \frac{\pi I_m^2}{2}$$

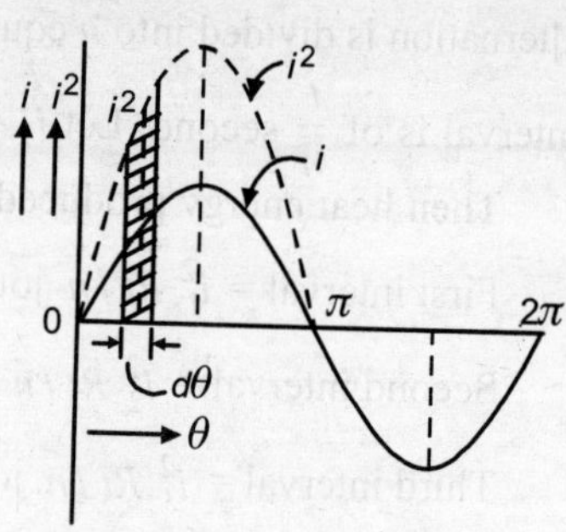

Fig.2.11: Squared wave of a sinusoidal altenating current

Base $= 0$ to $\pi = \pi - 0 = \pi$

Effective or r.m.s. value,

$$I_{r.m.s.} = \sqrt{\frac{\text{Area of first half of squared wave}}{\text{base}}} = \sqrt{\frac{\pi I_m^2}{2\pi}} = \sqrt{\frac{I_m^2}{2}} = \frac{I_m}{\sqrt{2}} = 0{\cdot}707\, I_m$$

Usually r.m.s. value of alternating current is simply represented by I instead of $I_{r.m.s.}$ Similarly, *r.m.s.* value of alternating voltage is represented by E or V.

2.15 FORM FACTOR AND PEAK FACTOR

There exists a definite relation among the average value, r.m.s. value and peak value of an alternating quantity. The relationship is expressed by the two factors namely, form factor and peak factor.

(i) **Form factor:** The ratio of r.m.s. value to average value of an alternating quantity is called *form factor*.

Mathematically, Form factor $= \dfrac{I_{r.m.s.}}{I_{av}}$ or $\dfrac{E_{r.m.s.}}{E_{av}}$

For the current varying sinusoidally;

$$\text{Form factor} = \frac{I_{r.m.s.}}{I_{av}} = \frac{I_m/\sqrt{2}}{2I_m/\pi} = \frac{\pi I_m}{2\sqrt{2}\,I_m} = \mathbf{1{\cdot}11}$$

(ii) **Peak factor:** The ratio of maximum value to r.m.s. value of an alternating quantity is called *peak factor*.

Mathematically, Peak factor $= \dfrac{I_m}{I_{r.m.s.}}$ or $\dfrac{E_m}{E_{r.m.s.}}$

For current varying sinusoidally:

$$\text{Peak factor} = \frac{I_m}{I_{r.m.s.}} = \frac{I_m}{I_m/\sqrt{2}} = \sqrt{2} = \mathbf{1{\cdot}4142}$$

Example 2.1. *An alternating voltage is given by v =* 282.8 sin 377*t*. Find

(*i*) *Frequency*

(*ii*) *r.m.s. value*

(*iii*) *average value*

(*iv*) *the instantaneous value of voltage when 't' is 3 m sec.*

(*v*) *the time taken for the voltage to reach* 200V *for the first time after passing through zero value.*

Solution: (*i*) Given that, $v = 282.8 \sin 314\, t$

Comparing above equation with

$$v = V_{in} \sin \omega t$$

$$\omega = 377 \text{ rad/sec.}$$

$$\therefore \quad \text{Frequency, } f = \frac{377}{2\pi} = \frac{377}{2 \times 3.14} = \mathbf{60\ Hz}\ (Ans.)$$

(*ii*) RMS value of the voltage,

$$V_{rms} = \frac{V_m}{\sqrt{2}} = \frac{282.8}{\sqrt{2}} = \mathbf{200\ V}\ (Ans.)$$

(*iii*) Average value,

$$V_{av} = \frac{2V_m}{\pi} = \frac{2 \times 282.8}{\pi} = \mathbf{180\ V}\ (Ans.)$$

(*iv*) $v = 282.8 \sin 377\, t.$ At $t = 3 \text{ m s} = 3 \times 10^{-3}$ s

$$v = 282.8 \sin 377 \times 3 \times 10^{-3} = \mathbf{255.9\ V}\ (Ans.)$$

(*v*) $v = 282.8 \sin 377\, t$ or $200 = 282.8 \sin 377\, t$

or

$$t = \frac{1}{377} \sin^{-1}\left(\frac{200}{282.8}\right) = 2{\cdot}08 \times 10^{-3} \text{ s} = \mathbf{2{\cdot}08\ m\ s}\ (Ans.)$$

Example 2.2. *Find the rms value, average value and form factor of the voltage waveform shown in fig.*2.12.

Solution: Average value of voltage over one cycle

$$= \frac{2\int_0^{\pi} V_m \sin \theta\, d\theta}{2\pi} = \frac{V_m}{\pi}\int_0^{\pi} \sin \theta\, d\theta$$

$$= \frac{2V_m}{\pi} = \frac{2}{\pi} \times 200 = \mathbf{127.32\ V}\ (Ans.)$$

Fig.2.12: Full-wave rectified wave

RMS value of voltage over one cycle

$$= \sqrt{\frac{2\int_{2}^{\pi} V_m^2 \sin^2 \theta d\theta}{2\pi}} = \sqrt{\frac{V_m^2}{2\pi}\int_0^{\pi} 2\sin^2\theta d\theta} = \sqrt{\frac{V_m^2}{2\pi}\int_0^{\pi}(1-\cos 2\theta)\, d\theta}$$

$$= \sqrt{\frac{V_m^2}{2\pi} \times \pi} = \frac{V_m}{\sqrt{2}} = \frac{200}{\sqrt{2}} = \mathbf{141.42\ V}\ (Ans.)$$

$$\text{Form factor} = \frac{\text{RMS value}}{\text{Average value}} = \frac{141.42}{127.32} = \mathbf{1{\cdot}11}\ (Ans.)$$

Example 2.3. *Find the average and rms values of the current, I(t) = 10 + 10 sin 314t.*

Solution: The wave diagram of given current $i(t) = 10 + 10 \sin 314\, t$ is shown in fig.2.13.

Average value of the given current,

$$I_{av} = I_{dc} + I_{av} \text{ of ac component.}$$

$$= 10 + \frac{2 I_m}{\pi} = 10 + \frac{2 \times 10}{\pi}$$

$$= 10 + 6{\cdot}366 = \mathbf{16{\cdot}366\ A}\ (Ans.)$$

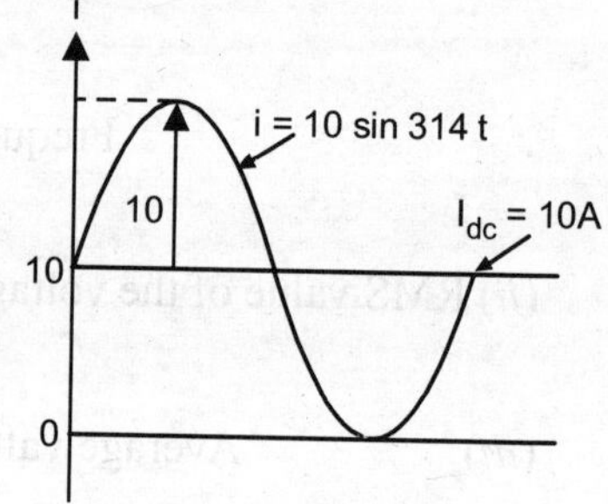

Fig.2.13: Wave diagram

RMS value of the given current

$$I_{rms} = I_{dc} + I_{rms} \text{ of ac component}$$

$$= 10 + \frac{I_m}{\sqrt{2}} = 10 + \frac{10}{\sqrt{2}}$$

$$= 10 + 7{\cdot}071 = \mathbf{17{\cdot}071\ A}\ (Ans.)$$

Example 2.4. *Calculate the average value, r.m.s. value, form factor and peak factor of a periodic current wave having values for equal time interval changing suddenly from one value to next: 0, 30, 45, 60, 90, 60, 45, 30, 0, – 30, –45, –60 etc. ampere. What would be the average and the r.m.s. value of a sine wave having the same peak value ?*

Solution: The periodic wave form of the alternating current is shown in fig.2.14.

Average value of current,

$$I_{av} = \frac{i_1 + i_2 + ... + i_n}{n} \quad (\text{where } n = 8)$$

$$= \frac{0 + 30 + 45 + 60 + 90 + 60 + 45 + 30}{8}$$

$$= \mathbf{45\ A}\ (Ans.)$$

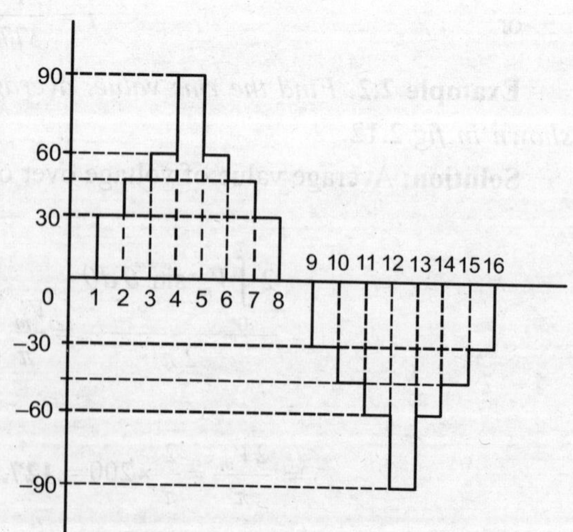

Fig.2.14: Wave shape of periodic current

R.M.S. value of current,

$$I_{r.m.s.} = \sqrt{\frac{i_1^2 + i_2^2 + ... i_n^2}{n}} = \sqrt{\frac{0 + (30)^2 + (45)^2 + (60)^2 + (90)^2 + (60)^2 + (45)^2 + (30)^2}{8}}$$

$$= \sqrt{\frac{21150}{8}} = \mathbf{51.42\ A}\ (Ans.)$$

$$\text{Form factor} = \frac{I_{r.m.s.}}{I_{av}} = \frac{51.42}{45} = \mathbf{1{\cdot}1427}\ (Ans.);$$

$$\text{Peak factor} = \frac{I_m}{I_{r.m.s.}} = \frac{90}{51.42} = \mathbf{1{\cdot}75}\ (Ans.)$$

Average value of sinusoidal a.c. having peak value of 90 A;

$$I_{av} = 0{\cdot}637\ I_m = 0{\cdot}637 \times 90 = \mathbf{57{\cdot}33\ A}\ (Ans.)$$

R.M.S. value,

$$I_{r.m.s.} = \frac{I_m}{\sqrt{2}} = \frac{90}{\sqrt{2}} = \mathbf{63{\cdot}64\ A}\ (Ans.)$$

PRACTICE EXERCISE

Q.1 *An alternating voltage is given by* $v = 141.4 \sin 314t$. *Find*

(i) *Frequency* (ii) *r.m.s. value* (iii) *average value*

(iv) *the instantaneous value of voltage when 't' is 3 m sec.*

(v) *the time taken for the voltage to reach 200V for the first time after passing through zero value.*

(*Ans, 50Hz, 100V, 90V, 114.4V, 2.5ms*)

Q.2 *Find the rms value, average value and form factor of a rectified simusoidal ac voltage having maximum value of 100V.* (*Ans, 63.66V, 70.71V, 1.11*)

Q.3 *Find the average and rms values of the current,* $I(t) = 15 + 15 \sin 314t$ (*Ans, 24.55A, 25.6A*).

Q.4 *Calculate the average value, r.m.s. value, form factor and peak factor of a periodic current wave having values for equal time interval changing suddenly from one value to next: 0, 20, 45, 50, 80, 50, 45, 20, 0, – 20, –45, –50 etc. ampere. What would be the average and the r.m.s. value of a sine wave having the same peak value ?* (*Ans, 38.75A, 45.07A, 1.163, 1.775, 50.96A, 56.56A*)

2.16 PHASOR REPRESENTATION OF SINUSOIDAL QUANTITY

It has been seen that an alternating quantity (varying sinusoidally) can be represented in the form of wave and equation. The wave form shows the graphical representation, whereas, the equation represents the mathematical expression of the instantaneous value of an alternating quantity.

The same alternating quantity can also be represented by a line of definite length (representing its maximum value) rotating in counter-clockwise direction at a constant velocity (ω radians/second). Such a rotating line is called a *phasor*.

Thus, an alternating quantity can be represented by a phasor which shows its magnitude and direction at that instant.

For instant, consider an alternating quantity (current) represented by the equation $i = I_m \sin \omega t$. Take a line OA to represent the maximum value of current I_m to scale. Imagine this line is rotating in counter-clockwise direction at an angular velocity of ω radian/s about point O. After t seconds the line is rotated through an angle $\theta (\theta = \omega t)$, from its horizontal position as shown in fig.2.15. The projection of line OA on the Y-axis is OB.

$$OB = OA \sin \theta = I_m \sin \omega t$$
$$= i \text{ (the value of current at that instant)}$$

Hence, the projection of the phasor OA on the Y-axis (*i.e.* OB) at any instant gives the value of current at that instant.

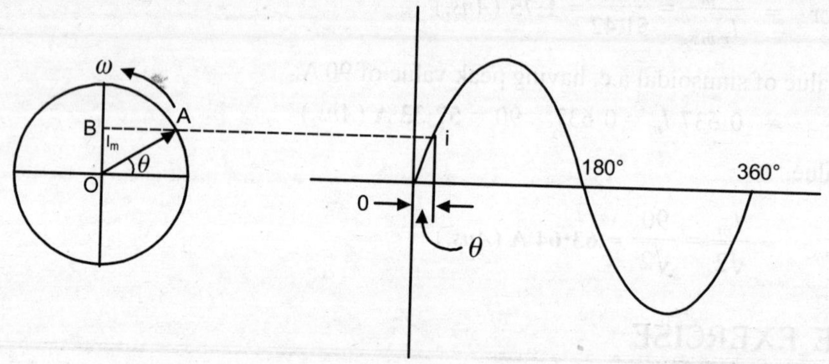

Fig.2.15: Plotting of sine wave

Thus, a sinusoidal alternating quantity is represented by a phasor (vector) of length to scale equal to its maximum value rotated through an angle θ with the axis of reference (*i.e.* X-axis).

The phasor representation of an alternating quantity enables us to understand its magnitude and position on the axis. The alternating quantities can be added and subtracted with a fair degree of ease by representing them vectorially (phasor diagram).

2.17 PHASE AND PHASE DIFFERENCE

The phase of an alternating quantity (current or voltage) at an instant is defined as the fractional part of a cycle through which the quantity has advanced from a selected origin (see fig.2.16). In actual practice, we are more concerned with the phase difference between the two alternating quantities rather than their *absolute phase*.

The two alternating quantities having same frequency, when attain their zero value at different instants, the quantities are said to have a phase difference. This angle between zero points (and are becoming positive) of two alternating quantities is called *angle of phase difference*.

In fig.2.17, two alternating currents of magnitude I_{m1} and I_{m2} are shown vectorially. Both the vectors are rotating at same angular velocity of ω radian per second. The zero values are obtained by the two currents at different instants. Therefore, they are said to have a phase difference of angle ϕ.

In other words, **the phase difference** *may be defined as the angular displacement between the maximum positive value of the two alternating quantities having the same frequency.*

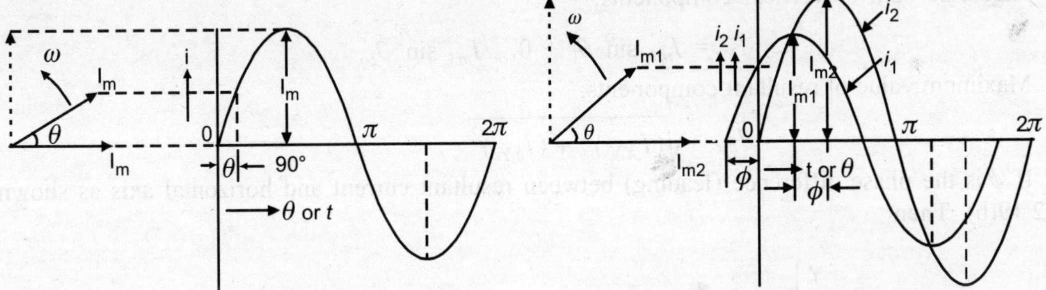

Fig.2.16: Phasor and wave diagram (single quantity) **Fig.2.17:** Phasor and wave diagram (two quantities)

The quantity which attains its +ve maximum value prior to the other is called a *leading* quantity, whereas, the quantity which attains its +ve maximum value after the other is called a *lagging* quantity. In this case current I_{m1} is leading current *w.r.t.* I_{m2} or is other words current I_{m2} is the lagging current *w.r.t.* I_{m1}.

2.18 ADDITION AND SUBTRACTION OF ALTERNATING QUANTITIES

In a.c. circuits, sometimes it is required to add or subtract the alternating quantities. In such cases, we proceed as follows:

Addition of alternating quantities

The given alternating quantities are represented as phasor and then they are added in the same manner as forces are added. Only phasors of the similar quantities are added *i.e.* either all the currents are added or all the voltages are added. Voltages and currents are never added with each other. For addition, the following method is most suitable and simple:

Method of components

In this method each phasor is resolved into horizontal and vertical components. The horizontal components are added algebraically to obtain the resultant horizontal component I_{XX}. Similarly, vertical components are summed up algebraically to obtain the resultant vertical component I_{YY}.

Consider an a.c. parallel circuit consisting of three branches each carrying a current of i_1, i_2 and i_3 respectively as shown in fig.2.18. Let the three currents be represented by;

$$i_1 = I_{m1} \sin (\omega t + \theta_1);$$

$$i_2 = I_{m2} \sin \omega t$$

and

$$i_3 = I_{m3} \sin (\omega t - \theta_2)$$

The maximum values of the three currents I_{m1}, I_{m2} and I_{m3} are represented by the phasors as shown in fig.2.19(a). Resolving the components horizontally and vertically.

i1
i2
i3
ir
A.C. SUPPLY

Fig.2.18: Circuit diagram

Algebraic sum of horizontal components;

$$I_{XX} = I_{m1} \cos \theta_1 + I_{m2} + I_{m3} \cos \theta_2$$

Algebraic sum of vertical components;

$$I_{YY} = I_{m1}\ \sin\ \theta_1 + 0 - I_{m3}\ \sin\ \theta_2$$

Maximum value of resultant components;

$$I_{mr} = \sqrt{(I_{XX})^2 + (I_{YY})^2}$$

If ϕ is the phase difference (leading) between resultant current and horizontal axis as shown in fig.2.19(b). Then

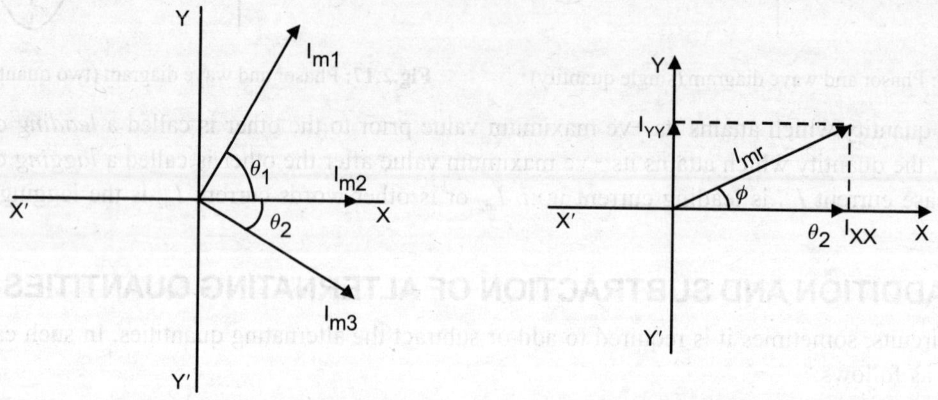

(a) Phasor position of three currents (b) Resultant phasor diagram

Fig.2.19: Phasor diagram.

$$\phi = \tan^{-1} \frac{I_{YY}}{I_{XX}}$$

The instantaneous value of the resultant current is given by the relation;

$$i_r = I_{mr}\ \sin\ (\omega\ t + \phi)$$

However, if I_{YY} comes out to be negative, the angle of phase difference will be lagging (*i.e.*$-\phi$). Then the instantaneous value of the resultant current will be given by the relation

$$i_r = I_{mr}\ \sin\ (\omega\ t - \phi)$$

Subtraction of alternating quantities

The methods explained above (*i.e.* parallelogram method and method of components) are also applied for the subtraction of an alternating quantity. The only difference is that in this case, the phasor of the alternating quantity which is to be subtracted is reversed or represented 180° out of phase. Then it is added with the other alternating quantity (or quantities) as usual.

Example 2.5. *Draw a phasor diagram showing the following voltages.*

$$v_1 = 100\ sin\ 314\ t; \qquad v_2 = 200\ sin\ (314\ t + \pi/3)$$

$$v_3 = -\ 50\ cos\ 314\ t; \qquad v_4 = 150\ sin\ (314\ t - \pi/4)$$

Find RMS value of resultant voltage.

Solution: $v_1 = 100 \sin 500\, t$

$$v_2 = 200 \sin\left(500t + \frac{\pi}{3}\right)$$

$$v_3 = -\,50 \cos 500\, t$$

$$= -50 \sin\left(\frac{\pi}{2} - 500\,t\right)$$

$$= 50 \sin (500\, t - \pi/2)$$

$$v_4 = 150 \sin (150\, t - \pi/4)$$

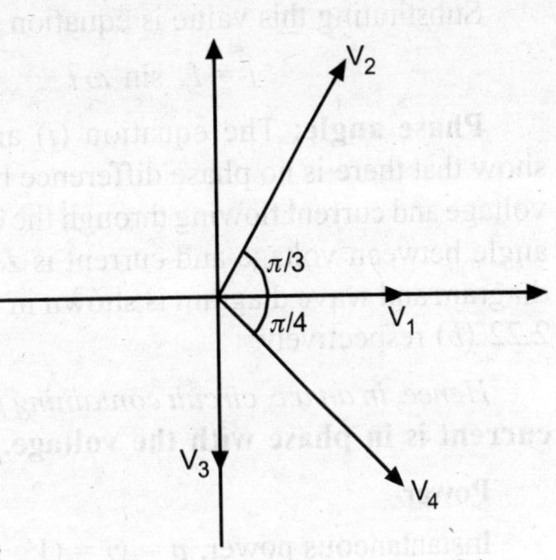

Fig.2.20: Phasor representation of three voltages

All the four voltages are shown vectorially in fig.2.20.

Resolving the phasors in horizontal axis;

$$V_{xx} = V_1 \cos 0 + V_2 \cos \frac{\pi}{3} + V_3 \cos \frac{\pi}{2} + V_4 \cos \frac{\pi}{4}$$

$$= 100 \times 1 + 200 \times 0{\cdot}5 + 50 \times 0 + 150 \times 0{\cdot}707 = 306{\cdot}05 \text{ V}$$

Resolving the phasors in vertical axis;

$$V_{yy} = V_1 \sin 0 + V_2 \sin \frac{\pi}{3} - V_3 \sin \frac{\pi}{2} - V_4 \sin \frac{\pi}{4}$$

$$= 100 \times 0 + 200 \times 0{\cdot}866 - 50 \times 1 - 150 \times 0{\cdot}707 = 17{\cdot}15 \text{ V}$$

Maximum value of resultant voltage;

$$V_{mr} = \sqrt{V_{xx}^2 + V_{yy}^2} = \sqrt{(306 \cdot 05)^2 + (17 \cdot 15)^2} = 306{\cdot}53 \text{ V}$$

RMS value of resultant voltage, $V_{rms(r)} = \dfrac{V_{mr}}{\sqrt{2}} = \dfrac{306 \cdot 53}{\sqrt{2}} = \mathbf{216{\cdot}75\ V}$ *(Ans.)*

2.19 A.C CIRCUIT CONTAINING ONLY RESISTANCE (R)

The circuit containing a pure resistance (R) ohm is shown in fig.2.21.

The alternating voltage applied across the circuit is given by the equation;

$$v = V_m \sin \omega\, t \qquad \ldots(i)$$

The instantaneous value of current flowing through the resistor will be;

$$i = \frac{v}{R} = \frac{V_m}{R} \sin \omega t \qquad \ldots(ii)$$

The value of current will be maximum when

$$\omega\, t = 90° \text{ or } \sin \omega\, t = 1$$

$$I_m = V_m/R$$

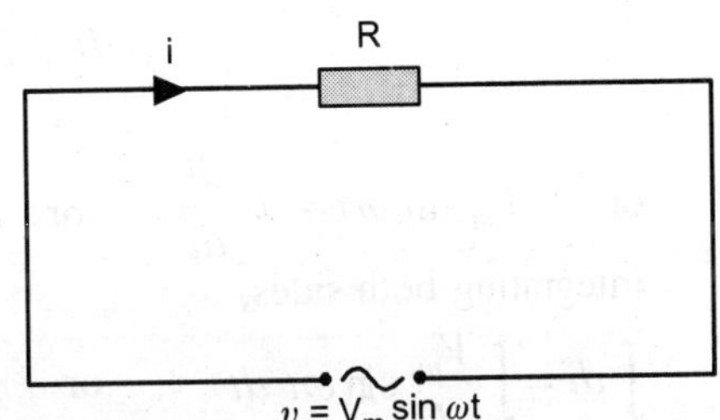

Fig.2.21: Circuit containing only resistance

Substituting this value is equation (*ii*), we get,

$$i = I_m \sin \omega t \qquad ....(iii)$$

Phase angle: The equation (*i*) and (*iii*), clearly show that there is no phase difference between applied voltage and current flowing through the circuit *i.e.* phase angle between voltage and current is zero. The phasor diagram and wave diagram is shown in fig.2.22 (*a*) and 2.22 (*b*) respectively.

(a) Phasor diagram (a) Wave diagram

Fig.2.22: Phasor and Wave diagram

Hence, in an a.c. circuit containing pure resistance, **current is in phase with the voltage.**

Power

Instantaneous power, $p = vi = (V_m \sin \omega t)\,(I_m \sin \omega t)$

$$= \frac{V_m I_m}{2}\, 2 \sin^2 \omega t$$

$$= \frac{V_m}{\sqrt{2}} \frac{I_m}{\sqrt{2}} (1 - \cos 2\,\omega t) = \frac{V_m}{\sqrt{2}} \frac{I_m}{\sqrt{2}} - \frac{V_m}{\sqrt{2}} \frac{I_m}{\sqrt{2}} \cos 2\,\omega t$$

Average power consumed in the circuit over a complete cycle.

$$P = \text{average of } \frac{V_m}{\sqrt{2}} \frac{I_m}{\sqrt{2}} - \text{average of } \frac{V_m}{\sqrt{2}} \frac{I_m}{\sqrt{2}} \cos \omega t$$

or $$P = V_{r.m.s.}\; I_{r.m.s.} - \text{zero} \quad \text{or} \quad P = VI$$

Power curve: Fig.2.22. (*b*) shows the power curve for a pure resistive circuit. Points on the power curve are obtained from the product of the corresponding instantaneous values of voltage and current.

2.20 A.C CIRCUIT CONTAINING PURE INDUCTANCE (L)

The circuit containing a pure inductance of *L* henry is shown in fig.2.23.

The alternating voltage applied across the circuit is given by the equation;

$$v = V_m \sin \omega t \qquad ...(i)$$

As a result, an alternating current *i* flows through the inductance which induces an e.m.f. in it, given by the relation;

$$e = -L \frac{di}{dt}$$

This induced e.m.f. is equal and opposite to the applied voltage.

$$\therefore \qquad v = -e = -\left(-L \frac{di}{dt}\right)$$

or $$V_m \sin \omega t = L \frac{di}{dt} \quad \text{or} \quad di = \frac{V_m}{L} \sin \omega t\, dt$$

Integrating both sides,

$$\int di = \int \frac{V_m}{L} \sin \omega t\, dt \quad \text{or} \quad i = \frac{V_m}{\omega L} (-\cos \omega t)$$

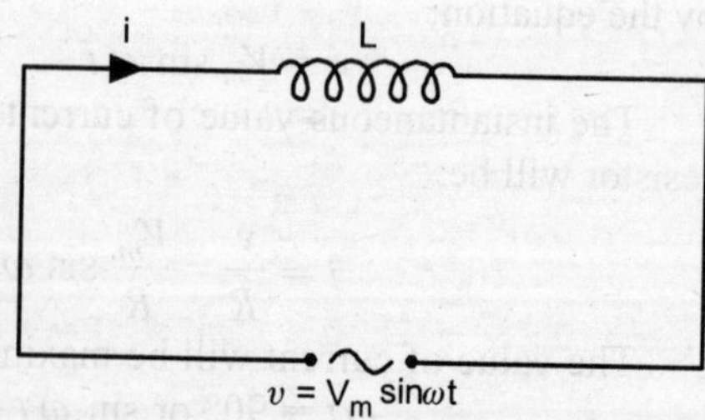

Fig.2.23: Circuit containing pure inductance

or $$i = \frac{V_m}{\omega L} \sin(\omega t - \pi/2) = \frac{V_m}{X_L} \sin(\omega t - \pi/2) \qquad ...(ii)$$

where $X_L = \omega L$ is the opposition offered to the flow of alternating current by a pure inductance and is called **inductive reactance.**

The value of current will be maximum when $\sin(\omega t - \pi/2) = 1$ *i.e.* $I_m = \frac{V_m}{X_L}$

$$\therefore \qquad i = I_m \sin(\omega t - \pi/2) \qquad ...(iii)$$

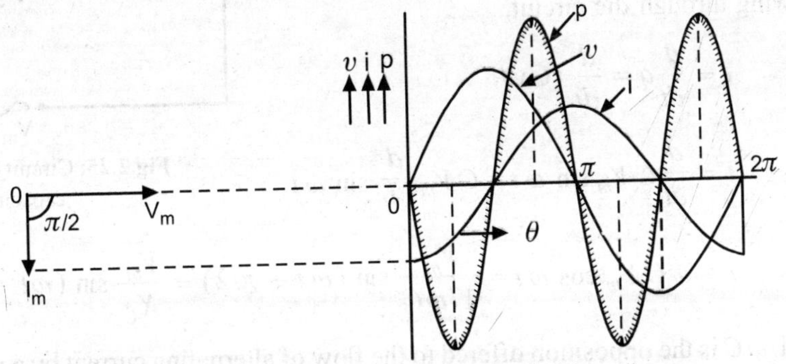

(a) Phasor diagram

(b) Wave diagram

Fig.2.24: Phasor and Wave diagram

Phase angle: The equation (*i*) and (*iii*), clearly show current flowing through a pure inductive circuit lags behind the applied voltage *v* by 90°. The phasor diagram is shown in fig.2.24 (*a*) and 2.24 (*b*) respectively.

Hence, in an a.c. circuit containing pure inductance, **current lags behind the voltage by 90°.**

Power: Instantaneous power, $p = vi = V_m \sin \omega t \times I_m \sin(\omega t - \pi/2)$

$$= V_m I_m \sin \omega t \cos \omega t = \frac{V_m I_m}{2} 2 \sin \omega t \cos \omega t$$

$$= \frac{V_m}{\sqrt{2}} \frac{I_m}{\sqrt{2}} \sin 2\omega t$$

Average power consumed in the circuit over a complete cycle,

$$P = \text{average } \frac{V_m}{\sqrt{2}} \cdot \frac{I_m}{\sqrt{2}} \sin 2\omega t = \text{zero}$$

Hence, average power consumed in a pure inductive circuit is **zero.**

Power curve: The power curve for a pure inductive circuit is shown in fig.3.4 (*b*). It is very clear that average power in a half cycle (one alternation) is zero, as the negative and positive loop area under power curve is the same.

It is interesting to note that during the first quarter cycle, what so ever power (or energy) is supplied by the source to the inductance (or coil) is stored in the magnetic field set-up around it. However, in the next quarter cycle, the magnetic field collapses and the power (or energy) stored in the field is returned to the source. This process is repeated in each and every alternation. **Hence, no power or energy is consumed in this circuit.**

2.21 A.C CIRCUIT CONTAINING PURE CAPACITANCE (C)

The circuit containing a pure capacitor of capacitance C farad is shown in fig.2.25.

The alternating voltage applied across the circuit is given by the equation;

$$v = V_m \sin \omega t \qquad ...(i)$$

Charge on the capacitor at any instant,

$$q = C v$$

Current flowing through the circuit,

$$i = \frac{d}{dt} q = \frac{d}{dt}(Cv)$$

or

$$i = \frac{d}{dt} C V_m \sin \omega t = C V_m \frac{d}{dt} \sin \omega t$$

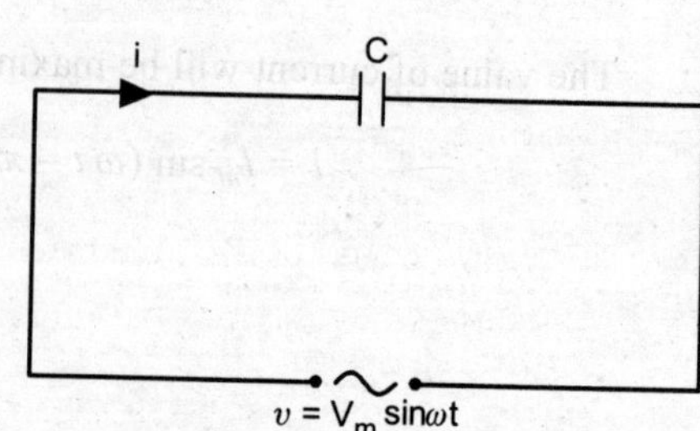

Fig.2.25: Circuit containing pure capacitance

or

$$i = \omega C V_m \cos \omega t = \frac{V_m}{1/\omega C} \sin(\omega t + \pi/2) = \frac{V_m}{X_C} \sin(\omega t + \pi/2) \qquad ...(ii)$$

where $X_C = 1/\omega C$ is the opposition offered to the flow of alternating current by a pure capacitor and is called *capacitive reactance*.

The value of current will be maximum when $\sin(\omega t + \pi/2) = 1$

i.e. $I_m = V_m / X_C$

Substituting this value is equation (*ii*), we get,

$$i = I_m \sin(\omega t + \pi/2) \qquad ...(iii)$$

Phase angle: The equation (*i*) and (*iii*), clearly show that current flowing through pure capacitive circuit leads the applied voltage by 90°. the phasor diagram and wave diagram is shown in fig.2.26 (*a*) and 2.26 (*b*) respectively.

Hence, in an a.c. circuit containing pure capacitance **current leads the voltage by 90°.**

Power: Instantaneous power, $p = v\,i = V_m \sin \omega t \times I_m \sin(\omega t + \pi/2)$

$$= V_m I_m \sin \omega t \cos \omega t = \frac{V_m}{\sqrt{2}} \frac{I_m}{\sqrt{2}} \sin 2\,\omega t$$

or Average power over a complete cycle,

$$P = \text{zero}$$

Hence, average power consumed in a pure capacitive circuit is **zero.**

Power curve: The power curve for a pure capacitive circuit is shown in fig.2.26 (*b*). It is very clear from the curve that average power in a half cycle (one alternation) is zero since the positive and negative loop area under power curve is the same.

It is interesting to note that, during the first quarter cycle, what so ever power (or energy) is supplied by the source to the capacitor is stored in the electric field set-up between the capacitor plates. In the next quarter cycle, the electric field collapses and the power (or energy) stored in the field is returned to the source. This process is repeated in each alternation. **Hence, no power is consumed by this circuit.**

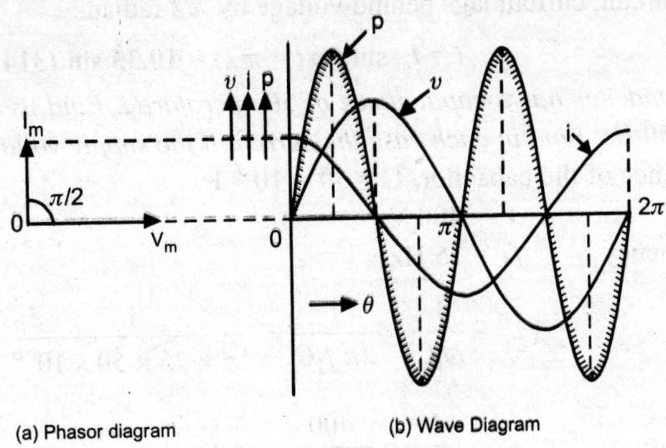

(a) Phasor diagram (b) Wave Diagram

Fig.2.26: Phasor and wave diagram

Example 2.6. *An a.c. circuit consists of a pure resistance of 8 ohm and is connected across an a.c. supply of 240 V, 50 Hz. Calculate (i) current; (ii) power consumed and (iii) write down the equation for voltage and current.*

Solution: (*i*) Current in the circuit, $I = \frac{V}{R} = \frac{240}{8} = \mathbf{30\ A}$ (*Ans.*)

(*ii*) Power consumed, $P = VI = 240 \times 30 = \mathbf{7200\ W}$ (*Ans.*)

(*iii*) Maximum value of applied voltage, $V_m = \sqrt{2}\ V = \sqrt{2} \times 240 = 339.4$ V

Maximum value of current, $I_m = \sqrt{2} \times 30 = 42.42$ A

Angular velocity, $\omega = 2\ \pi f = 2\ \pi \times 50 = 314{\cdot}16$ rad/s

Equation for applied voltage;

$$v = V_m \sin \omega t = \mathbf{339.4 \sin 314{\cdot}16\ t} \ (Ans.)$$

As in a pure resistive circuit, voltage and current are in phase with each other, therefore, current is given by the equation;

$$i = I_m \sin \omega t = \mathbf{42.42 \sin 314{\cdot}16\ t} \ (Ans.)$$

Example 2.7. *An inductive coil having negligible resistance and 0·1 henry inductance is connected across 230 V, 50 Hz supply. Find (i) the inductive reactance (ii) r.m.s. value of current (iii) power and (iv) equations for voltage and current.*

Solution: Inductive reactance, $X_L = 2\ \pi L = 2\ \pi \times 50 \times 0{\cdot}1 = \mathbf{31{\cdot}416\ \Omega}$ (*Ans.*)

Current, $I = 230/X_L = 200/31{\cdot}416 = \mathbf{7.32\ A}$ (*Ans.*)

Power, P = **Zero** (*Ans.*)

Now, $V_m = \sqrt{2}\ V = \sqrt{2} \times 230 = 325.27$ V;

$I_m = \sqrt{2}\ I = \sqrt{2} \times 7{\cdot}32 = 10.35$ A

and $\omega = 2\ \pi f = 314$ rad/s

$\therefore$ $v = V_m \sin \omega t = \mathbf{325.27 \sin 314\ t}$ (*Ans.*)

In pure inductive circuit, current lags behind voltage by $\pi/2$ radian.

$$\therefore \quad i = I_m \sin(\omega t - \pi/2) = \mathbf{10.35 \sin(314\, t - \pi/2)} \text{ (Ans.)}$$

Example 2.8. *A capacitor has a capacitance of 50 microfarad. Find its capacitive reactance for frequencies of 25 and 50 Hz. Find in each case the current if the supply voltage is 400 V.*

Solution: Capacitance of the capacitor, $C = 50 \times 10^{-6}$ F

Supply voltage, $V = 400$ V

When supply frequency, $f_1 = 25$ Hz

Capacitive reactance, $$X_{C1} = \frac{1}{\omega_1 C} = \frac{1}{2\pi f_1 C} = \frac{1}{2\pi \times 25 \times 50 \times 10^{-6}} = \mathbf{127.32\ \Omega} \text{ (Ans.)}$$

Current in the circuit, $$I_1 = \frac{V}{X_{C1}} = \frac{400}{127.32} = \mathbf{3.14\ A} \text{ (Ans.)}$$

When supply frequency, $f_2 = 50$ Hz

Capacitive reactance, $$X_{C2} = \frac{1}{\omega_2 C} = \frac{1}{2\pi f_2 C} = \frac{1}{2\pi \times 25 \times 50 \times 10^{-6}} = \mathbf{63.66\ \Omega} \text{ (Ans.)}$$

Current in the circuit, $$I_2 = \frac{V}{X_{C2}} = \frac{400}{63.66} = \mathbf{6{\cdot}28\ A} \text{ (Ans.)}$$

PRACTICE EXERCISE

Q.1 *An a.c. circuit consists of a pure resistance of 16 ohm and is connected across an a.c. supply of 230 V, 50 Hz. Calculate (i) current; (ii) power consumed and (iii) write down the equation for voltage and current.*

(***Ans.*** *23 A, 5290 W; 325.27 Sin 314.16 t; 32.53 sin 314.16 t*)

Q.2 *An inductive coil having negligible resistance and 0·1 henry inductance is connected across 200 V, 50 Hz supply. Find (i) the inductive reactance (ii) r.m.s. value of current (iii) power and (iv) equations for voltage and current.*

(***Ans.*** *31.416 W; 6.366 A; 0 W, 282.84 sin 314 t, 9 sin (314t-π/2)*)

Q.3 *A capacitor has a capacitance of 30 microfarad. Find its capacitive reactance for frequencies of 25 and 50 Hz. Find in each case the current if the supply voltage is 440 V.*

(***Ans.*** *212.2 Ω, 2.073 A; 106.1 Ω; 4.146 A*)

2.22 A.C SERIES CIRCUITS

An ac circuit may have combination of resistance, inductance and capacitance. In actual practice, a.c. circuits contain two or more than two such components connected in series or parallel. A series circuit is a circuit in which each component carries the same current. An a.c. series circuit may be;

(i) $R - L$ series circuit *(ii)* $R - C$ series circuit *(iii)* $R - L - C$ series circuit.

2.23 R – L SERIES CIRCUIT

An *R-L* series circuit contains resistance (*R*) and inductance (*L*) connected in series as shown in fig 2.27.

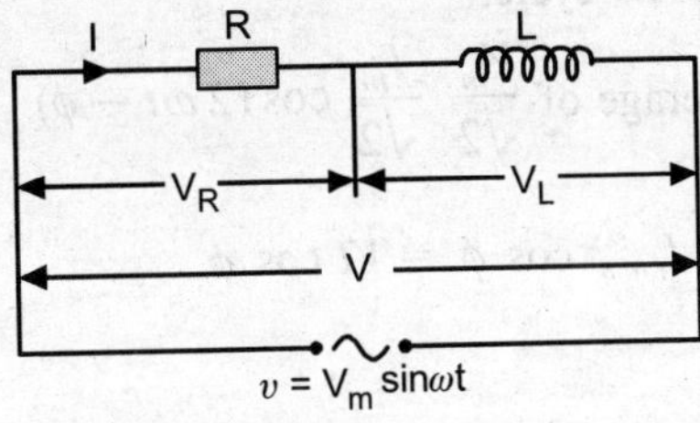

Fig.2.27: R-L series circuit

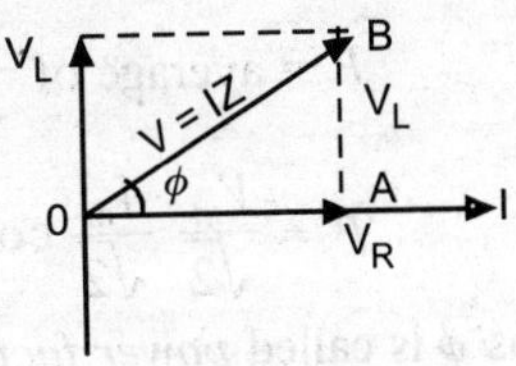

Fig.2.28: Phasor diagram

Its phasor diagram is shown in fig.2.28. To draw the phasor diagram, current I (r.m.s. value) is taken as the reference vector. Voltage drop in resistance V_R (= IR) is taken in phase with current vector, whereas, voltage drop in inductive reactance V_L (= IX_L) is taken 90° ahead of the current vector (since current lags behind the voltage by 90° in pure inductive circuit). The vector sum of these two voltages (drops) is equal to the applied voltage V (r.m.s. value).

Now, $V_R = IR$ and $V_L = IX_L$ (where $X_L = 2\,\pi f L$)

In right angle triangle OAB

$$V = \sqrt{(V_R)^2 + (V_L)^2} = \sqrt{(IR)^2 + (IX_L)^2} = I\sqrt{R^2 + X_L^2}$$

or $$I = \frac{V}{\sqrt{R^2 + X_L^2}} = \frac{V}{Z}$$

where $Z = \sqrt{R^2 + X_L^2}$ is the total opposition offered to the flow of alternating current by an $R - L$ series circuit and is called *impedance* of the circuit. It is measured in ohm.

Phase angle

From the phasor diagram shown in fig.2.28, it is clear that current, in this circuit, lags behind the applied voltage by an angle *f*, called *phase angle*.

From phasor diagram; $\tan\phi = \frac{V_L}{V_R} = \frac{IX_L}{IR} = \frac{X_L}{R}$ or $\phi = \tan^{-1} X_L/R$

Power: If the alternating voltage applied across the circuit is given by the equation.

$$v = V_m \sin \omega t$$

Then, $$i = I_m \sin(\omega t - \phi)$$

$\therefore$ Instantaneous power, $$p = v\,i$$

$$= V_m \sin \omega t \,.\, I_m \sin(\omega t - \phi) = \frac{V_m I_m}{2}\, 2 \sin \omega t \sin(\omega t - \phi)$$

$$= \frac{V_m}{\sqrt{2}} \frac{I_m}{\sqrt{2}} [\cos\phi - \cos(2\,\omega t - \phi)]$$

$$= \frac{V_m}{\sqrt{2}} \frac{I_m}{\sqrt{2}} \cos\phi - \frac{V_m}{\sqrt{2}} \frac{I_m}{\sqrt{2}} \cos(2\,\omega t - \phi)$$

Average power consumed in the circuit over a complete cycle,

$$P = \text{average of } \frac{V_m}{\sqrt{2}} \frac{I_m}{\sqrt{2}} \cos\phi - \text{average of } \frac{V_m}{\sqrt{2}} \frac{I_m}{\sqrt{2}} \cos(2\omega t - \phi)$$

or

$$P = \frac{V_m}{\sqrt{2}} \frac{I_m}{\sqrt{2}} \cos\phi - \text{zero} = V_{r.m.s.} I_{r.m.s.} \cos\phi = VI\cos\phi$$

where $\cos\phi$ is called *power factor* of the circuit.

From phasor diagram; $\cos\phi = \frac{V_R}{V} = \frac{IR}{IZ} = \frac{R}{Z}$

Thus, **power factor** *is defined as the cosine of the angle between voltage and current in an a.c. circuit. It may also be defined as the ratio of resistance to impedence of an a.c. circuit.*

Alternatively: Power, $P = VI\cos\phi = IZ \cdot I \cdot \frac{R}{Z} = I^2 R$

This shows that power is actually consumed in resistance only; inductance does not consume any power.

Power curve: The phasor diagram and wave diagram for voltage and current are shown in fig.2.29 (*a*) and 2.29 (*b*) respectively where applied voltage ($v = V_m \sin wt$) is taken as reference quantity. The power curve for $R - L$ series circuit is also shown in fig.2.29 (*b*). The points on the power curve are obtained from the product of the corresponding instantaneous values of voltage and current. It is clear that power is negative between angle 0 and *f* and between 180° and (180 + *f*). During rest of the cycle the power is positive. Since the area under the positive loops is greater than that under the negative loops, the net power over a complete cycle is positive. Hence, a definite quantity of power is utilised or consumed by this circuit.

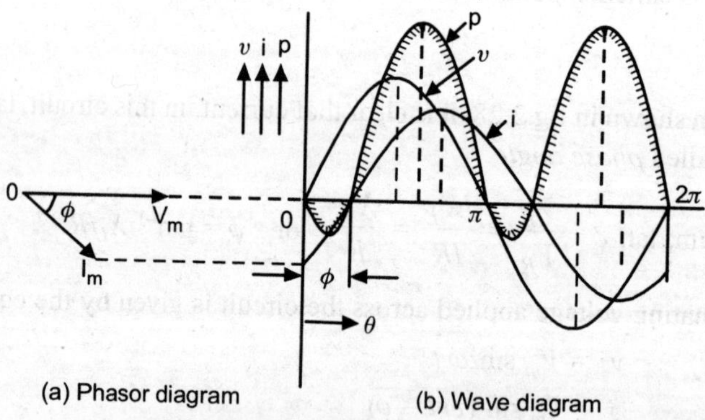

Fig.2.29: Phasor and wave diagram.

2.24 IMPEDENCE TRIANGLE

The simplified phasor diagram of $R - L$ series circuit is shown in fig.2.30. When each side of this phasor diagram is divided by a common factor I, we get another right angled triangle, as shown in fig.2.31, whose sides represent R, X_L and Z. Such a triangle is known as *impedance triangle.*

Thus, a right angled triangle whose base represents circuit resistance, perpendicular represents circuit reactance and hypotenuse represents circuit impedance is called an **impedance triangle.**

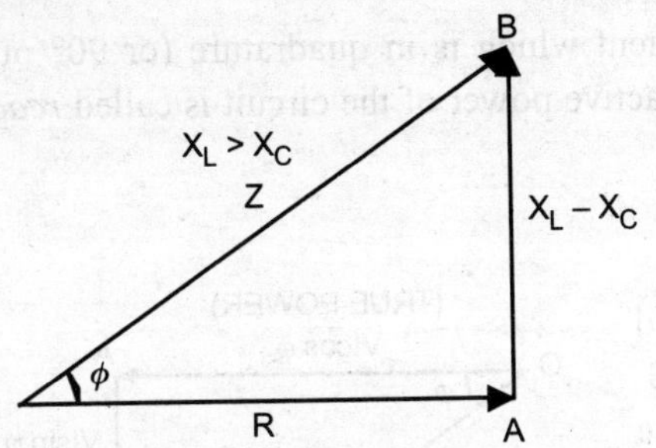

Fig.2.30: Voltage triangle for an R-L series circuit

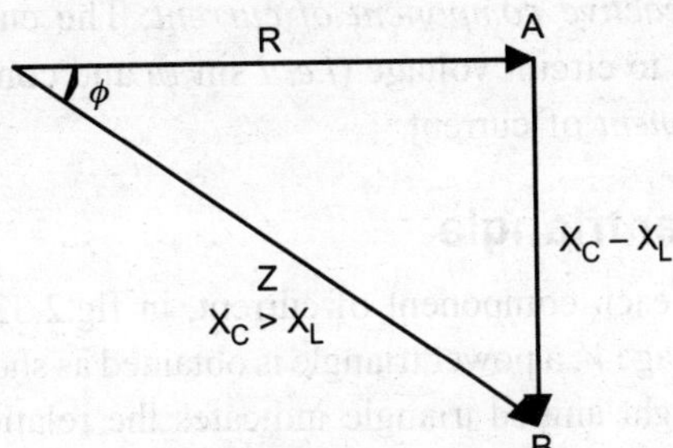

Fig.2.31: Impedance triangle

The concept of impedance triangle is useful since it enables us to calculate:

(*i*) the impedance of the circuit, $Z = \sqrt{R^2 + X^2}$

(*ii*) the power factor of the circuit, $\cos\phi = R/Z$

(*iii*) phase angle, $\phi = \tan^{-1} X_L/R$.

2.25 TRUE POWER AND REACTIVE POWER

The power which is actually consumed or utilised in an a.c. circuit is called *true power* or *active power* or *real power*. We have seen that power is consumed only in resistance. A pure inductor and a pure capacitor do not consume any power, since in a half cycle what so ever power is received from the source by these components the same is returned to the source. This power which flows back and forth (*i.e.* in both directions in the circuit) or reacts upon itself is called *reactive power*. It does not do any useful work in the circuit. It has been seen that in pure resistive circuit current is in phase with the applied voltage, whereas, in pure inductive and capacitive circuit, current is 90° out of phase. Thus, it is concluded that the current in phase with the voltage produces true or active power, whereas, the current 90° out of the phase with the voltage contributes to reactive power. Hence,

True power = voltage × current in phase with voltage

Reactive power = voltage × current 90° out of phase with voltage.

The phasor diagram for an inductive circuit is shown in fig.2.32. where current I lags behind the voltage V by an angle $\phi°$. The current I can be resolved into two rectangular components *i.e.* (*i*) $I\cos\phi$, in phase with voltage V and (*ii*) $I\sin\phi$, which is 90° out of phase with voltage V.

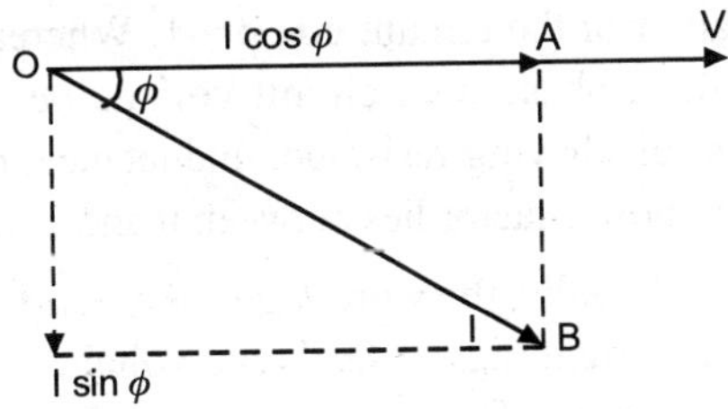

Fig.2.32: Active and reactive component of current

$\therefore$ True power, $P = V \times I\cos\phi = VI\cos\phi$ watt

Reactive power, $P_r = V \times I\sin\phi = VI\sin\phi$ VAR

Apparent power, $P_a = V \times I = VI$ VA

The bigger units of true power, reactive power and apparent power are *kW* (or *MW*), *kVAR* (or *MVAR*) and *kVA* (or *MVA*) respectively.

Active component of current: The current component which is in phase with circuit voltage (*i.e.* $I\cos\phi$) and contributes to active or true power of the circuit is called *active component* or *wattfull component* or *in-phase component* of current.

Reactive component of current: The current component which is in quadrature (or 90° out of phase) to circuit voltage (*i.e.* $I \sin \phi$) and contributes to reactive power of the circuit is called *reactive component* of current.

Power triangle

When each component of current, in fig.2.32, is multiplied by voltage V, a power triangle is obtained as shown in fig.2.33. This right angled triangle indicates the relation among true power, reactive power and apparent power.

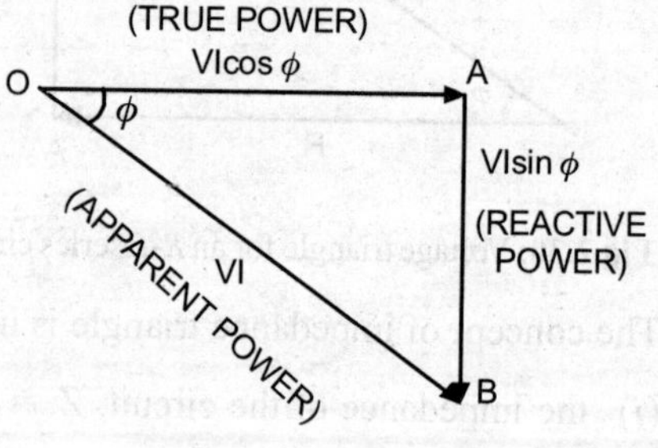

Fig.2.33: Power triangle

In the above discussion, the following points are worth noting:

(*i*) When an active component of current is multiplied with circuit voltage, it results in active or true power. It is this power which produces torque in motors, heat in heaters, light in lamps etc. wattmeters indicate this power.

(*ii*) When the reactive component of current is multiplied with circuit voltage, it results in reactive power. It is this power which merely flows back and forth without doing any work. This power determines the power factor of the circuit.

(*iii*) When the circuit current is multiplied with circuit voltage, it results in apparent power. It is so called because it appears that product of voltage and current is power. But is a.c. circuits (except pure resistive circuit) there is usually phase difference between voltage and current so that VI does not give real power. To avoid confusion, it is measured in volt-ampere.

(*iv*) From power triangle shown in fig.2.33, the power factor may also be determined by taking ratio of true power to apparent power, *i.e.* power factor, $\cos \phi$ = true power/apparent power.

2.26 POWER FACTOR AND ITS IMPORTANCE

In a.c. circuits, the power factor may be expressed as:

$$p.f. = \cos \phi = R/Z = \text{true power/ apparent power}$$

In case of pure resistive circuit, current is in phase with circuit voltage *i.e.* $\phi = 0$. Therefore, power factor of the circuit, $\cos \phi = 1$. Whereas, in case of pure inductive or capacitive circuit, current is 90° out of phase with circuit voltage *i.e.* $\phi = 90°$. Therefore, power factor of the circuit $\cos \phi = 0$. For circuits having resistance-inductance; resistance-capacitance or resistance-inductance and capacitance, the power factor lies between 0 and 1. *It may be noted that the value of p.f. can never be* **more than one.**

Usually, the word *lagging* or *leading* is attached with the numerical value of p.f. to signify whether the current lags or leads the voltage. In inductive circuits, current always lags behind the voltage and their power factors are mentioned as lagging p.f. Whereas, for capacitive circuits, the power factor is mentioned as leading p.f. since in these cases current always leads the voltage vector.

Importance of power factor

The power factor of an a.c. circuit plays an important role in the power system. Since power of an a.c. circuit is given by the relation;

$$P = VI \cos \phi = \text{or } I = \frac{P}{V \cos \phi}$$

From the above relation, it is clear that for fixed power at constant voltage, the current drawn by the circuit increases with decrease in p.f.

Thus at low p.f., a.c. circuits draw more current from their mains and results in the following disadvantages:

1. **Greater conductor size:** At low p.f., the conductors are to carry more current for the same ower therefore, they require larger area of cross-section.
2. **Poor efficiency:** At low power factors, the conductors have to carry larger current which increases copper losses (I^2R) and results in poor efficiency.
3. **Larger voltage drop:** At lower power factors, the conductors have to carry larger current which increases voltage drop (IR) in the system and results in poor regulation.
4. **Larger kVA rating of equipment:** The kVA rating of electrical machines and equipment connected in the power system such as alternators, transformers, switch gears etc. will be more at lower power factors since it is inversely proportional to power factor (*i.e.* kVA = kW/cos ϕ).

To improve the power factor of an a.c. circuit **a capacitor** *is connected across the circuit i.e. parallel to the circuit.*

2.27 Q-FACTOR OF A COIL

Reciprocal of power factor of a coil is known as its *Q-factor*. It is also called quality factor or figure of merit of the coil.

$$\text{Mathematically, } Q\text{-factor} = \frac{1}{p.f.} = \frac{1}{\cos\phi} = \frac{Z}{R}$$

If the value of R is very small in comparison to its inductive reactance X_L then,

$$Q\text{-factor} = \frac{X_L}{R} = \frac{\omega L}{R}$$

$$\text{Also, } Q = 2\pi \times \frac{\text{maximum energy stored}}{\text{energy dissipated per cycle}}$$

Example 2.9. *A coil having a resistance of 15 Ω and an inductance of 63.66 mH is connected across a 250 V, 50 Hz supply. Calculate (i) the reactance and impedance of the coil, (ii) the current, (iii) the phase difference between current and the applied voltage and (iv) the power factor. Draw also the phasor diagram showing voltage and current.*

Solution: The circuit is shown in fig.2.34.

(*i*) Reactance, $X_L = 2\pi f L = 2\pi\, 50 \times 63.66 \times 10^{-3} = \mathbf{20\ \Omega}$ (*Ans.*)

Impedance, $Z = \sqrt{R^2 + X_L^2} = \sqrt{(15)^2 + (20)^2} = \mathbf{25\ \Omega}$ (*Ans.*)

(*ii*) Current, $I = \frac{V}{Z} = \frac{250}{25} = \mathbf{10\ A}$ (*Ans.*)

(*iii*) Phase difference, $\phi = \tan^{1} \frac{X_L}{R} = \tan^{-1} \frac{20}{15} = \tan^{-1} 1.333 = \mathbf{53.13°}$ (*Ans.*)

(*iv*) Power factor, cos ϕ = **0·6 lag** (*Ans.*)

The phasor diagram for the circuit is shown in fig.2.35.

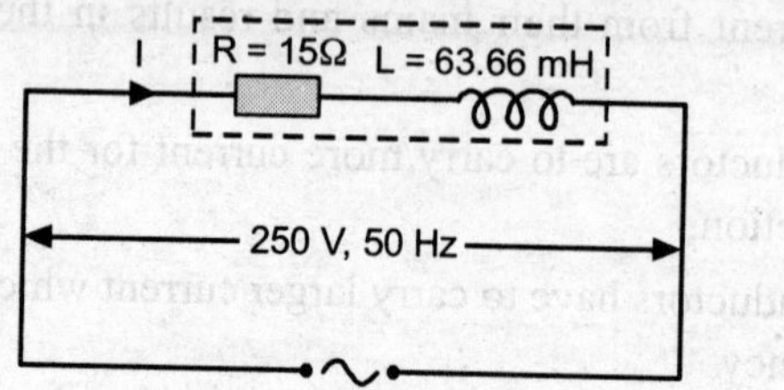

Fig.2.34: Circuit diagram

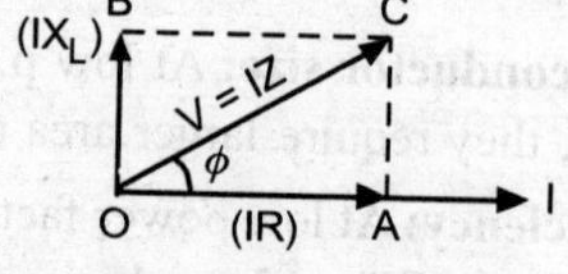

Fig.2.35: Phasor diagram

Example 2.10. *A coil connected to* 50 V *d.c. supply ,draws* 5 A *and the same coil when connected to* 50 V, *a.c. voltage of frequency* 50 Hz *draws* 2.5A. *Calculate the parameters of the coil and power factor.*

Solution: Let the resistance and inductance of the coil be R ohm and L henry respectively. When coil in connected to *d.c.* supply, the opposition is only resistance of the coil,

$$\therefore \quad \text{Resistance of the coil, } R = \frac{V_{d.c.}}{I_{d.c.}} = \frac{50}{5} = \mathbf{10\Omega}\ (Ans.)$$

When coil is connected across *a.c.* supply of 50 V, 50 Hz, the opposition is impedance of the coil,

$$\therefore \quad \text{Impedance of the coil, } Z = \frac{V_{ac}}{I_{ac}} = \frac{50}{2.5} = 20\ \Omega$$

$$\text{Now, } Z = \sqrt{R^2 + X_L^2} \quad \text{or} \quad Z^2 = R^2 + X_L^2$$

or

$$X_L = \sqrt{Z^2 - R^2} = \sqrt{(20)^2 - (10)^2} = \sqrt{300}$$

and

$$L = \frac{X_L}{2\pi f} = \frac{\sqrt{300}}{2\pi \times 50} = \mathbf{55{\cdot}13\ mH}\ (Ans.)$$

$$\therefore \quad \text{Parameters are } R = 10\ \Omega \text{ and } L = 55{\cdot}13 \text{ mA}$$

$$\bullet \quad \text{Power factor, } \cos\phi = \frac{R}{Z} = \frac{10}{20} = \mathbf{0{\cdot}5\ lagging}\ (Ans.)$$

Example 2.11. *A coil when connected to 230 V d.c. supply dissipates 2645 watt of power. When connected across 230 V a.c. supply of frequency 50 Hz., it dissipates 1058 watt of power. Calculate the value of resistance and inductance of the coil.*

Solution: When d.c. is applied across the coil, the opposition is only resistance of the coil,

$$\therefore R = \frac{V_{dc}^2}{P_{dc}} = \frac{(230)^2}{2645} = \mathbf{20\ \Omega}\ (Ans.)$$

When a.c. is applied across the coil;

$$P_{ac} = I_{ac}^2 R \text{ or } I_{ac} = \sqrt{P_{ac}/R} = \sqrt{1058/20} = 7.273 \text{ A}$$

$$\text{Impedance, } Z = \frac{V_{ac}}{I_{ac}} = \frac{230}{7.273} = 31{\cdot}623\ \Omega$$

$$X_L = \sqrt{Z^2 - R^2} = \sqrt{(31{\cdot}623)^2 - (20)^2} = 24{\cdot}495\ \Omega$$

$$\text{Inductance, } L = \frac{X_L}{2\pi f} = \frac{24{\cdot}495}{2\pi \times 50} = \mathbf{0{\cdot}078\ H}\ (Ans.)$$

Example 2.12. *The voltage and current through a circuit element are*

$$v = 100 \sin(314\,t + 55°) \text{ volt}$$
$$i = 10 \sin(314\,t + 325°) \text{ ampere}$$

Find the value of power drawn by the element.

Solution: Given

$$v = 100 \sin(314\,t + 55°) \text{ V}$$
$$i = 10 \sin(314\,t + 325°) \text{ A}$$

or

$$i = 10 \sin(314\,t - 35°) \text{ A}$$

Now their phasor representation is shown in fig.2.36.

∴ Phase difference between voltage and current is 90°.

Now power drawn by the circuit, $P = VI \cos\phi$

$$= \frac{100}{\sqrt{2}} \times \frac{10}{\sqrt{2}} \times \cos 90°$$

$$= \frac{1000}{2} \times 0° = \mathbf{0\ W}\ (Ans.)$$

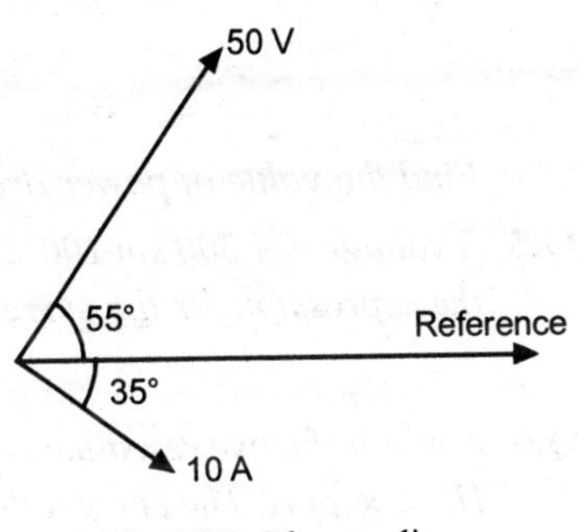

Fig.2.36: Phasor diagram

which indicates that element is pure inductive.

Example 2.13. *A coil of resistance 1·5 ohm and impedance 6 ohm is placed in series with a second coil of resistance 2 ohm. When a voltage of 230 V, 50 Hz is applied to the circuit, the current flowing through the circuit is 7 A. Find the inductance of the second coil.*

Solution: The circuit is shown fig.2.37.

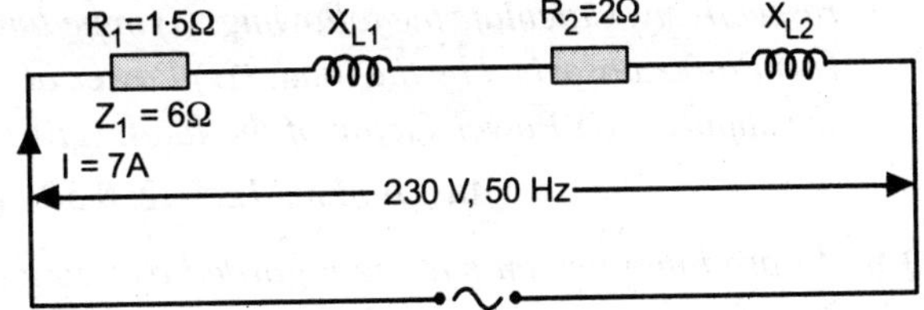

Fig.2.37: Circuit as per given data

Impedance of whole circuit,

$Z = V/I = 230/7 = 32{\cdot}86\ \Omega$

Resistance of whole circuit,

$R = R_1 + R_2 = 1{\cdot}5 + 2 = 3{\cdot}5\ \Omega$

Inductive reactance,

$$X_L = \sqrt{Z^2 - R^2} = \sqrt{(32{\cdot}86)^2 - (3{\cdot}5)^2} = 32{\cdot}67\ \Omega$$

Inductive reactance of I coil, $X_{L1} = \sqrt{Z_1^2 - R_1^2} = \sqrt{(6)^2 - (1{\cdot}5)^2} = 5{\cdot}81\ \Omega$

Inductive reactance of II coil, $X_{L2} = X_L - X_{L1} = 32{\cdot}67 - 5{\cdot}81 = 26{\cdot}86\ \Omega$

∴ Inductance, $L_2 = X_{L2}/2\pi f = 26{\cdot}86/2\pi \times 50 =$ **85·5 m H** *(Ans.)*

PRACTICE EXERCISE

Q.1 *A coil having a resistance of 12 Ω and an inductance of 0.1 H is connected across a 100 V, 50 Hz supply. Calculate (i) the reactance and impedance of the coil, (ii) the current, (iii) the phase difference between current and the applied voltage and (iv) the power factor. Draw also the phasor diagram showing voltage and current.* (**Ans.** 31.416 W; 33.63 W; 2.97 A; 69.1°; 0.3568 lag)

Q.2 *A coil connected to* 100 V *d.c. supply draws* 10 *and the same coil when connected to* 100 V, *a.c. voltage of frequency* 50 *Hz draws* 5*A. Calculate the parameters of the coil and power factor.* (**Ans.** *10 Ω; 55.13 mH; 0.5 lagging*)

Q.3 *A coil when connected to 230 V d.c. supply dissipates 2000 watt of power. When connected across 230 V a.c. supply of frequency 50 Hz., it dissipates 800 watt of power. Calculate the value of resistance and inductance of the coil.* (**Ans.** *20 W, 0.078 H*)

Q.4 *The voltage and current through a circuit element are*

$$v = 50 \sin (314\,t + 70°) \text{ volt}$$

$$i = 10 \sin (314\,t + 340°) \text{ ampere}$$

Find the value of power drawn by the element. (**Ans.** *0 Ω*)

Q.5 *A voltage $e = 200 \sin 100\,\pi t$ is applied to a coil having R = 200 ohm and L = 638 millihenry. Find the expression for the current and also determine the power taken by the coil.* (**Ans.** *0.706 sin (100 π t – 45.06°) 50 Ω*)

Q.6 *A non-inductive resistance of* 10 *ohm is connected in series with an inductive coil across* 200 *V,* 50 *Hz ac supply. The current drawn by the series combination is* 10 *ampere. The resistance of the coil is* 2 *ohm. Determine (i) inductance of the coil (ii) Power factor (iii) voltage across the coil.* (**Ans.** *50.93 mH; 0.6 lag, 161.24 V*)

Q.7 *An inductive load is connected in series with a non-inductive resistance of 8 ohms. The combination is connected across an a.c. supply of 100 volt, 50 Hz. A voltmeter connected across the non-inductive resistor and then across the inductive load gives the reading of 64 volt and 48 volt respectively. Calculate the following: (i) impedance of the load; (ii) Impedance of the combination; (iii) Power absorbed by the load; (iv) Power absorbed by the resistor; (v) Total power taken from the supply; (vi) Power factor of the load; (vii) Power factor of the whole circuit.* (**Ans.** *6 ohm; 12.5 Ω; 0.586 lag; 225 W; 512 W; 737 W; 0.586 lag; 0.92 lag*)

Q.8 *An arc lamp (which may be regarded as being non-inductive) takes 10 A at 50 V. Calculate the impedance of choke of 1ohm resistance to be placed in series with it in order that it may be operated at 200 V, 50 Hz supply. Find also the total power used and the power factor.* (**Ans.** *19.105 Ω; 0.3 lag; 600 W*)

2.28 R – C SERIES CIRCUIT

A circuit that contains a pure resistance *R* ohm connected in series with a pure capacitor of capacitance *C* farad is known as *R* – *C* series circuit.

An *R* – *C* series circuit and its phasor diagram is shown in fig.2.38 and 2.39 respectively. To draw, the phasor diagram, current *I* (*r.m.s.* value) is taken as the reference vector. Voltage drop in resistance V_R

(= IR) is taken in phase with current vector, whereas, voltage drop in capacitive reactance V_C (= IX_C) is taken 90° behind the current vector (since current leads the voltage by 90° in pure capacitive circuit). The vector sum of these two voltage drops is equal to the applied voltage V (r.m.s. value).

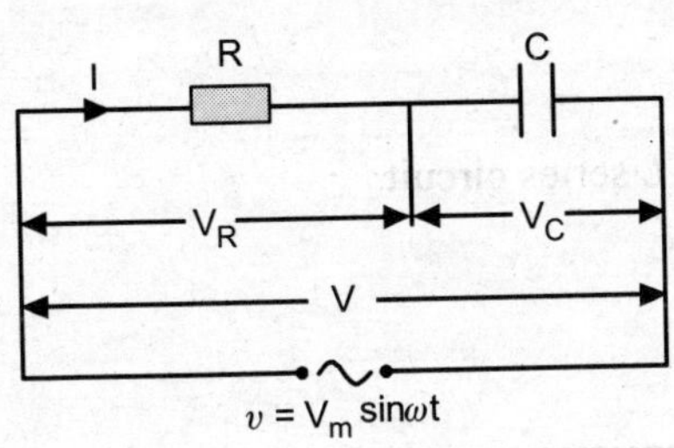

Fig.2.38: R-C series circuit

Fig.2.39: Phasor diagram

Now $V_R = IR$ and $V_C = IX_C$ (where $X_C = 1/2\pi f C$)

In right angled triangle OAB

$$V = \sqrt{(V_R)^2 + (V_C)^2} = \sqrt{(IR)^2 + (IX_C)^2} = I\sqrt{R^2 + X_C^2}$$

or
$$I = \frac{V}{\sqrt{R^2 + X_C^2}} = \frac{V}{Z}$$

where $Z = \sqrt{R^2 + X_C^2}$ is the total opposition offered to the flow of alternating current by an $R - C$ series circuit and is called *impedance* of the circuit. It is measured in ohm.

Phase angle: From the phasor diagram it is clear that current in this circuit leads the applied voltage by an angle ϕ called *phase angle.*

From the phasor diagram shown in fig.2.38;

$$\tan\phi = \frac{V_C}{V_R} = \frac{IX_C}{IR} = \frac{X_C}{R} \quad \text{or} \quad \phi = \tan^{-1} X_C/R$$

Power: If the alternating voltage applied across the circuit is given by the equation:

$$v = V_m \sin\omega t \quad ...(i)$$

Then,
$$i = I_m \sin(\omega t + \phi) \quad ...(ii)$$

∴ Instantaneous power,

$$p = vi = V_m \sin\omega t\, I_m \sin(\omega t + \phi) = \frac{V_m I_m}{2} 2\sin(\omega t + \phi)\sin\omega t$$

$$= \frac{V_m}{\sqrt{2}}\frac{I_m}{\sqrt{2}}[\cos\phi - \cos(2\omega t + \phi)] = \frac{V_m}{\sqrt{2}}\frac{I_m}{\sqrt{2}}\cos\phi + \frac{V_m}{\sqrt{2}}\frac{I_m}{\sqrt{2}}\cos(2\omega t + \phi)$$

Average power consumed in the circuit over a complete cycle,

$$P = \text{average of } \frac{V_m}{\sqrt{2}}\frac{I_m}{\sqrt{2}}\cos\phi - \text{average of } \frac{V_m}{\sqrt{2}}\frac{I_m}{\sqrt{2}}\cos(2\omega t + \phi)$$

or $$P = \frac{V_m}{\sqrt{2}} \cdot \frac{I_m}{\sqrt{2}} \cos\phi - \text{zero} = V_{r.m.s.}\, I_{r.m.s.} \cos\phi = VI\cos\phi$$

where $\cos\phi$ is called *power factor* of the circuit.

From phasor diagram

$$\cos\phi = \frac{V_R}{V} = \frac{IR}{IZ} = \frac{R}{Z} \text{ same as in } R - L \text{ series circuit.}$$

Alternatively; Power,

$$P = VI\cos\phi = IZ \cdot I \cdot \frac{R}{Z} = I^2R$$

This shows that power is actually consumed in resistance only; capacitor does not consume any power.

Power curve: The phasor diagram and wave diagram for voltage and current are shown in fig.2.40 (*a*) and 2.40 (*b*) respectively, where applied voltage ($v = V_m \sin \omega t$) is taken as reference quantity. The power curve for $R - C$ circuit is also shown in 2.40 (*b*). The points on the power curve are obtained from the product of the corresponding instantaneous values of voltage and current. It is clear that power is negative between angle ($180° - \phi$) and 180° and between ($360° - \phi$) and 360°. During rest of the cycle, the power is positive. Since the area under the positive loops is greater than that under the negative loops, the net power over a complete cycle is positive. Hence, a definite quantity of power is utilised or consumed by this circuit.

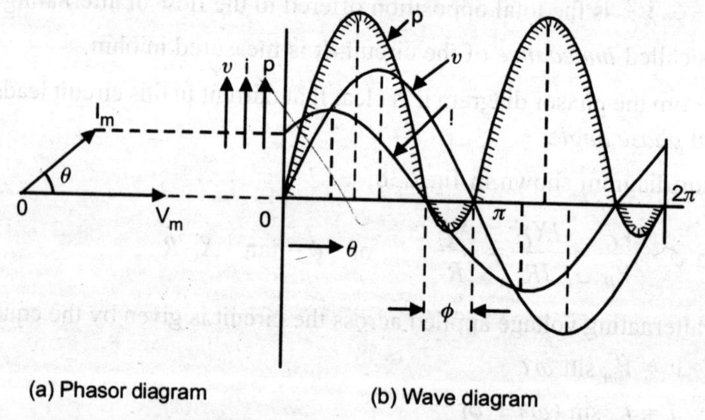

Fig.2.40: Phasor and wave diagram

Impedance triangle

When each side of the simplified phasor diagram shown in fig.2.41 is divided by a common factor I, we get another right angled triangle (shown in fig.2.42 known as *impedance triangle*.

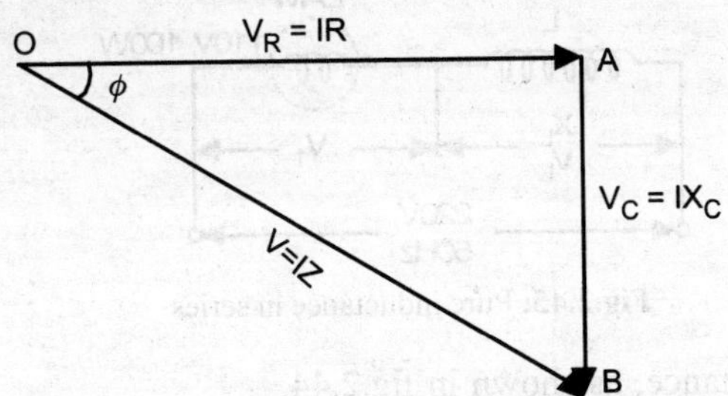

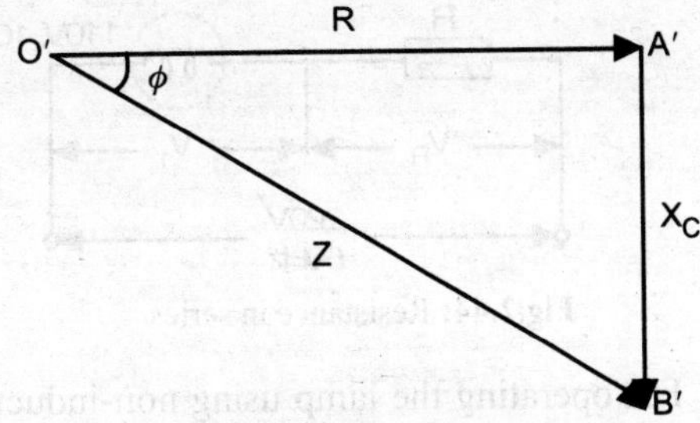

Fig.2.41: Voltage Triangle **Fig.2.42:** Impedance triangle

Example 2.14. *A resistance of 20 ohm and capacitor of 212.2 μF capacitance are connected in series across a 250 V, 50 Hz supply. Calculate (i) impedance of the circuit; (ii) current; (iii) power factor and phase angle; (iv) power consumed in the circuit.*

Solution: The circuit is shown in fig.2.43.

Impedance, $Z = \sqrt{R^2 + X_C^2}$

where, $X_C = 1/2\,\pi f C = 1/2\,\pi \times 50 \times 212.2 \times 10^{-6}$

$= 15\ \Omega;$

$R = 20\ \Omega$

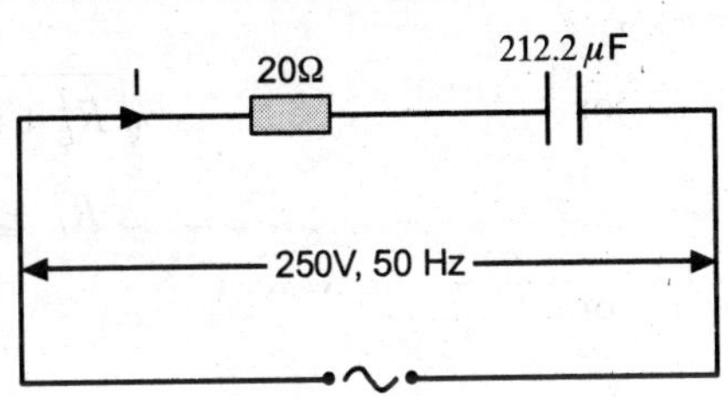

Fig.2.43: Circuit diagram

(*i*) $\therefore Z = \sqrt{(20)^2 + (15)^2} = \mathbf{25\ \Omega}$ (*Ans.*)

(*ii*) Current, $I = \dfrac{V}{Z} = \dfrac{250}{25} = \mathbf{10\ A}$ (*Ans.*)

(*iii*) Power factor, $\cos\phi = \dfrac{R}{Z} = \dfrac{20}{25} = \mathbf{0{\cdot}8}$ **leading** (*Ans.*)

Phase angle, $\phi = \cos^{-1} 0{\cdot}8 = \mathbf{36.87^\circ}$ (*Ans.*)

(*iv*) Power, $P = V I \cos\phi = 250 \times 10 \times 0{\cdot}8 = \mathbf{2000\ W}$ (*Ans.*)

Example 2.15. *A 110 V, 100 W lamp is to be operated on 220 V, 50 Hz supply mains. In order that lamp should operate in correct voltage. Calculate value of:*

(i) non inductive resistance *(ii) pure inductance* *(iii) pure capacitance*

Ans. Lamp's rating: 110 V, 100 W

Supply voltage, $V_S = 220$ V and Frequency, $f = 50$ Hz

Resistance of the lamp, $R_L = \dfrac{(110)^2}{100} = 121\ \Omega$

Operating current, $I = \dfrac{100}{110} = 0{\bullet}9091$ A

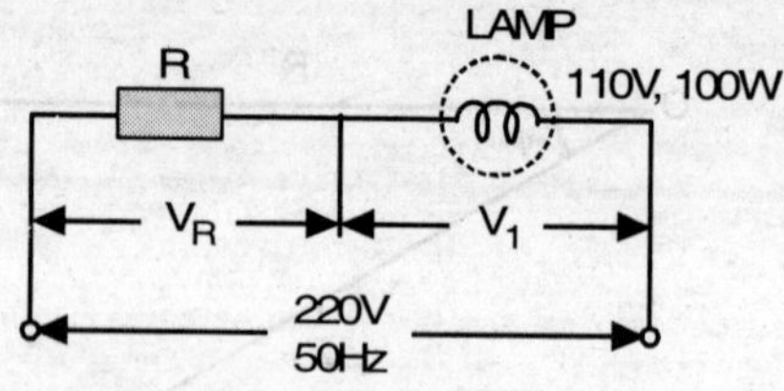

Fig.2.44: Resistance in series

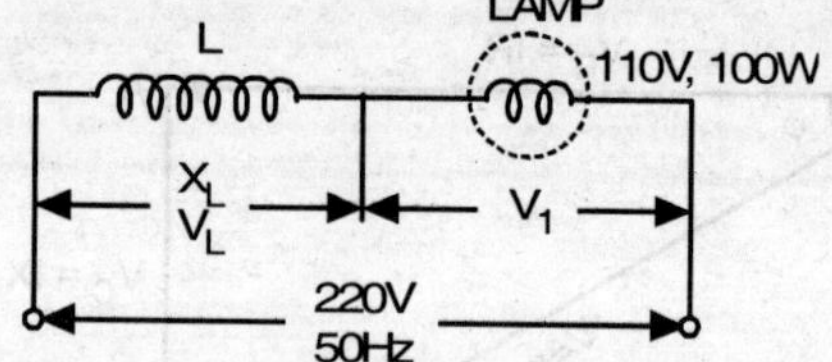

Fig.2.45: Pure inductance in series

(i) For operating the lamp using non-inductive resistance, as shown in fig.2.44.
Let the value of resistance be $R\ \Omega$.

$\therefore \quad I\,(R + R_L) = V$

or $\quad R + R_L = \dfrac{220}{0.9091} = 242 \text{ ohm}$

or $\quad R = 242 - R_L = 242 - 121 = \mathbf{121\ \Omega}$ (*Ans.*)

(ii) For operating the lamp using pure inductance as shown in Fig.2.45.
Let the value of inductance be L henry and $X_L = 2\,\pi f L$,

$\therefore \quad I\,Z = V \quad$ or $\quad \dfrac{V}{I} = \dfrac{220}{0.9091} = 242\ \Omega$

or $\quad \sqrt{R_L^2 + X_L^2} = 242$

$R_L^2 + X_L^2 = (242)^2 \quad$ or $\quad X_L^2 = 242^2 - 121^2$

or $\quad X_L = \sqrt{242^2 - 121^2} = 209.58\ \Omega$

or $\quad 2\,\pi f L = 209.58 \quad$ or $\quad L = \dfrac{209.58}{2\,\pi \times 50} = \mathbf{0.667\ H}$ (*Ans.*)

(iii) For operating the lamp using pure capacitor, as shown in fig.2.46.
Let C be the capacitance of the capacitor.

$$X_C = \frac{1}{2\,\pi f C}$$

And $\quad I\,Z = V \quad$ or $\quad Z = \dfrac{V}{I} = \dfrac{220}{0.9091} = 242\ \Omega$

$\sqrt{R_L^2 + X_c^2} = Z \quad$ or $\quad R_L^2 + X_C^2 = Z^2$

or $\quad X_C = \sqrt{Z^2 - R_L^2} = \sqrt{(242)^2 + (121)^2} = 209.58\ \Omega$

or $\quad \dfrac{1}{2\pi f C} = X_c \quad$ or $\quad C = \dfrac{1}{2\,\pi \times 50 \times 209.58} = \mathbf{15.188\ \mu F}$ (*Ans*)

Fig.2.46: Pure capacitor in series

Example 2.16. *A 400 Hz generator has an induced e.m.f. of 100 V and an internal impedance (5 + j0) ohm. If it supplies an impedance consisting of a 40 ohm capacitive reactance in series with 10* *ohm resistance, what is the magnitude of the current passing through it ? Determine also the voltage at the terminals of the generator.*

Solution: Simple circuit is shown in fig.2.47.

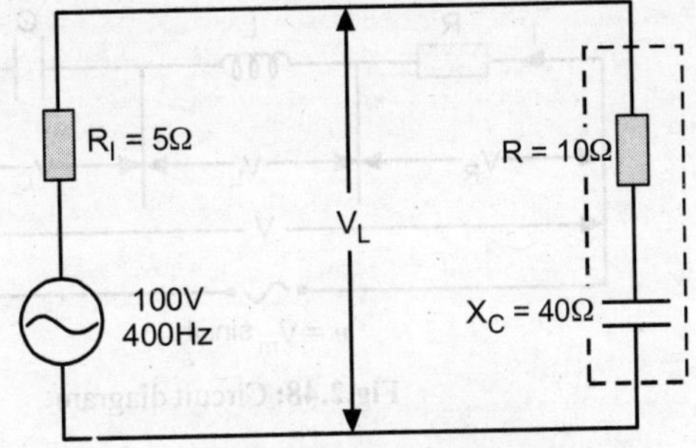

Fig.2.47: Circuit diagram

Total impedance of the circuit.

$$Z_T = \sqrt{(R_i + R)^2 + (X_C)^2}$$

$$= \sqrt{(5+10)^2 + (40)^2} = 42{\cdot}72\ \Omega$$

Circuit current, $I = V/Z = 100/42{\cdot}72 =$ **2·341 A** (*Ans.*)

Load impedance, $Z_L = \sqrt{R^2 + X_C^2} = \sqrt{(10)^2 + (40)^2} = 41{\cdot}231\ \Omega$

Terminal voltage, $V_L = IZ_L = 2{\cdot}341 \times 41{\cdot}231 =$ **96·52 V** (*Ans.*)

PRACTICE EXERCISE

Q.1 *A resistance of 15 ohm and capacitor of 150 m F capacitance are connected in series across a 230 V, 50 Hz supply. Calculate (i) impedance of the circuit; (ii) current; (iii) power factor and phase angle; (iv) power consumed in the circuit.*

(***Ans.*** 25·987 Ω; 8·85 A; 0·577 leading, 54.75°; 1174·9 W)

Q.2 *A 120 V, 60 W lamp is to be operated on 220 V, 50 Hz supply mains. In order that lamp should operate in correct voltage. Calculate value of: (i) non inductive resistance, (ii) pure inductance, (iii)* Pure capacitance (***Ans.*** 200 Ω; 1·174 H; 8.63mF Ω)

Q.3 *A voltage of 125 V at 50 Hz is applied across a non inductive resistor connected in series with a condenser. The current in the circuit is 2·2 A. The power loss in the resistor is 96·8 W and that in the condenser is negligible. Calculate the resistance and the capacitance.*

(***Ans.*** 20 ohm; 53.18 ohm; 59·85 μ F)

Q.4 *A resistor R in series with a capacitor C is connected to a 50 Hz, 240 V supply. Find the value of C so that R absorbs, 300 W at 100 V.* (***Ans.*** 43·77 μF)

2.29 R – L – C SERIES CIRCUIT

A circuit that contains a pure resistance of R ohm, a pure inductance of L henry and a pure capacitor of capacitance C farad; all connected in series is known as $R - L - C$ series circuit.

An $R - L - C$ series circuit is shown in fig.2.48.

Here $X_L = 2\ \pi f L$ and $X_C = 1/2\ \pi f C$

When a resulting current I (r.m.s. value) flows through the circuit, the voltage across each component will be

$V_R = IR$ *i.e.* voltage across R .... in phase with I;
$V_L = IX_L$ *i.e.* voltage across L .... leads I by 90°;
$V_C = IX_C$ *i.e.* voltage across C .... lags I by 90°;

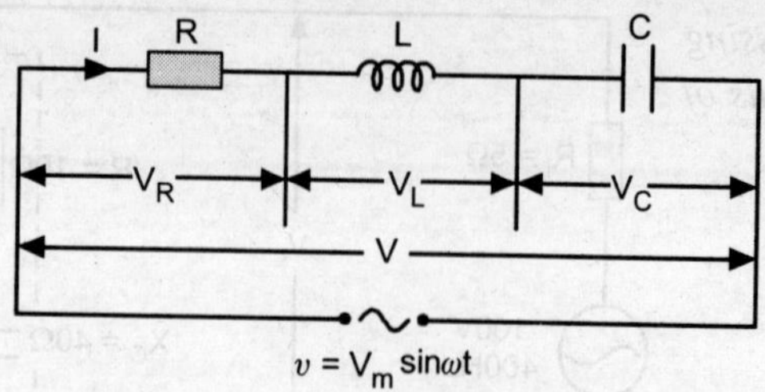

Fig.2.48: Circuit diagram

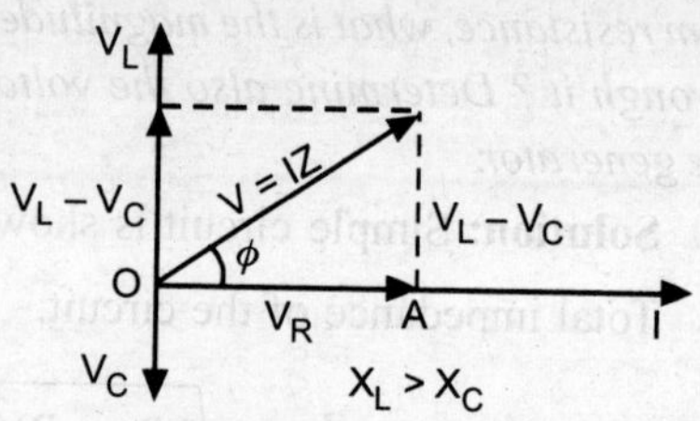

Fig.2.49: Phasor diagram

The phasor diagram is shown in fig.2.49 where current is taken as the reference phasor. Since voltage across inductance V_L leads the current vector I by 90° and voltage across capacitance V_C lags the current vector I by 90°, they act opposite to each other. If $V_L > V_C$, in effect, the circuit behaves as an inductive circuit but when $V_L < V_C$, the circuit behaves as a capacitive circuit. Here, the phasor diagram is drawn for an inductive circuit (*i.e.* when $V_L > V_C$).

$$V = \sqrt{(V_R)^2 + (V_L - V_C)^2} = \sqrt{(IR)^2 + (IX_L - IX_C)^2}$$

or $$V = I\sqrt{R^2 + (X_L - X_C)^2}$$

or $$I = \frac{V}{\sqrt{(R)^2 + (X_L - X_C)^2}} = \frac{V}{Z}$$

where $Z = \sqrt{R^2 + (X_L - X_C)^2}$ is the total opposition offered to the flow of alternating current by an $R - L - C$ series circuit and is called *impedance* of the circuit.

Phase angle: From phasor diagram: $\tan\phi = \frac{V_L - V_C}{V_R} = \frac{X_L - X_C}{R}$

or $$\phi = \tan^{-1}\frac{X_L - X_C}{R}$$

Power: Average power, $P = VI\cos\phi = I^2 R$

Power, factor, $\cos\phi = \frac{V_R}{V} = \frac{R}{Z}$

If the alternating voltage applied across the circuit is given by the equation

$$v = V_m \sin\omega t$$

the circuit current is represented by the equation as per the constants or parameters explained below:

Three cases of $R - L - C$ series circuit

(*i*) When $X_L > X_C$, the phase angle ϕ is positive. In effect, the circuit behaves as an $R - L$ series circuit. The circuit current lags behind the applied voltage and p.f. is lagging. The current is given by the equation.

$$i = I_m \sin(\omega t - \phi)$$

(*ii*) When $X_L < X_C$, the phase angle ϕ is negative. In effect, the circuit behaves as an $R - C$ series circuit. The circuit current leads the applied voltage and p.f. is leading. The current is given by the equation.

$$i = I_m \sin(\omega t + \phi)$$

(*iii*) When $X_L = X_C$, the phase angle ϕ is zero. In effect, the circuit behaves like a pure resistive circuit. The circuit current is in phase with applied voltage and p.f. is unity. The current is given by the equation.

$$i = I_m \sin \omega t$$

Impedance triangle

Fig.2.50 shows the impedance triangle of the circuit when $X_L > X_C$, whereas, fig.2.51 shows the impedance triangle of the circuit when $X_L < X_C$.

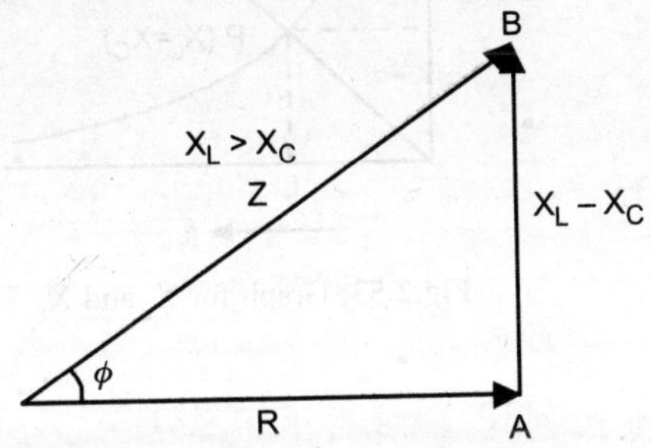

Fig.2.50: Impedance triangle $X_L > X_C$

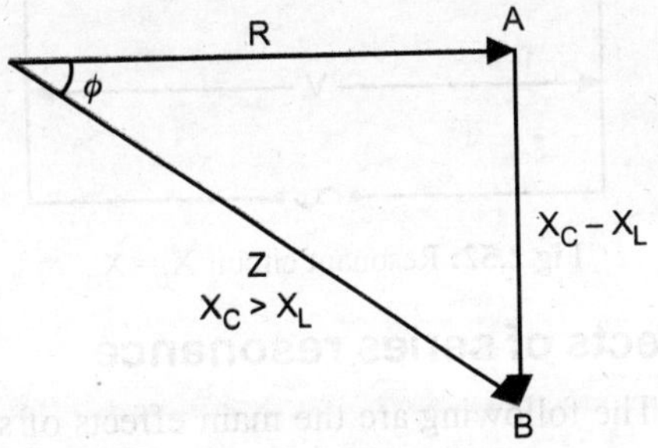

Fig.2.51: Impedance triangle $X_L < X_C$

2.30 SERIES RESONANCE

In an $R - L - C$ series circuit, when circuit current is in phase with the applied voltage, the circuit is said to be in series resonance. This condition is obtained in an $R - L - C$ circuit, shown in fig.2.52,

when $X_L = X_C$ (or $X_L - X_C = 0$)

At resonance; $X_L - X_C = 0$ or $X_L = X_C$

$$\text{Impedance, } Z_r = \sqrt{R^2 + (X_L - X_C)^2} = R$$

$$\text{Current, } I_r = \frac{V}{Z_r} = \frac{V}{R}$$

Since, at resonance, the opposition to the flow of current is only resistance (R) of the circuit, the circuit draws *maximum current* under this conditions.

Resonant frequency: The value of X_L (= $2\pi f L$) and X_C (= $1/2\pi f C$) can be changed by changing the supply frequency. When frequency increases, the value of X_L, increases, whereas, the value of X_C decreases, and *vice-versa.* Thus to obtain series resonance, the frequency is adjusted to f_r so that $X_L = X_C$, the condition at point P shown in fig.2.53.

$\therefore$ At series resonance, $X_L = X_C$

$$2\,\phi f_r\, L = \frac{1}{2\pi f_r C} \quad \text{or} \quad f_r = \frac{1}{2\pi\sqrt{LC}}$$

where f_r is the *resonant frequency* in Hz when L and C are measured in henry and farad respectively.

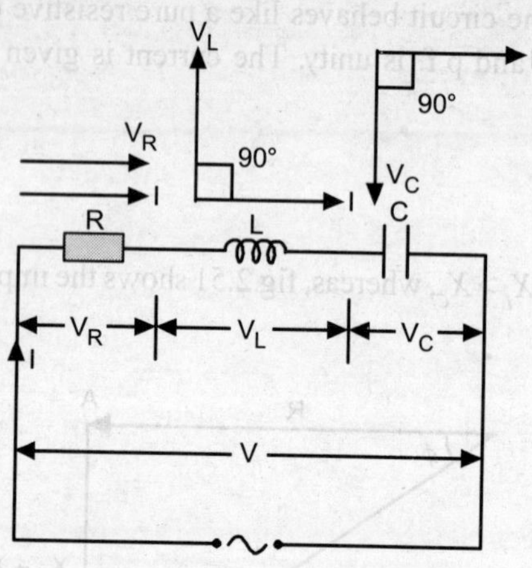

Fig.2.52: Resonant circuit $X_L = X_C$

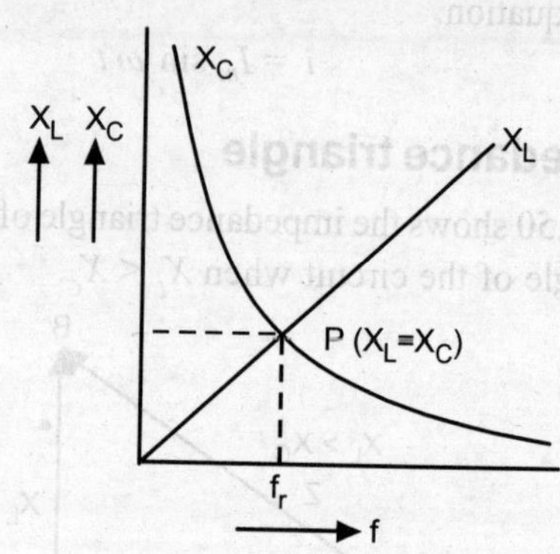

Fig.2.53: Graph for X_L and X_C

Effects of series resonance

The following are the main effects of series resonance:

(*i*) At resonance $X_L = X_C$, therefore, the impedance of the circuit is minimum and is reduced to the resistance of the circuit only, *i.e.*

$$Z_r = R$$

(*ii*) Since, impedance is minimum, the circuit current is maximum at resonance, *i.e.*

$$I_r = V/Z_r = V/R$$

(*iii*) Power taken by the circuit is maximum, as I_r is maximum, *i.e.*

$$P_r = I_r^2 R$$

(*iv*) As the current drawn by the circuit, at resonance, is very large (maximum), the voltage drop across L (*i.e.* $V_L = IX_L = I \times 2\pi f_r L$) and C (*i.e.* $V_C = IX_C = I \times 1/2\pi f_r C$) are also very large.

In power system, at resonance, the excessive voltage built up across the inductive and capacitive components (such as circuit breakers, reactors, etc.) may cause damage. Therefore, series resonance should be avoided in power system. However, in some of the electronic devices (such as antenna circuit of radio and *TV* receiver, tuning circuits etc.), the principle of series resonance is used to increase the signal voltage and current at a desired frequency (f_r).

Since a series resonant circuit has the capability to draw heavy current and power from the mains, it is often regarded as **acceptor circuit.**

2.31 RESONANCE CURVE

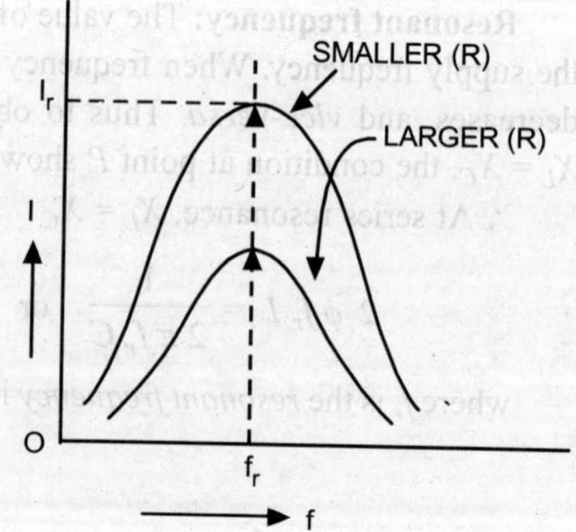

Fig.2.54: Resonance curve

The curve obtained by plotting a graph between current and frequency is known as resonance curve. A resonance curve of a typical $R - L - C$ series circuit is shown in fig.2.54. It may be noted that current reaches its maximum value at the resonant frequency (f_r), falling off rapidly on either side of that point. It is because when the value of frequency is lower than resonance frequency, $X_C > X_L$ and when the value of frequency is higher than f_r, $X_C < X_L$. In both the cases, impedance of the circuit increases ($Z > Z_r$) and the value of current decreases.

Note that resistance of the circuit also plays its own role. Smaller the resistance, the greater the current at resonance.

Bandwidth: The range of frequency over which circuit current is equal to or more than 70·7% of maximum value (*i.e.* I_r, current at resonance) is known as the *bandwidth* of a series resonant circuit.

Fig.2.55 shows a resonance curve of a typical $R - L - C$ circuit where the circuit current is equal to or greater then 70·7% of maximum current (*i.e.* $I_r = V/R$) between frequency range f_1 to f_2.

$\therefore$ Bandwidth, $BW = f_2 - f_1$

Here, the frequency f_1 is called the *lower cut-off frequency* and the frequency f_2 is called the *upper cur-off frequency*. The bandwidth represents the frequency range at which the circuit offers low impedance to circuit current.

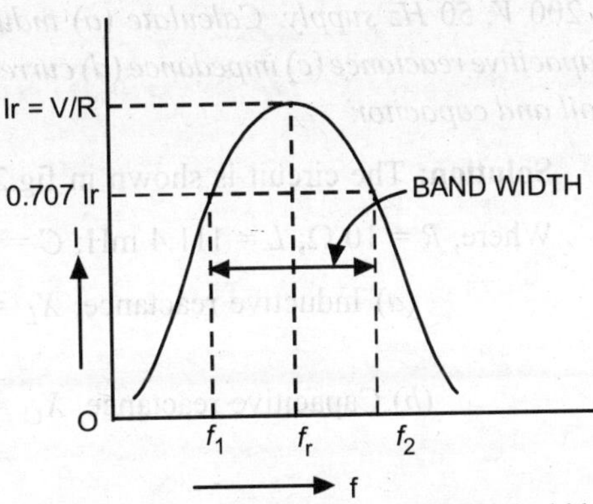

Fig.2.55: Resonance curve showing band width.

The following points may be noted here:

(*i*) If the resonant frequency is not located at the centre of upper and lower cur-off frequency, then

$$f_r = \sqrt{f_1 f_2}$$

(*ii*) When the resonant frequency is located sufficiently near to the centre of the two cut-off frequencies and Q of the circuit is ³ 10, then

$$f_1 = f_r - \frac{BW}{2} \text{ and } f_2 = f_r + \frac{BW}{2}$$

Selectivity: From the resonance curve, it is clear that for smaller resistance the resonance curve is sharp and flat for the larger resistance. A sharper resonance curve provides smaller band of frequencies to give reasonable response and hence provides better selectivity. It also shows that selectivity is reciprocal of bandwidth.

2.32 Q-FACTOR OF SERIES RESONANT CIRCUIT

We have seen that at series resonance, the circuit draws largest current from the mains, this produces a heavy voltage across L or C. The factor by which the *p.d.* across L or C rises to that of the applied voltage is called the *Q-factor* of the series resonant circuit.

$$\therefore \quad Q\text{-factor} = \frac{\text{Voltage across } L \text{ or } C}{\text{Applied voltage}} = \frac{I_r X_L}{I_r R} = \frac{X_L}{R} = \frac{\omega_r L}{R}$$

where, $$\omega_r = 2\pi f_r = 2\pi \frac{1}{2\pi\sqrt{LC}} = \frac{1}{\sqrt{LC}}$$

$$\therefore \quad Q\text{-factor} = \frac{L}{R} \times \frac{1}{\sqrt{LC}} = \frac{1}{R}\sqrt{\frac{L}{C}}$$

The value of Q-factor depends entirely upon the design of coil (*i.e.* $R - L$ which is a part of $R - L - C$ circuit).

Example 2.17. *A coil resistance 10Ω and inductance 111.4 mH is connected in series with a capacitor of 159.16 μF across a 200 V, 50 Hz supply. Calculate (a) inductive reactance (b) capacitive reactance (c) impedance (d) current (e) voltage across coil and capacitor.*

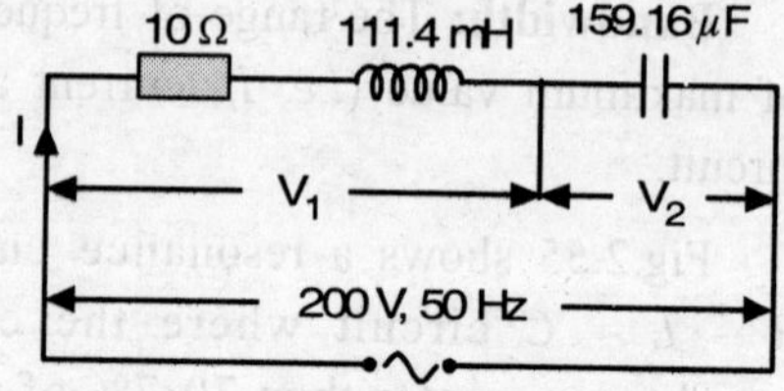

Fig.2.56: Circuit as per deta given

Solution: The circuit is shown in fig.2.56.

Where, $R = 10\ \Omega$; $L = 111.4$ mH; $C = 159.16\ \mu\text{F} = 159.16 \times 10^{-6}$ F; $V = 200$ V; $f = 50$ Hz

(*a*) Inductive reactance, $X_L = 2\pi f L = 2\pi \times 50 \times 111.14 \times 10^{-3} = \mathbf{35\ \Omega}$ (*Ans.*)

(*b*) Capacitive reactance, $X_C = \frac{1}{2\pi f C}$

$$= \frac{1}{2\pi \times 50 \times 159.16 \times 10^{-6}}$$

$$= \mathbf{20\ \Omega}\ (Ans.)$$

(*c*) Impedance, $Z = \sqrt{R^2 + (X_L - X_C)^2} = \sqrt{10^2 + (35 - 20^2)}$

$= \mathbf{25\ \Omega}$ (*Ans.*)

(*d*) Current, $I = \frac{V}{Z} = \frac{200}{25} = \mathbf{8\ A}$ (*Ans.*)

(*e*) Voltage across coil, $V_1 = I\,Z_{\text{COIL}} = I\sqrt{R^2 + X_L^2}$

$= 8\sqrt{10^2 + 35^2} = \mathbf{291.2\ V}$ (*Ans.*)

Voltage across capacitor, $V_2 = I X_C = 8 \times 20 = \mathbf{160\ V}$ (*Ans.*)

Example 2.18. *Find applied voltage and power loss in the circuit shown in fig.2.57.*

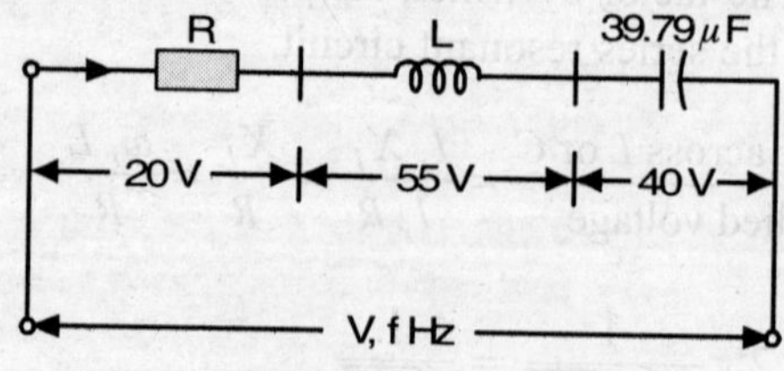

Fig.2.57: Given circuit

Solution: Here, $C = 39.79$ mF; $X_C = \dfrac{1}{2\pi fC} = \dfrac{1}{2\pi \times 50 \times 39.79 \times 10^{-6}} = 80$ W

Applied voltage, $V = \sqrt{20^2 + (55 - 40)^2} = \mathbf{25\ V}$ *(Ans.)*

Current, $I = \dfrac{V_C}{X_C} = \dfrac{40}{80} = 0.5\text{A}$

Power loss, $P = V_R \times I = 20 \times 0{\cdot}5 = \mathbf{10\ watt}$ *(Ans.)*

Example 2.19. *A coil of resistance 12 Ω and inductance 0·051 H, a non-inductive resistance of 20 Ω resistance and a loss-free 212.2 μF capacitor are connected across a 240 V, 50 Hz sinusoidal supply, Calculate (i) the current and (ii) the power factor of the circuit.*

Fig.2.58: Circuit as per data

Solution: The circuit is shown in fig.2.58.

Here, $R_1 = 12\ \Omega$; $L = 0{\cdot}051$ H; $R_2 = 20\ \Omega$;
$C = 212.2\ \mu\text{F} = 212.2 \times 10^{-6}$ F; $V = 240$ V; $f = 50$ Hz

Resistance of the whole circuit, $R = R_1 + R_2$

$= 12 + 20 = 32\ \Omega$

Inductive reactance, $X_L = 2\pi f L$

$= 2\pi \times 50 \times 0{\cdot}051 = 16\ \Omega$

Capacitive reactance, $X_C = \dfrac{1}{2\pi f C}$

$= \dfrac{1}{2\pi \times 50 \times 212.2 \times 10^{-6}} = 15\ \Omega$

Impedance, $Z = \sqrt{R^2 + (X_L - X_C)^2}$

$= \sqrt{32^2 + (16 - 15)^2} = 32.016\ \Omega$

(i) Circuit current, $I = \dfrac{V}{Z} = \dfrac{240}{32.016} = \mathbf{7.496\ A}$ *(Ans.)*

(ii) Power factor of the circuit, $\cos\phi = \dfrac{R}{Z} = \dfrac{32}{32.016} =$ **0.9995 (lagging)** *(Ans.)*

(since $X_L > X_C$)

Example 2.20. *A series R-L-C circuit consisting of a resistance of* 20 W, *inductance* 0·2 H *and capacitance of* 150 *m F is connected across a 230 V, 50 Hz source. Calculate (i) the impedance (ii) the current (iii) the magnitude and nature of the power factor (iv) the frequency of supply to be adjusted to make power factor unity.*

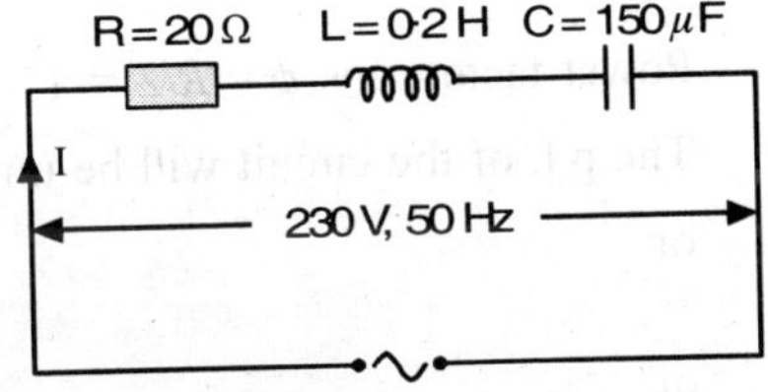

Fig.2.59: Circuit diagram

Solution: The circuit is shown in fig.2.59

Here, $R = 20\ \Omega;\ L = 0{\cdot}2\ \text{H};\ C = 150\ \mu\text{F} = 150 \times 10^{-6}\ \text{F}$

Inductive, reactance, $X_L = 2\pi f L = 2\ \pi \times 50 \times 0{\cdot}2 = 62{\cdot}83\ \Omega$

Capacitance reactance, $X_C = \dfrac{1}{2\pi f C} = \dfrac{1}{2\pi \times 50 \times 150 \times 10^{-6}} = 21{\cdot}22\ \Omega$

(*i*) Impedance, $Z = \sqrt{R^2 + (X_L - X_C)^2}$

$= \sqrt{20^2 + (62\cdot 83 - 21\cdot 22)^2}$

$=$ **46·17 Ω (*Ans.*)**

(*ii*) Circuit current, $I = \dfrac{V}{Z} = \dfrac{230}{46\cdot 17} =$ **4·98 A (*Ans.*)**

(*iii*) Power factor $= \cos\phi = \dfrac{R}{Z} = \dfrac{20}{46\cdot 167} =$ **0·433 (*Ans.*)**

The power factor, is **lagging** because inductive reactance is more than capacitive reactance *i.e.* the circuit behaves as an inductive circuit.

Power factor will be unity when

$$X_L = X_C \quad \text{or} \quad 2\,\pi f_r L = \frac{1}{2\,\pi f_r C}$$

or $$f_r = \frac{1}{2\pi}\frac{1}{\sqrt{LC}} = \frac{1}{2\,\pi\sqrt{0\cdot 2 \times 150 \times 10^{-6}}} = \mathbf{29{\cdot}06\ Hz}\ (\textit{Ans.})$$

Example 2.21. *An inductive coil takes 10 A and dissipates 1500 W when connected to a 250 V, 25 Hz supply. Calculate the following, (i) the impedance; (ii) the effective resistance; (iii) the reactance; (iv) the value of the capacitance required to be connected in series with coil to make the power factor of the circuit unity; and (vi) what is now the current taken by the coil ? Also draw the phasor diagram of the two cases.*

Solution: The circuit is shown in fig.2.60.

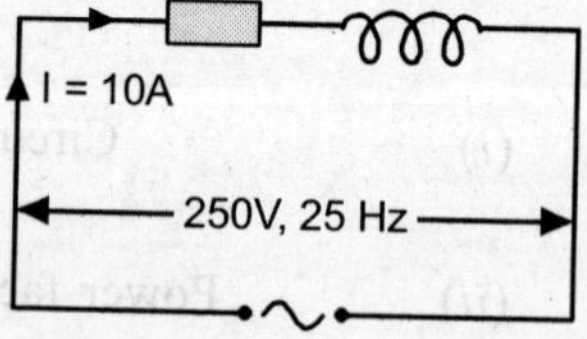

Fig.2.60: Circuit diagram

Power dissipated in the coil, $I^2 R = 1500$ W

∴ Resistance of the coil, $R = 1500/(10)^2 = 15\ \Omega$ (*Ans.*)

Impedance of the coil, $Z = V/I = 250/10 = 25\ \Omega$ (*Ans.*)

Inductive reactance of the coil, $X_L = \sqrt{Z^2 - R^2} = \sqrt{(25)^2 - (15)^2}$

$=$ **20 Ω** (*Ans.*)

Power factor, $\cos\phi = R/Z = 15/25 =$ **0·6 lag** (*Ans.*)

The p.f. of the circuit will be unity when, $X_C = X_L$

or $1/2\ \pi f C = 20$

or $$C = \frac{1}{2\pi \times 25 \times 20} = \mathbf{159.2\ \mu\ F}\ (\textit{Ans.})$$

Now, current, $I = \frac{V}{R} = \frac{250}{15} = \mathbf{16.67\ A}$ (*Ans.*)

The phasor diagram for the two cases is drawn in fig.2.61 (*a*) and (*b*) respectively.

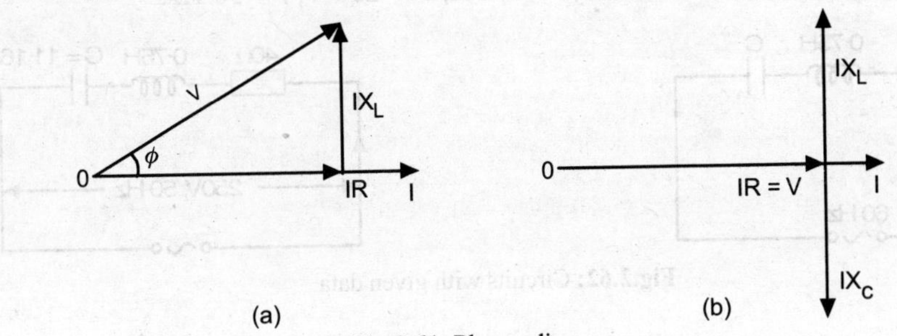

Fig.2.61: Phasor diagram

Example 2.22 *A choke coil is connected series with a 100 μF capacitor. With a constant supply voltage of 250 V, it is found that the circuit takes its maximum current of 50 A when the supply frequency is 100 Hz. Determine (i) resistance and inductance of the choke coil; (ii) voltage across the capacitor and (iii) Q-factor of the circuit.*

Solution: At resonance; Current, $I_r = V/R$

or $\quad R = V/I_r = 250/50 = \mathbf{5\ \Omega}$ (*Ans.*)

Also $\quad 2\,\pi f_r L = 1/2\,\pi f_r C$

or
$$L = \frac{1}{(2\pi f_r)^2 C} = \frac{1}{(2\pi \times 100)^2 \times 100 \times 10^{-6}} = \mathbf{25.33\ mH}\ (Ans.)$$

Voltage across capacitor,

$$V_C = I_r X_C = 50 \times 1/2\,\pi \times 100 \times 100 \times 10^{-6} = \mathbf{795.8\ V}\ (Ans.)$$

$$Q\text{-}\ factor = \frac{V_L}{V_R} = \frac{X_L}{R} = \frac{2\pi f_r L}{R} = \frac{2\,\pi \times 100 \times 25.33 \times 10^{-3}}{5} = \mathbf{3.183}\ (Ans.)$$

Example 2.23. *A series resonant circuit has a Q-factor of 150; an inductance of 0·1 H and a capacitance of 0·1 μF. Calculate the band width of the circuit.*

Solution: At resonance, frequency

$$f_r = \frac{1}{2\,\pi\sqrt{LC}} = \frac{1}{2\,\pi\sqrt{0.1 \times 0.1 \times 10^{-6}}} = \frac{10^4}{2\,\pi}\text{Hz}$$

Now, $\quad Q = \frac{f_r}{BW}$

∴ $\quad$ Bandwidth, $BW = \frac{f_r}{Q} = \frac{10^4}{2\,\pi \times 150} = \mathbf{10 \cdot 61\ Hz}$ (*Ans.*)

Q-factor $= V_C/V = 397{\cdot}88/250 = \mathbf{1{\cdot}59}$ (*Ans.*)

Example 2.24. *A coil of resistance 40Ω and inductance 0·75 H are in a series circuit with a capacitor C. The resonant frequency is 60 Hz. If supply is 250 V, 50 Hz find*

(i) line current, *(ii) power factor,* *(iii)* power consumed.

Solution: Here, $R = 40\ \Omega$; $L = 0{\cdot}75$ H; $f_r = 60$ Hz; $V = 250$ V; $f = 50$ Hz.

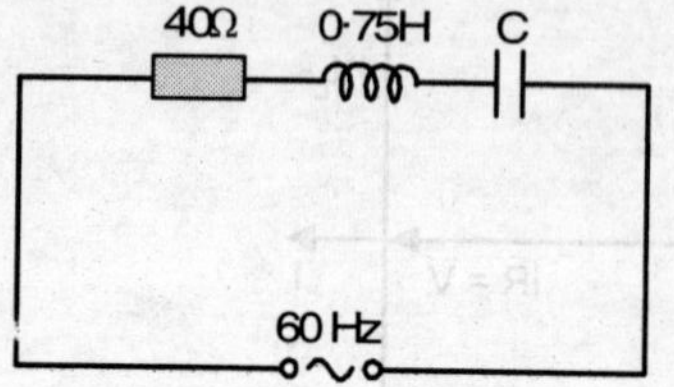

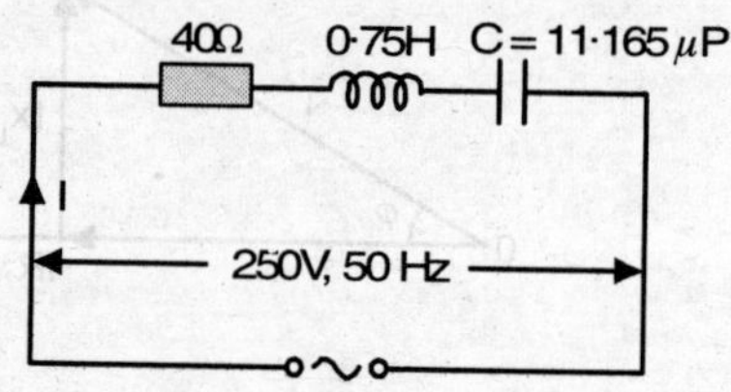

Fig.2.62: Circuits with given data

At resonance, $X_L = X_C$ or $2\pi f_r L = \dfrac{1}{2\pi f_r C}$

or $$C = \frac{1}{(2\pi f_r)^2 L} = \frac{1}{(2\pi \times 60)^2 \times 0.75} = 9.38\ \mu\text{F}$$

At 50 Hz, $X_L = 2\pi f L = 2\pi \times 50 \times 0{\cdot}75 = 235{\cdot}6\ \Omega$

$$X_C = \frac{1}{2\pi f C} = \frac{1}{2\pi \times 50 \times 9.38 \times 10^{-6}} = 339.35\ \Omega$$

Impedance, $Z = \sqrt{R^2 + (X_L - X_C)^2} = \sqrt{(40)^2 + (235.6 - 339.35)^2} = 111.2\ \Omega$

(i) Line current, $I = \dfrac{V}{Z} = \dfrac{250}{111.2} =$ **2·248 A** *(Ans.)*

(ii) Power factor, $\cos\phi = \dfrac{R}{Z} = \dfrac{40}{111.2} =$ **0·36 leading** *(Ans.)*

(iii) Power consumed, $P = VI\cos\phi = 250 \times 2.248 \times 0{\cdot}36 =$ **202.32 W** *(Ans.)*

PRACTICE EXERCISE

Q.1 *A coil of resistance* 8W *and inductance* 0·12 *henry is connected in series with a condenser of capacitance* 140 *microfarad across a* 230 *volt, 50 Hz supply. Determine:*

(i) Impedance of the entire circuit,

(ii) Current flowing through the condenser,

(iii) Power factor of the circuit,

(iv) Voltage across the condenser.

(*Ans.* 17 W; 13·53 A; 0·471 (lagging); 307·7 V)

Q.2 *A coil of resistance 10 ohms and inductance 0.1 H is connected in series with a condenser of capacitance 150 micro-farad across a 200 V, 50 Hz supply. Determine*

(i) impedance

(ii) *current*

(iii) *power factor*

(iv) *voltage across the coil*

(v) *voltage across the condenser*

(**Ans.** 14·28 Ω; 14 A;0·7 lag; 461·57 V; 297·08 V)

Q.3. *A series R-L-C circuit with* $R = 10\Omega$; $L = 0{\cdot}02\ H$, $C = 2\ \mu F$ *is connected to* 100 *V variable frequency source. Find the frequency for which the current is maximum.* (**Ans.** 795·77 Hz)

Q.4 *Determine the parameters of an R – L – C series circuit that will resonate at 1000 Hz, has a bandwidth of 100 Hz and draws 16 W from a 200 V generator operating at the resonant frequency of the circuit.* (**Ans.** 2500 W; 3·98 H, 6.3 ′ 10⁻⁶F)

Q.5 *A* 10 *mH coil is connected in series with a loss free capacitor to a variable frequency source of* 20 *V. The current in the circuit has maximum value of* 0·2 *A at a frequency of* 100 *kHz. Calculate:*

(i) *the value of capacitance*

(ii) *the Q factor of the coil*

(iii) *the half power frequencies.*

(**Ans.** 253·3 pF; 62·8; 99·204 kHz; 100·796 kHz)

2.33 A.C PARALLEL CIRCUITS

The a.c. circuits in which number of branches are connected in such a manner so that voltage across each branch is the same but current flowing through them is different, are called *a.c. parallel circuits*. The parallel circuits are used more frequently in a.c. system because of the following reasons:

(*i*) Almost all the electrical appliances (or devices) of different ratings are operated at the same supply voltage and are connected in parallel.

(*ii*) Each device is required to be operated independently (with a switch) without disturbing the operation of other devices. Hence, connected in parallel.

2.34 METHODS OF SOLVING PARALLEL A.C CIRCUITS

In parallel circuits, number of branches are connected in parallel. Each branch, generally, contains number of components like resistance, inductance and capacitance forming series circuits. Therefore each branch is analysed separately as a series circuit and then the effects of separate branches are combined together. While carrying out circuit calculations, the magnitudes and phase angles of voltages and currents are taken into account. The following methods may be applied for solving a.c. parallel circuits.

(*i*) Phasor (or vector) method.

(*ii*) Admittance method.

(*iii*) Method of phasor algebra (or symbolic method or J-method).

The method to be applied for the solution, depends upon the conditions of the problem. However, in general, the method which yields quick results is applied.

2.35 PHASOR (OR VECTOR) METHOD

To solved parallel a.c. circuits by this method, proceeds as follows:

Step I: Draw the circuit as per the given problem as shown in fig.2.63. (Here, for illustration, we have considered two branches connected in parallel. One branch contains resistance and inductance in series, whereas second branch contains resistance and capacitance in series. The supply voltage is V volts).

Step II: Find the impedance of each branch of the circuit separately, *i.e.*

$$Z_1 = \sqrt{R_1^2 + X_{L1}^2} \text{ where } X_{L1} = 2\pi f L_1$$

$$Z_2 = \sqrt{R_2^2 + X_{C2}^2} \text{ where } X_{C2} = 1/2\pi f C_2$$

Step III: Determine the magnitude of current and phase angle with the voltage in each branch.

$$I_1 = \frac{V}{Z_1};\ \phi_1 = \tan^{-1}\frac{X_{L1}}{R_1} \quad (lagging) \text{ [for inductive branch]}$$

$$I_2 = \frac{V}{Z_2};\ \phi_2 = \tan^{-1}\frac{X_{C2}}{R_2} \quad (lagging) \text{ [for capacitive branch]}$$

Step IV: Draw the phasor diagram taking voltage as the reference phasor. Represent the branch currents on it as shown in fig.2.64.

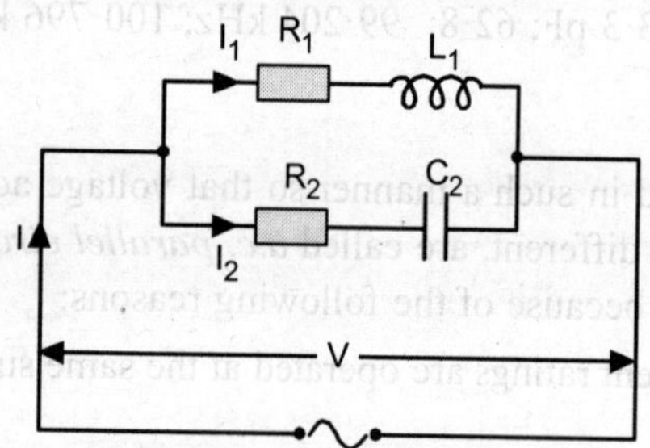

Fig; 2.63: Parallel ac circuit

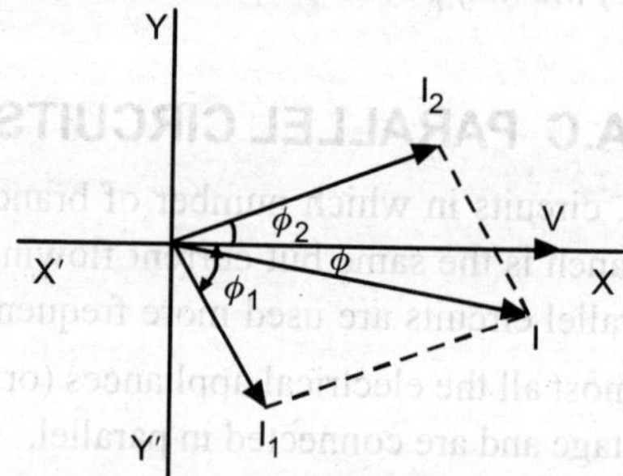

Fig.2.64: Phasor diagram

Step V: Find the phasor sum of branch currents by the method of components.

$$I_{XX} = I_1 \cos\phi_1 + I_2 \cos\phi_2$$

$$I_{YY} = -I_1 \sin\phi_1 + I_2 \sin\phi_2 \quad (negative)$$

$$I = \sqrt{(I_{XX})^2 + (I_{YY})^2}$$

Step VI: Find the phase angle ϕ between the total current I and circuit voltage V.

$$\phi = \tan^{-1}\frac{I_{YY}}{I_{XX}} \text{ lagging (since, } I_{YY} \text{ is negative)}$$

Power factor of the circuit = cos ϕ (*lagging*)

or Power factor $= \frac{I_{XX}}{I}$ (*lagging*)

Example 2.25. *The circuits A and B are connected in parallel to a 250 V, 50 Hz supply. Circuit A consists of resistance 15 ohm in series with an inductive reactance of 15 ohm and circuit B consists of resistance 30 ohm in series with a capacitive reactance of 15 ohm. Determine (i) the current drawn by each circuit; (ii) total current drawn from the mains.*

Solution: The circuit is shown in fig.2.65.

Impedance, $Z_1 = \sqrt{R_1^2 + X_{L1}^2} = \sqrt{(15)^2 + (15)^2} = 21.21\ \Omega$

Impedance, $Z_2 = \sqrt{R_2^2 + X_{C2}^2} = \sqrt{(30)^2 + (15)^2} = 33.54\ \Omega$

Current, $I_1 = \dfrac{V}{Z_1} = \dfrac{250}{21.21} =$ **11.78 A (*Ans.*)**

Current, $I_2 = \dfrac{V}{Z_2} = \dfrac{250}{33.54} =$ **7.45 A (*Ans.*)**

Phase angle, $\phi_1 = \tan^{-1}\dfrac{X_{L1}}{R_1} = \tan^{-1}\dfrac{15}{15} = 45°$ lag

Phase angle, $\phi_2 = \tan^{-1}\dfrac{X_{C2}}{R_2} = \tan^{-1}\dfrac{15}{30} = 26{\cdot}56°$ leading

The phasor diagram representing voltage as reference phasor is shown in fig.2.66.

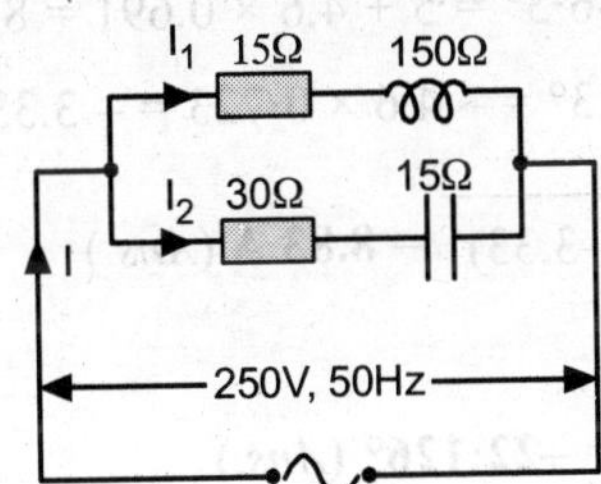

Fig.2.65: Circuit diagram

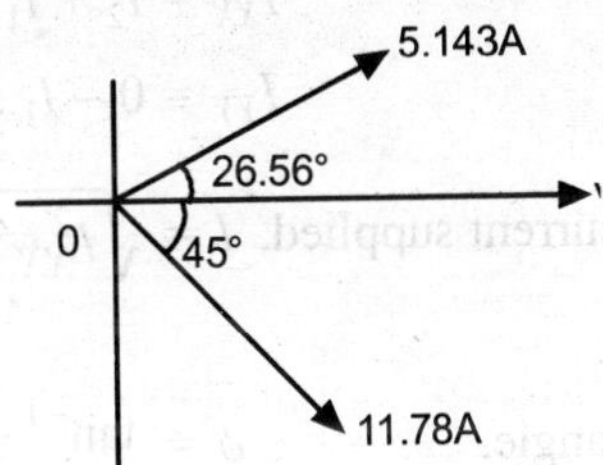

Fig.2.66: Phasor diagram

Resolving the currents horizontally and vertically, we get,

$$I_{XX} = I_1 \cos\phi_1 + I_2 \cos\phi_2 = 11.78 \cos 45° + 7.45 \times \cos 26{\cdot}56° = 15\ \text{A}$$

$$I_{YY} = -I_1 \sin\phi_1 + I_2 \sin\phi_2$$

$$= -11.78 \times \sin 45° + 7.45 \sin 26{\cdot}56° = -5\ \text{A}$$

Current drawn from the mains, $I = \sqrt{I_{XX}{}^2 + I_{YY}{}^2} = \sqrt{(15)^2 + (5)^2} =$ **15.81 A (*Ans.*)**

Example 2.26. *A coil of resistance 30 ohm and inductance 0·1 H is connected in parallel with a non-inductive resistance of 40 ohm. Find (i) current in each branch of the circuit, (ii) total current supplied, (iii) phase angle and p.f. of combination when a voltage of 200 volt at 50 Hz is applied (iv) power consumed in the circuit.*

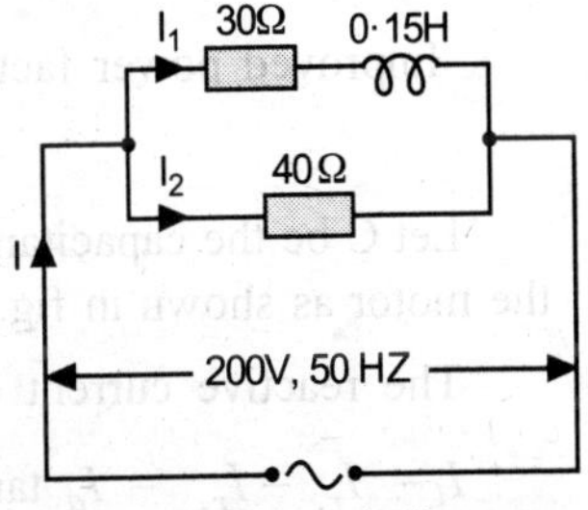

Fig.2.67: Circuit diagram

Solution: The circuit is shown in fig.2.67.

Applied voltage, $V = 200$ V; Supply frequency, $f = 50$ Hz

Branch *I*

$$R_2 = 30\ \Omega;\ X_{L1} = 2\pi f L_1 = 2\pi \times 50 \times 0{\cdot}1 = 31{\cdot}4\ \Omega$$

Impedence, $Z_1 = \sqrt{R_1^2 + X_L^2} = \sqrt{(30)^2 + (31.4)^2} = 43.43$ ohm

Current in the coil, $I_1 = \frac{V}{Z_1} = \frac{200}{43.43} = \mathbf{4.6\ A}$ (*Ans.*)

Phase angle, $\phi_1 = \tan^{-1}\frac{X_{L1}}{R_1} = \tan^{-1}\frac{31.4}{30} = 46{\cdot}3°$ lagging

Branch II

Resistance, $R_2 = 40$ ohm

Branch current, $I_2 = \frac{V}{R_2} = \frac{200}{40} = \mathbf{5\ A}$ (*Ans.*)

Phase angle, $\phi_2 = 0$ (I_2 is in phase with V)

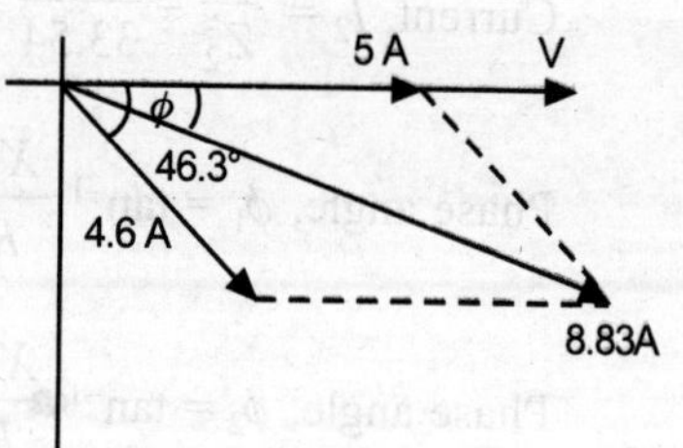

Fig.2.68: Phasor diagram

The two currents are shown vectorially in fig.2.68. Resolving the currents horizontally and vertically,

$$I_{XX} = I_2 + I_1 \cos\phi_1 = 5 + 4.6 \cos 46{\cdot}3° = 5 + 4.6 \times 0.691 = 8.18\ A$$

$$I_{YY} = 0 - I_1 \sin\phi_1 = 0 - 4.6 \sin 46{\cdot}3° = -4.6 \times 0{\cdot}723 = -3.33\ A$$

Total current supplied, $I = \sqrt{I_{XX}^2 + I_{YY}^2} = \sqrt{(8.18)^2 + (-3.33)^2} = \mathbf{8.83\ A}$ (*Ans.*)

Phase angle, $\phi = \tan^{-1}\frac{I_{YY}}{I_{XX}} = \tan^{-1}\left(\frac{-3{\cdot}33}{8.18}\right) = \mathbf{-22{\cdot}126°}$ (*Ans.*)

Power factor of the circuit, $\cos\phi = \cos(-22{\cdot}126°) = \mathbf{0{\cdot}9264}$ **(lagging)** (*Ans.*)

Power, $P = VI\cos\phi = 200 \times 8.83 \times 0{\cdot}9264 = \mathbf{1635.7\ W}$ (*Ans.*)

Example 2.27. *A single phase motor takes* 30 *A at a pf of* 0·6 *lagging from* 230 *V,* 50 *Hz supply. What value of capacitance must a shunt capacitor have to raise the overall power factor to* 0·9 ?

Solution: Active component of current drawn by motor, $I_a = I\cos\phi_1 = 50 \times 0{\cdot}6 = 30$ A

Initial power factor, $\cos\phi_1 = 0{\cdot}6$ lagging

$$\tan\phi_1 = \tan\cos^{-1} 0{\cdot}6 = 1{\cdot}333$$

Improved power factor, $\cos\phi_2 = 0{\cdot}9$ lagging

$$\tan\phi_2 = \tan\cos^{-1} 0{\cdot}9 = 0{\cdot}4843$$

Let C be the capacitance of capacitor connected across the motor as shown in fig.2.69

The reactive current drawn by capacitor,

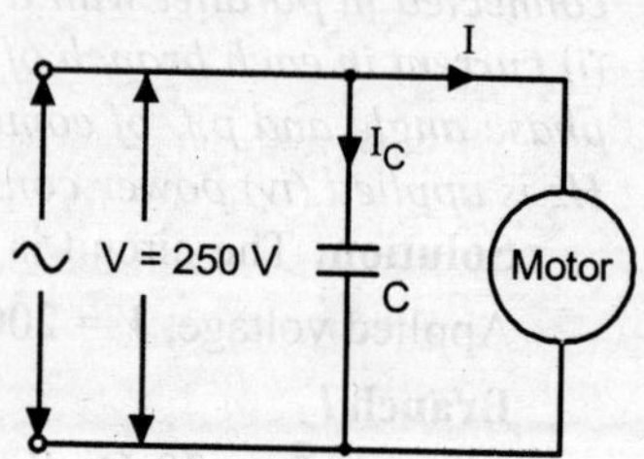

Fig.2.69: Circuit diagram

$$I_C = I_{r_1} - I_{r_2} = I_a \tan\phi_1 - I_a \tan\phi_2$$

$$I_C = I_a(\tan\phi_1 - \tan\phi_2) = 30\,(1{\cdot}333 - 0{\cdot}4843) = 25{\cdot}47\ A$$

The value of capacitance required, $C = \dfrac{I_C}{2\pi f V} = \dfrac{25 \cdot 47}{2\pi \times 50 \times 250} = \mathbf{324 \cdot 3\ \mu F}$ (*Ans.*)

PRACTICE EXERCISE

Q.1 *A coil of resistance 15 ohm and inductance 0·05 H is connected in parallel with a non-inductive resistance of 20 ohm. Find:*

(i) current in each branch of the circuit

(ii) total current supplied

(iii) phase angle and p.f. of combination when a voltage of 200 volt at 50 Hz is applied

(iv) power consumed in the circuit.

(*Ans.* 9·2 A; 10 A; 17·656 A; -22·126°; 0·9264 (lagging); 3271·3 W)

Q.2 *A series ac circuit has a resistance of 15 Ω and inductive reactance of 10 Ω. Calculate the value of capacitor which is connected across this series combination so that system has unity power factor. The frequency of ac supply is 50 Hz.* (*Ans.* 97.94mF)

Q.3 *The circuits A and B are connected in parallel to a 230 V, 50 Hz supply. Circuit A consists of resistance 20 ohm in series with an inductive reactance of 20 ohms and circuit B consists of resistance 40 ohm in series with a capacitive reactance of 20 ohm. Determine:*

(i) the current drawn by each circuit

(ii) total current drawn from the mains. (*Ans.* 8·13 A; 5·143 A; 10·91 A)

Q.4 *A single phase motor takes 5 A current at 230 V, 50 Hz supply at a p.f. 0·707 lagging. It is required to improve the p.f. of the motor to 0·9 by connecting a capacitor in parallel with it. Determine the capacitance of capacitor.* (*Ans.* *25·23 m F*)

Q.5 *A single phase motor takes 50 A at a pf of 0·6 lagging from 250 V, 50 Hz supply What value of capacitance must a shunt capacitor have to raise the overall power factor to 0·9 ?* (*Ans.*324·3 mF)

2.36 PARALLEL RESONANCE

An a.c. circuit containing an inductor and capacitor in parallel is said to be in parallel resonance when the circuit current is in phase with the applied voltage. Consider an inductor of L henry having some resistance R ohm connected in parallel with a capacitor of capacitance C farad across a supply voltage of V volt as shown in fig.2.70. The phasor diagram of the circuit is shown in fig.2.71.

The circuit current I_r will only be in phase with the supply voltage when

$$I_C = I_L \sin \phi_L$$

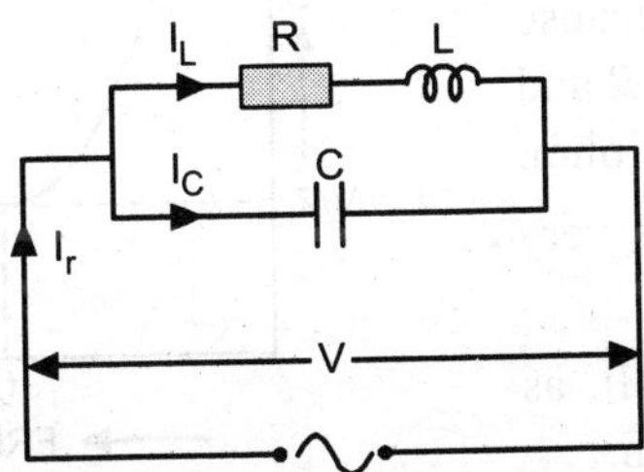

Fig.2.70: Parallel resonance circuit

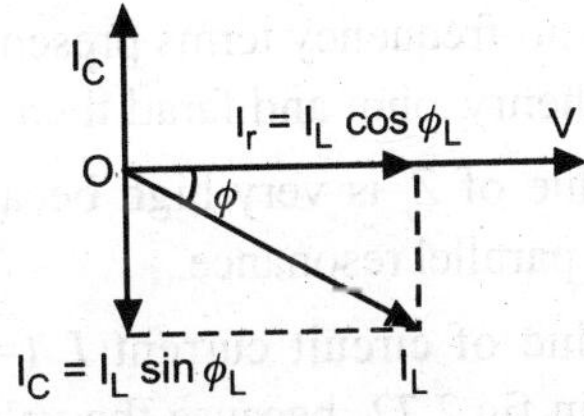

Fig.2.71: Phasor diagram

Since, at resonance the reactive component of current is suppressed, the circuit draws *minimum current* under this condition.

Resonant frequency

The value of X_L (= $2pfL$) and X_C (= $1/2pfC$) can be changed by changing the supply frequency. When frequency increases, the value of X_L and consequently the value of Z_L increases. This decreases the magnitude of current I_L which also lags behind the voltage V by a progressively greater angle. On the other hand, the value of X_C decreases and consequently the value of I_C increases. At some frequency f_r (called resonance frequency), $I_C = I_L \sin f_L$ and resonance occurs.

$\therefore$ At parallel resonance, $I_C = I_L \sin \phi_L$

where, $$I_L = \frac{V}{Z_L};\ \sin \phi_L = \frac{X_L}{Z_L} \text{ and } I_C = \frac{V}{X_C}$$

$$\therefore \quad \frac{V}{X_C} = \frac{V}{Z_L} \times \frac{X_L}{Z_L} \quad \text{or} \quad X_L X_C = Z_L^2$$

or $$\frac{\omega L}{\omega C} = Z_L^2 = \left(R^2 + X_L^2\right) \qquad ...(i)$$

or $$\frac{L}{C} = R^2 + (2\pi f_r L)^2$$

or $$2\pi f_r L = \sqrt{\frac{L}{C} - R^2}$$

or $$f_r = \frac{1}{2\pi L}\sqrt{\frac{L}{C} - R^2} = \frac{1}{2\pi}\sqrt{\frac{1}{LC} - \frac{R^2}{L^2}} \qquad ...(ii)$$

If R is very small as compared to L, then

Resonance frequency, $$f_r = \frac{1}{2\pi\sqrt{LC}}$$

Effect of parallel resonance

At parallel resonance; Line current, $I_r = I_L \cos \phi$

or $$\frac{V}{Z_r} = \frac{V}{Z_L} \times \frac{R}{Z_L} \text{ or } \frac{1}{Z_r} = \frac{R}{Z_L^2} \text{ or } \frac{1}{Z_r} = \frac{R}{L/C} = \frac{CR}{L}$$ (since $Z_L^2 = L/C$ from exp. *i.*)

$\therefore$ Circuit impedance, $Z_r = L/CR$

It shows that:

(*i*) Circuit impedance Z_r (= L/CR) is a pure resistive because there is no frequency terms present. If the value of L, R and C is in henry, ohm and farad then the value of Z_r is in ohm.

(*ii*) The value of Z_r is very high because the ratio L/C is very large at parallel resonance.

(*iii*) The value of circuit current I_r (= V/Z_r) is very small, as shown in fig.2.72, because the value of Z_r is very high.

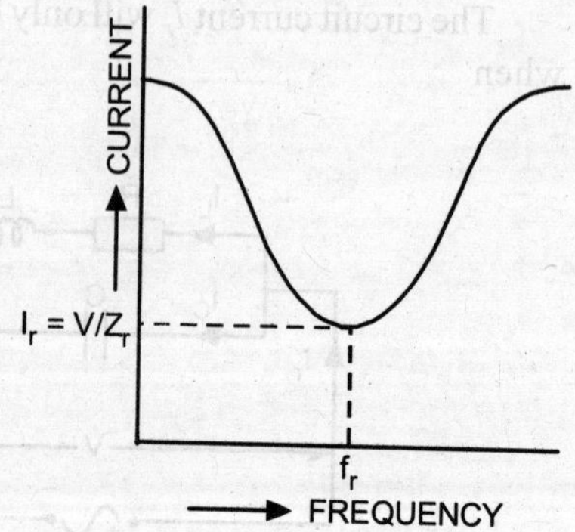

Fig.2.72: Curve between f and I.

(*iv*) The current flowing through the capacitor and coil is much greater than the line current because the impedance of each branch is quite low than circuit impedance Z_r.

Since a parallel resonant circuit can draw a very small current and power from the mains, it is often regarded as **rejected circuit.**

Resonance curve: A current-frequency curve for a typical parallel resonant circuit is shown in fig.2.72. The value of line current I_r (= V_r/Z_r) is minimum at resonance.

2.37 Q-FACTOR OF A PARALLEL RESONANT CIRCUIT

We have seen that at parallel resonance, the current circulating between the two branches is many times greater than the line current drawn from the mains. This current simplification produced by the resonance is called the *Q-factor* of the parallel resonant circuit.

$$\therefore \; Q\text{-factor} = \frac{\text{Current circulating between } L \text{ and } C}{\text{Line current}} = \frac{I_C}{I_r}$$

Now, $$I_C = V/X_C = 2\,\pi f_r\, CV \text{ and } I_r = \frac{V}{L/CR}$$

$$\therefore \; Q\text{-factor} = \frac{2\,\pi f_r\, CV}{V} \times \frac{L}{CR} = \frac{2\,\pi f_r\, L}{R} \quad \text{...same as for series circuit}$$

$$\text{or } Q\text{-factor} = \frac{2\,\pi L}{R} \cdot \frac{1}{2\,\pi\sqrt{LC}} \quad \left(\because f_r = \frac{1}{2\,\pi\sqrt{LC}} \text{ neglecting } R\right)$$

$$\text{or } Q\text{-factor} = \frac{1}{R}\sqrt{\frac{L}{C}} \quad \text{or} \quad Q\text{-factor} = \sqrt{\frac{L}{C}} \text{ (neglecting } R)$$

The value of Q-factor is same as for series resonance.

2.38 COMPARISON OF SERIES AND PARALLEL RESONANT CIRCUITS

The comparison of series and parallel resonant circuit is given below:

S.No.	Particulars	Series Circuit	Parallel Circuit
1.	*Impedance,*	Minimum *i.e.* $Z_r = R$	Maximum *i.e.* $Z_r = L/CR$
2.	*Current*	Maximum *i.e.* $I_r = V/R$	Minimum *i.e.* $I_r = V/Z_r$
3.	*Resonant frequency*	$f_r = \frac{1}{2\,\pi\sqrt{LC}}$	$f_r = \frac{1}{2\,\pi}\sqrt{\frac{1}{LC} - \frac{R^2}{L^2}}$
4.	*Power factor*	Unity	Unity
5.	*Q-factor*	X_L/R	X_L/R
6.	*Amplification*	It amplifies voltage	It amplifies current

Example 2.28. *A parallel circuit consists of a coil having 25 Ω resistance and 300 mH inductance in parallel with a capacitor of capacitance 4μF. Determine (i) the resonant frequency (ii) dynamic impedance of the circuit and (iii) Q-factor of the circuit at resonance.*

Solution: (*i*) Resonant frequency, $f_r = \frac{1}{2\pi}\sqrt{\frac{1}{LC} - \frac{R^2}{L^2}}$

or $$f_r = \frac{1}{2\pi}\sqrt{\frac{1}{0.3 \times 4 \times 10^{-6}} - \frac{(25)^2}{(0.3)^2}} = \textbf{144·68 Hz } (Ans.)$$

(*ii*) Dynamic impedance, $Z_r = \frac{L}{CR} = \frac{0.3}{4 \times 10^{-60} \times 25} = \textbf{3000 }\Omega \text{ } (Ans.)$

(*iii*) $Q\text{-factor} = \frac{2\pi f_r L}{R} = \frac{2\pi \times 145.27 \times 0.3}{25} = \textbf{10.953 } (Ans.)$

Example 2.29. *A coil of 15Ω resistance and 0·1 H inductance is connected in parallel with a capacitor of 15 μ F. Calculate the frequency at which the circuit will act as a non-inductive resistance of R ohm. Find also the value of R, input current at resonant frequency and the ratio of the circulating current to the supply current at resonant frequency. The applied voltage is* 100 V (*rms*).

Solution: Capacitance, $C = 10\ \mu F = 10 \times 10^{-6}$ F; $R = 10\ \Omega$; $L = 0·1$ H; $V = 100$ V

The circuit will act as a resistive circuit at resonant frequency, *i.e.*

$$f_r = \frac{1}{2\pi}\sqrt{\frac{1}{LC} - \frac{R^2}{L^2}} = \frac{1}{2\pi}\sqrt{\frac{1}{0.1 \times 15 \times 10^{-6}} - \frac{(10)^2}{(0.1)^2}} = \textbf{127.74 Hz } (Ans.)$$

Dynamic resistance (or impedance) $= \frac{L}{CR} = \frac{0.1}{15 \times 10^{-6} \times 15} = \textbf{444.44 }\Omega \text{ } (Ans.)$

Input current at resonant frequency, $I = \frac{V}{L/CR} = \frac{100}{444.44} = \textbf{0·225 A } (Ans.)$

The ratio of circulating current to line current at resonant frequency,

$$\frac{I_C}{I} = \frac{2\pi f_r L}{R} = \frac{2\pi \times 127.74 \times 0.1}{15} = \textbf{5.35 } (Ans.)$$

Example 2.30. *An inductive circuit of resistance 2Ω and inductance 0·01 H is connected to a 250 V, 50 Hz supply. What value of capacitance is required to be placed in parallel with it to produce resonance ?*

Solution: Here, $R = 2\ \Omega$; $L = 0·014$ H; $V = 250$ V; $f = 50$ Hz

Inductive reactance, $X_L = 2\pi f L = 2\pi \times 50 \times 0·01 = 3·14\ \Omega$

Impedance, $Z = \sqrt{R^2 + X_L^2} = \sqrt{(2)^2 + (3.14)^2} = \textbf{3.72 }\Omega$

At resonance, $Z = \sqrt{\frac{L}{C}}$

or $$C = \frac{L}{Z^2} = \frac{0.01}{(3.72)^2} = \textbf{722.6 }\mu\textbf{F } (Ans.)$$

2.39 POLY-PHASE SYSTEM

Poly means many (more than one) and *phase* means windings or circuits, each of them having a single alternating voltage of the same magnitude and frequency. Hence, a polyphase system is essentially a combination of two or more than two voltages having same magnitude and frequency but displaced from one another by equal electrical angle. This angular displacement between the adjacent voltages is called phase difference and depends upon the number of phases.

$$\text{Phase difference} = \frac{360 \text{ electrical degrees}}{\text{Number of phases}}$$

$$\text{For three-phase system phase difference} = \frac{360°}{3} = 120° \text{ electrical.}$$

Thus, an a.c. system having a group of (two or more than two) equal voltages of same frequency arranged to have equal phase difference between them is called a **polyphase system.**

2.40 ADVANTAGES OF 3-PHASE SYSTEM OVER 1-PHASE SYSTEM

1. **Constant power:** In single phase circuits, the power delivered is pulsating, whereas, in poly-phase system, power delivered is almost constant when the loads are balanced.
2. **Higher rating:** The rating (output) of a 3-phase machine is nearly 1.5 times the rating (output) of a single-phase machine of the same size.
3. **Power transmission economics:** To transmit the same amount of power over a fixed distance at a given voltage, 3-phase system requires only 75% of the weight of conducting material as required by single-phase system.
4. **Superiority of 3-phase induction motors:** The 3-phase induction motors are considered to be superior because of the following reasons:
   (*i*) Three-phase induction motors are self starting, whereas 1-phase induction motors have no starting torque without auxiliary means.
   (*ii*) Three-phase induction motors have higher power factor and efficiency than that of single-phase induction motors.

2.41 GENERATION OF 3-PHASE E.M.FS.

In a 3-phase system, three are three equal voltages (or e.m.fs.) of the same frequency having a phase difference of 120°. These voltages can be produced by a three-phase, a.c. generator having three identical windings (or phases) displaced 120° electrical apart. When these windings are rotated in a stationary magnitude field [see fig.2.73 (*a*)] or when these windings are kept stationary and the magnetic field is rotated [see fig.2.73 (*b*)], an e.m.fs. is induced in each winding or phase. These e.m.fs. are of same magnitude and frequency but are displaced from one another by 120° electrical.

Consider three identical coils $a_1\ a_2$, $b_1\ b_2$ and c_1c_2 mounted as shown in fig.2.73 (*a* and *b*). Here, a_1, b_1 and c_1 are the start terminals, whereas, a_2, b_2 and c_2 are the finish terminals of the three coils. It may be noted that a phase difference of 120° electrical is maintained between the corresponding start terminals a_1, b_1 and c_1. Let the three coils mounted on the same axis be rotated (or the magnetic field system be rotated keeping coils stationary) is anticlockwise direction at ω radians/second, as shown in fig.2.73 (*a*) and 2.73 (*b*) respectively. Three e.m.fs. are induced in the three coils respectively. Their magnitude and direction, at this instant, are as mentioned below:

(*i*) The e.m.f. induced in coil a_1a_2 is zero (consider start terminal a_1) and is increasing in the positive direction as shown by wave e_{a1a2} in fig.2.73 (*c*).

(*ii*) The coil b_1b_2 is 120° (electrical) behind the coil a_1a_2. The e.m.f. induced in this coil is negative and is becoming maximum negative (consider start terminal b_1) as shown by the e_{b1b2} in fig.2.73 (*c*).

(*iii*) The coil c_1c_2 is 120° (electrical) behind b_1b_2 or 240° (electrical) behind a_1a_2. The e.m.f. induced in this coil is positive and is decreasing (consider start terminal c_1) as shown by wave e_{c1c2} in fig.2.73 (*c*).

Phasor diagram

The e.m.fs. induced in three coils are of the same magnitude and frequency but are displaced by 120° (electrical) from each other as shown by phasor diagram in fig.2.73 (*d*). These can be represented by the equations;

$$e_{a1a2} = E_m \sin \omega t;$$

$$e_{b1b2} = E_m \sin (\omega t - 2\pi/3)$$

$$e_{c1c2} = E_m \sin (\omega t - 4\pi/3) = E_m \sin (\omega t - 240°)$$

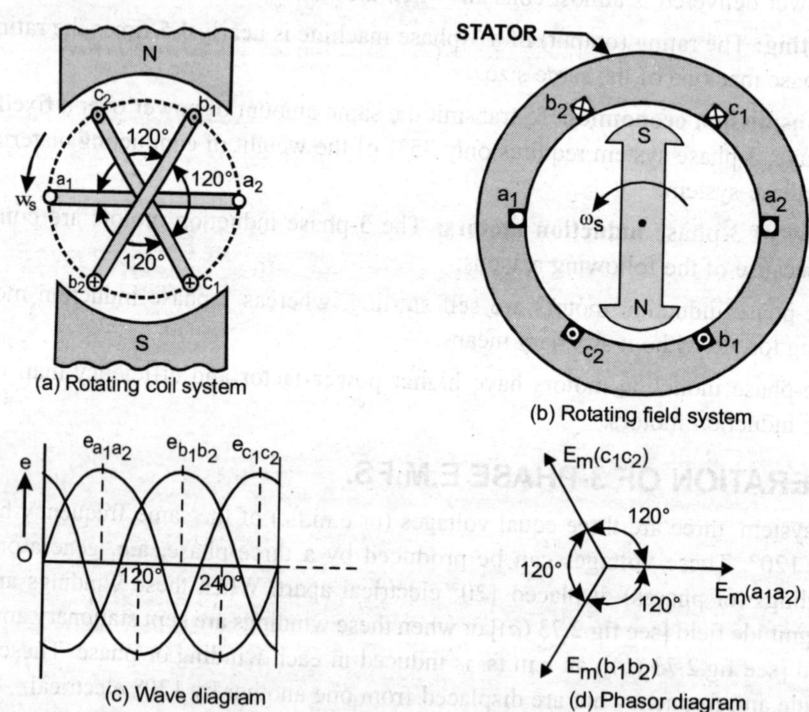

Fig.2.73: Generation of 3-phase *emfs*

2.42 NAMING THE PHASES

The three phases may be named out by numbers (1,2 and 3); by letters (*a*, *b* and *c*) or by colours (*Red*, *Yellow* and *Blue i.e. R Y B*). In India they are named by *R Y B i.e.* red, yellow and blue.

2.43 PHASE SEQUENCE

In a three phase system, there are three voltages having same magnitude and frequency displaced by an angle of 120° electrical. They are attaining their positive maximum value in a particular order.

The order in which the voltages (or e.m.fs.) in the three phases attain their maximum positive value is called the **phase sequence.**

In the above article, the e.m.fs. in the three phases attain their positive maximum vlaue in the order of a_1a_2, b_1b_2, c_1c_2, therefore, the phase sequence is *a, b, c*. However, if the coils, or phases are being named out as *R, Y, B* in place of *a, b, c* respectively, then the phase sequence will be *RYB*.

The sequence *RYB* (or *YBR* or *BRY*) is considered as positive phase sequence, whereas, the *RBY* (or *BYR* or *YRB*) is considered as negative phase sequence.

The sequence knowledge of phase sequence is essential in the following important applications:

(*i*) The direction of rotation of 3-phase induction motors depends upon the phase sequence of 3-phase supply. To reverse the direction of rotation, the phase sequence of the supply given to the motor has to be changed.

(*ii*) The parallel operation of 3-phase alternators and transformers is only possible if phase sequence is known.

2.44 INTERCONNECTION OF THREE-PHASES

In a 3-phase a.c. generator, three are three windings. Each winding has two terminals (start and finish). If a separate load is connected across each phase winding as shown in fig.2.74, then each phase

supplies an independent load through a pair of leads (wires). Thus, six wires will be required in this case to connect the load to generator. This will make the whole system complicated and expensive.

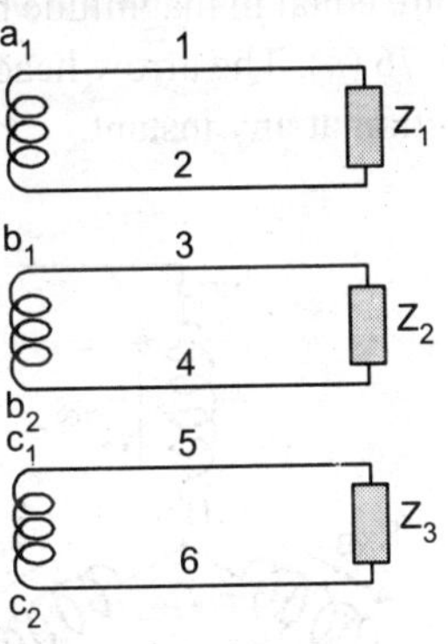

Fig.2.74: Three coils representing three phases loaded reparately.

In order to reduce the number of the conductors, the three phase windings of the a.c. generator are suitably inter-connected. The following are the two universally adopted methods of inter-connecting the three phases:

1. Star or wye (Y) connection; 2. Mesh or delta (D) connection.

2.45 STAR ORWYE (Y) CONNECTION

In star or wye (Y) connections, the similar ends (either start or finish) of the three windings are connected to a common point called star or neutral point. The three line conductors are run from the remaining three free terminals called line conductors. Ordinarily only three wires are carried to the external circuit giving 3-phase, 3-wire star connected system. However, sometimes a fourth wire is carried from the star point to the external circuit, called neutral wire, giving 3-phase, 4-wire star connected system.

As shown in fig.2.75, the finish terminals a_2, b_2 and c_2 of the three windings are connected to form a star or neutral point. From the remaining three free terminals three conductors are run, named *R*, *Y* and *B*. The current flowing through each phase is called phase current I_{ph} and current flowing through each line conductor is called line current I_L. Similarly, voltage across each phase is called phase voltage (E_{ph}) and voltage across two line conductors is called line voltage (E_L).

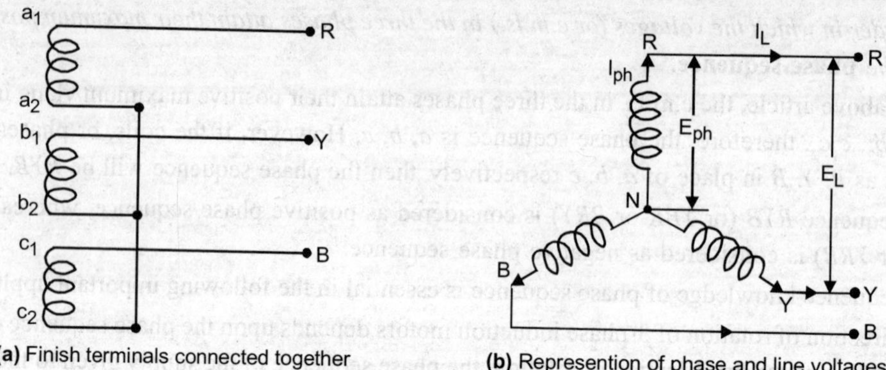

(a) Finish terminals connected together (b) Represention of phase and line voltages

Fig.2.75: Star connections

Relation between phase voltage and line voltage

The connections are shown in fig.2.76 (*a*). Since the system is balanced, the three voltages E_{NR}, E_{NY} and E_{NB} are equal in magnitude but displaced from one another by 120° electrical. Their phasor are shown in fig.2.76 (*b*). The arrow heads on e.m.fs. and currents indicate the positive direction and not their actual direction at any instant.

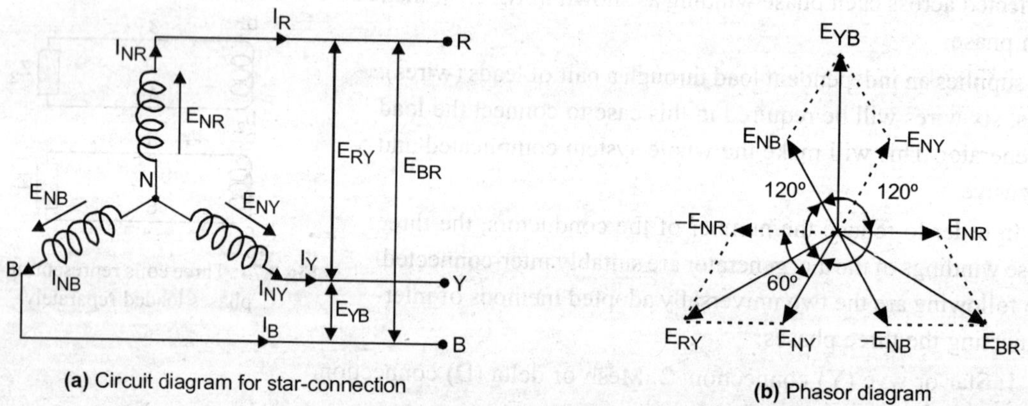

(a) Circuit diagram for star-connection (b) Phasor diagram

Fig.2.76: Relating between line and phase voltages in star connections

Now, $E_{NR} = E_{NY} = E_{NB} = E_{ph}$ *(in magnitude)*

It may be seen that between any two lines, there are two phase voltages.

Tracing the loop *N R Y N*, we get $\overline{E_{NR}} + \overline{E_{RY}} - \overline{E_{NY}} = 0$

or $\overline{E_{RY}} = \overline{E_{NY}} - \overline{E_{NR}}$ *(vector difference)*

To find the vector sum of E_{NY} and $-E_{NR}$, reverse the vector E_{NR} and add it vectroialy with E_{NY} as shown in fig.2.76(*b*).

$$\therefore \quad E_{RY} = \sqrt{E_{NY}^2 + E_{NR}^2 + 2E_{NY}\,E_{NR}\cos 60^\circ}$$

or $$E_L = \sqrt{E_{ph}^2 + E_{ph}^2 + 2E_{ph}\,E_{ph} \times 0\cdot 5} = \sqrt{3E_{ph}^2} = \sqrt{3}E_{ph} \quad \textit{(in magnitude)}$$

Similarly, $\overline{E_{YB}} = \overline{E_{NB}} - \overline{E_{NY}}$ or $E_L = \sqrt{3}\ E_{ph} = \overline{E_{NR}} - \overline{E_{NB}}$ or $E_L = \sqrt{3}\ E_{ph}$

Hence, in star connections; *Line voltage* = $\sqrt{3}$ × *Phase voltage*

Relation between phase current and line current

From fig.2.76(*a*), it is clear that same current flows through phase winding as well as the line conductor since line conductor is just connected in series with the phase winding.

$\therefore I_R = I_{NR}; I_Y = I_{NY}$ and $I_B = I_{NB}$

where, $I_{NR} = I_{NY} = I_{NB} = I_{ph}$ (phase current) and $I_R = I_Y = I_B = I_L$ (line current)

Hence, is star connections; *Line current* = *Phase current*

2.46 MESH OR DELTA (Δ) CONNECTION

In delta (Δ) or mesh connections, the finish terminal of one winding is connected to start terminal of the other winding and so on which forms a closed circuit. The three line conductors are run from three junctions of the mesh called line conductors, as shown in fig.2.77.

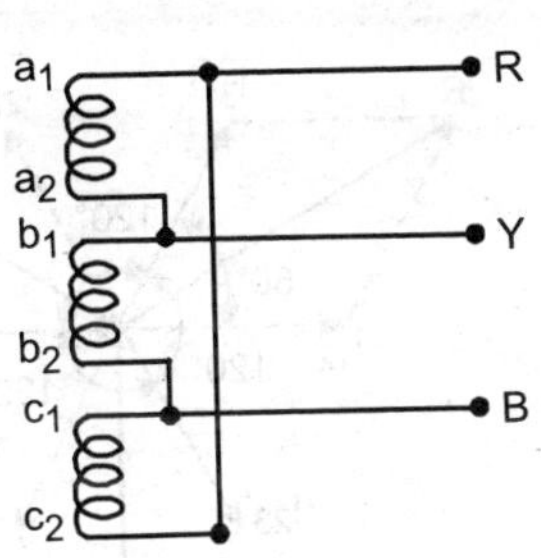

(a) Finish terminal of one phase connected to start of the next to from delta connections

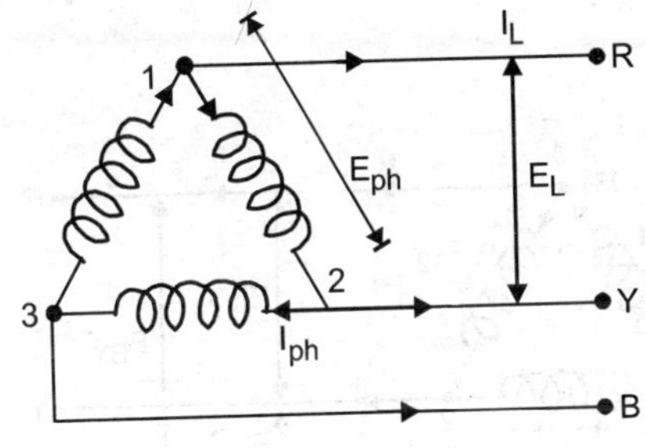

(b) Phase and line voltages in delta connections

Fig.2.77: Delta connection

To obtain delta connection a_2 is connected with b_1, b_2 is connected with c_1 and c_2 is connected with a_1 as shown in fig.2.77 (*a*). The three conductors *R*, *Y* and *B* are run from the three junctions called line conductors. The current flowing through each phase is called phase current (I_{ph}) and the current flowing through each line conductor is called line current (I_L) as shown in fig.2.77(*b*). Similarly, voltage across each phase is called phase voltage (E_{ph}) and voltage across two line conductors is called line voltage (E_L).

Relation between phase voltage and line voltage

From fig.2.78 (*a*), it is clear that voltage across terminals 1 and 2 is the same as across terminals *R* and *Y*.

$\therefore E_{12} = E_{RY}$; similarly $E_{23} = E_{YB}$ and $E_{31} = E_{BR}$

where $E_{12} = E_{23} = E_{31} = E_{ph}$ (phase voltage) and $E_{RY} = E_{YB} = E_{BR} = E_L$ (line voltage)

Hence, in delta connection; *Line voltage* = *Phase voltage.*

Relation between phase current and line current

Since the system is balanced, therefore, the three phase currents I_{12}, I_{23} and I_{31} are equal in magnitude but displaced from one another by 120° electrical. Their phasors are shown in fig.2.78 (*b*).

Thus $I_{12} = I_{23} = I_{31} = I_{ph}$ *(in magnitude)*

In fig.2.78 (*a*), it may be seen that current is divided at every junction 1, 2 and 3.

Applying Kirchhoff's first law at junction 1;

Incoming currents = outgoing currents

$\overline{I_{31}} = \overline{I_R} + \overline{I_{12}}$ or $\overline{I_R} = \overline{I_{31}} - \overline{I_{12}}$ *(vector difference)*

To find the vector sum of I_{31} and $-I_{12}$, reverse the vector I_{12} and add it vectorially with I_{31} as shown in fig.2.78 (*b*).

$$\therefore \quad I_R = \sqrt{I_{12}^2 + I_{12}^2 + 2I_{31}\, I_{12} \cos 60°} = \sqrt{I_{ph}^2 + I_{ph}^2 + 2I_{ph}\, I_{ph} \times 0 \cdot 5} \qquad (\because I_R = I_l)$$

$$\text{or} \quad I_L = \sqrt{3\, I_{ph}^2} = \sqrt{3}\, I_{ph} \quad \textit{(in magnitude)}$$

$$\text{Similarly,} \quad \overline{I_Y} = \overline{I_{12}} - \overline{I_{23}} \text{ or } I_L = \sqrt{3}\, I_{ph}$$

$$\text{and} \quad \overline{I_B} = \overline{I_{23}} - \overline{I_{31}} \text{ or } I_L = \sqrt{3}\, I_{ph}$$

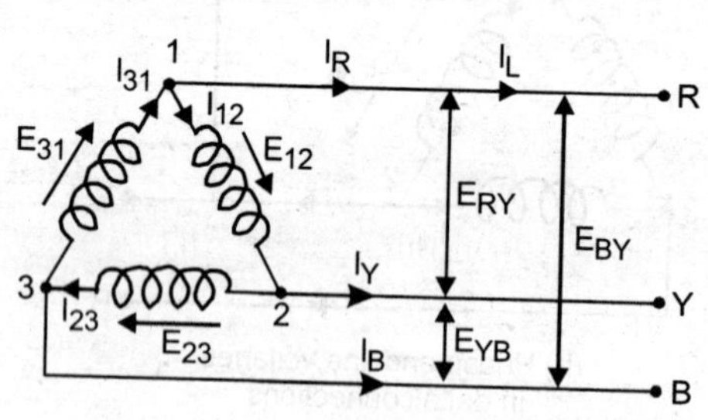

(a) Circuit diagram for delta connection

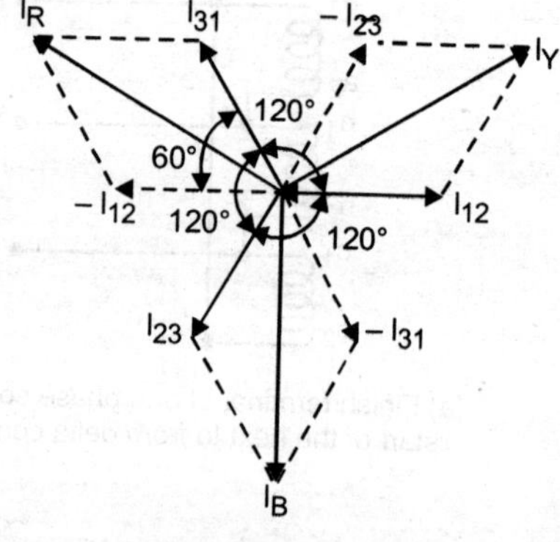

(b) Phasor diagram

Fig.2.78: Relation between line and phase currents in delta connection

2.47 CONNECTIONS OF 3-PHASE LOADS

Similar to 3-phase supply, the three-phase loads may also be connected in

(*i*) Star or (*ii*) Delta

The 3-phase loads connected in star and delta are shown in fig.2.79 (*a*) and (*b*) respectively.

The 3-phase loads may be balanced or unbalanced. If the three loads (impedances) Z_1, Z_2 and Z_3 are having same magnitude and phase angle, then the 3-phase load is said to be a balanced load. Under such connections, all the phase or line currents and all the phase or line voltages are equal in magnitude.

Throughout this book, balanced 3-phase system will be considered unless stated otherwise.

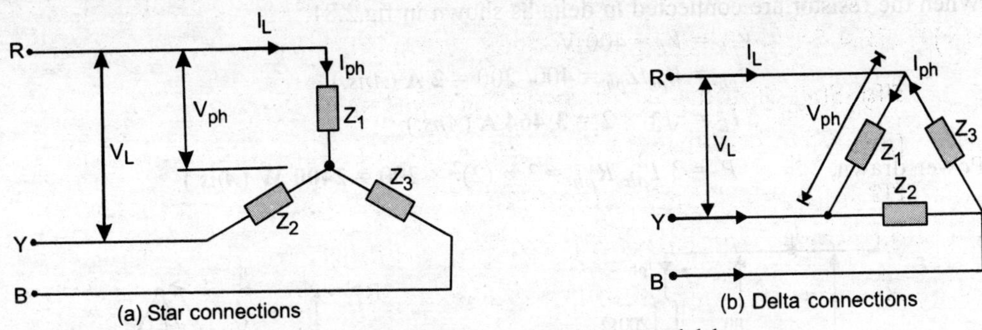

Fig.2.79: Three-phase load, connected in star and delta

2.48 POWER IN 3-PHASE CIRCUITS

Power in single-phase system or circuit is given by the relation;

$P = VI\cos\phi$, where, V = voltage of single-phase *i.e.* V_{ph}

I = current of single-phase *i.e.* I_{ph} and

$\cos\phi$ = power factor of the circuit

In 3-phase circuits (balanced load), the power is just the sum of powers in three phases, *i.e.*, $P = 3\,V_{ph}\,I_{ph}\cos\phi$.

In *star-connections*

$$P = 3\frac{V_L}{\sqrt{3}}\,I_L\cos\phi \text{ (since } V_{ph} = V_L/\sqrt{3} \text{ and } I_{ph} = I_L)$$

or $$P = \sqrt{3}\,V_L\,I_L\cos\phi$$

In *delta-connections*

$$P = 3\,V_L\,\frac{I_L}{\sqrt{3}}\cos\phi \text{ (since } V_{ph} = V_L \text{ and } I_{ph} = I_L/\sqrt{3})$$

or $$P = \sqrt{3}\,V_L\,I_L\cos\phi$$

Thus, the total power in a 3-phase balanced load, irrespective of connections (star or delta), is given by the relation $\sqrt{3}\,V_L\,I_L\cos\phi$. Its units are kW or W (watt).

Apparent power, $P_a = \sqrt{3}\,V_L\,I_L$ Its units are kVA or VA (volt-ampere)

Reactive power, $P_r = \sqrt{3}\,V_L\,I_L\sin\phi$ (volt-ampere-reactive).

Its units are kVAR or VAR

Example 2.31. *Three resistors each of 200 Ω are connected in star across 400 V, 3-phase supply. Calculate the line current, phase current and power taken from the source. What will be the same values if the resistors are connected in delta?*

Solution: When the resistors are connected in star as shown in fig.2.80.

Phase voltage, $V_{ph} = V_L/\sqrt{3} = 400/\sqrt{3} = 231$ V

Phase current, $I_{ph} = V_{ph}/Z_{ph} = 231/200 =$ **1.155 A** (*Ans.*)

Line current, $I_L = I_{ph} =$ **1.155 A** (*Ans.*)

Power drawn, $P = 3\,I_{ph}^2\,R_{ph} = 3 \times (1.155)^2 \times 200 =$ **800 W** (*Ans.*)

When the resistor are connected in delta as shown in fig.2.81.

$$V_{ph} = V_L = 400 \text{ V}$$

$$I_{ph} = V_{ph}/Z_{ph} = 400/\ 200 = \mathbf{2\ A}\ (Ans.)$$

$$I_L = \sqrt{3} \times 2 = \mathbf{3{\cdot}464\ A}\ (Ans.)$$

Power drawn, $P = 3\, I_{ph}^2\, R_{ph} = 3 \times (2)^2 \times 200 = \mathbf{2400\ W}\ (Ans.)$

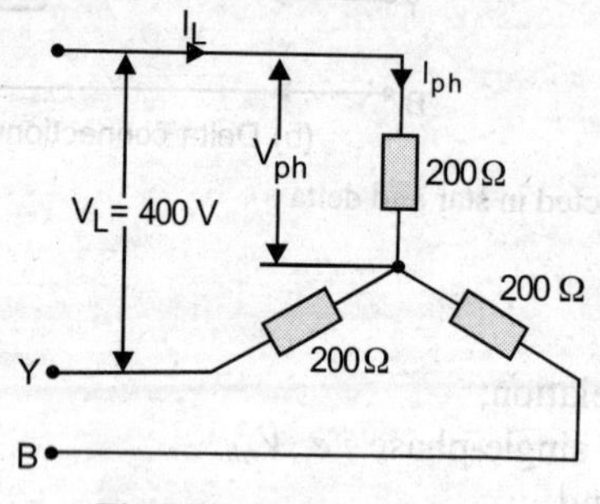

Fig.2.80: Star connections

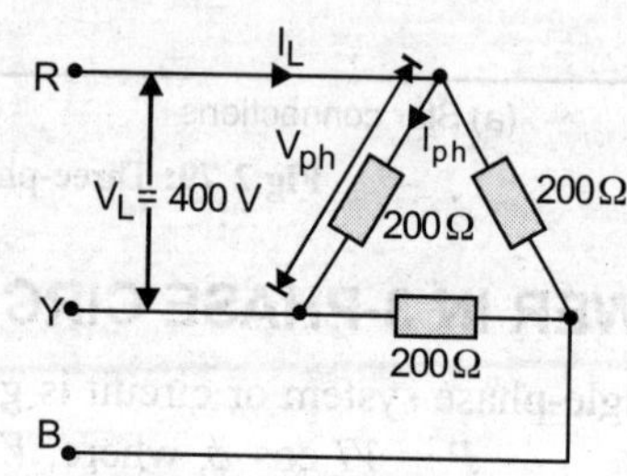

Fig.2.81: Delta connections

Example 2.32. *Three similar coils each having a resistance of* 6 *ohm and an inductance of* 0·0255 H *in are connected in star across a , three phase,* 400 V, 50 *Hz supply. Calculate the line current, power factor, power in kW, kVA and kVAR .*

Solution: The connection are shown in fig.2.82.

Here, $R = 6\ \Omega,\ \ L = 0{\cdot}0255 \text{ H},\ f = 50 \text{ Hz}$

$$X_L = 2\pi fL = 2\pi \times 50 \times 0{\cdot}0255 = 8\ \Omega$$

Line voltage, $V_L = 400$ V

Phase voltage, $V_{ph} = \dfrac{V_L}{\sqrt{3}} = \dfrac{400}{\sqrt{3}} = 231$ V

$$Z_{ph} = \sqrt{R^2 + X_L^2} = \sqrt{6^2 + 8^2} = \mathbf{10\Omega}$$

Phase current, $I_{ph} = \dfrac{V_{ph}}{Z_{ph}} = \dfrac{231}{10} = 23{\cdot}1$ A

Line current, $I_L = I_{ph} = \mathbf{23{\cdot}1\ A}$ *(Ans.)*

Power factor, $\cos\phi = \dfrac{R_{ph}}{Z_{ph}} = \dfrac{6}{10} = \mathbf{0{\cdot}6\ lagging}$ *(Ans.)*

Fig.2.82: Star connections

Power input, $P = \sqrt{3}\, V_L\, I_L \cos\phi = \sqrt{3} \times 400 \times 23.1 \times 0.6 = 9600 \text{ W} = \mathbf{9{\cdot}6\ kW}$ *(Ans.)*

Power in *KVA* $= \dfrac{kW}{\cos\theta} = \dfrac{9.6}{0.6} = \mathbf{16}$ *(Ans.)*

Power in kVAR = kVA × sin ϕ = kVA × sin $\text{Cos}^{-1}\phi$ = 16 × 0.8 = **12.8** *(Ans.)*

Example 2.33. *A 3-phase star connected balanced load is connected across a 3-phase 400V, 50Hz supply. The line current is* 20 *A and the power consumed by the load is* 12 *kW. Calculate the, phase current, power factor resistance and inductance of the load.*

Solution: The connection are shown in fig.2.83.

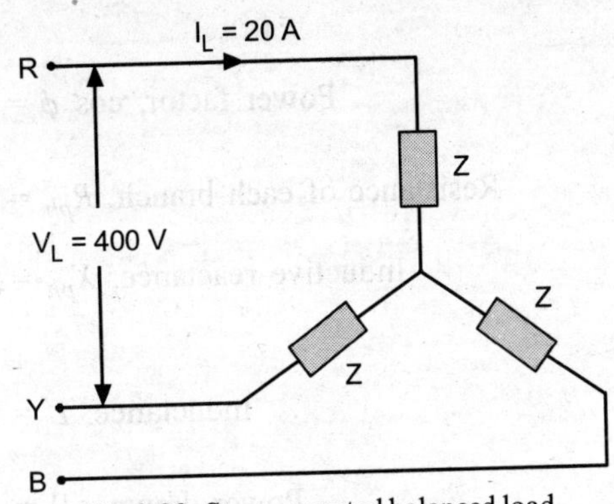

Fig.2.83: Star connected balanced load

Here, $P = 12\text{ kW} = 12000\text{ W};$

$V_L = 400\text{ V}; f = 50\text{ Hz}; I_L = 20\text{ A}$

$$\text{Phase voltage, } V_{ph} = \frac{V_L}{\sqrt{3}} = \frac{400}{\sqrt{3}} = 231\text{ V}$$

$$\text{Phage current, } I_{ph} = I_L = \mathbf{20\ A}\ (Ans.)$$

$$\text{Phase impedance, } Z_{ph} = \frac{V_{ph}}{I_{ph}} = \frac{231}{20} = 11{\cdot}55\ \Omega$$

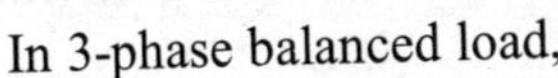

In 3-phase balanced load,

$$\text{Power } P = \sqrt{3}\, V_L\, I_L \cos\phi$$

$$\therefore \text{ Power factor, } \cos\phi = \frac{P}{\sqrt{3}\, V_L\, I_L} = \frac{12000}{\sqrt{3} \times 400 \times 20}$$

$$= \mathbf{0{\cdot}866}\ (Ans.)$$

$$\text{Resistance, } R = Z \cos\phi = 11.55 \times 0.866 = \mathbf{10\ \Omega}\ (Ans)$$

$$\text{Inductive reactance, } X_L = \sqrt{Z_{ph}^{\ 2} - R^2} = \sqrt{(11.55)^2 - (10)^2} = 5.78\ \Omega$$

$$\text{Inductance, } L = \frac{X_L}{2\pi f} = \frac{5.78}{2\pi \times 50} = \mathbf{0.0184\ H}\ (Ans.)$$

Example 2.34. *A 3-phase balanced delta connected load is connected to a 3-phase, 400 V, 50 Hz supply. It draws a line current of 17.32 A at 0·866 power factor lagging. Determine resistance and inductance of each branch. Also determine the power drawn by the load.*

Solution: The connection are shown in fig.2.84.

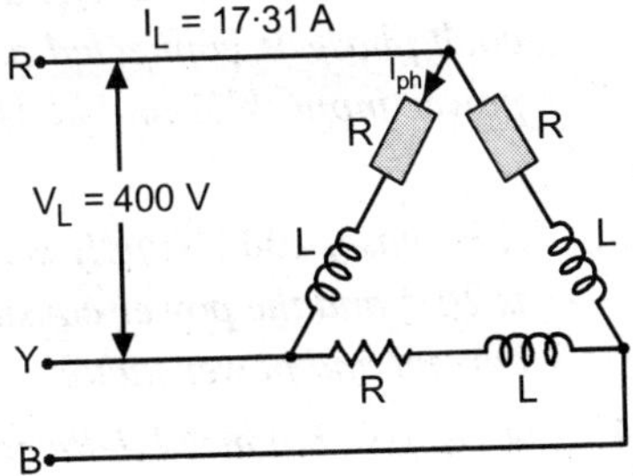

Fig.2.84: Delta connected load

Line voltage, $V_L = 400\text{ V}, \ f = 50\text{ Hz}$

Line current, $I_L = 17{\cdot}32\text{ A},$

power factor, $\cos\phi = 0{\cdot}866$ (lag).

$$\text{Phase current, } I_{ph} = \frac{I_L}{\sqrt{3}} = \frac{17.32}{\sqrt{3}} = 10\text{ A}$$

$$\cos\phi = 0{\cdot}866$$

$$\Rightarrow \quad \phi = \cos^{-1}(0{\cdot}866)$$

$$= 30°$$

$$\text{Phase voltage, } V_{ph} = V_L = 400\text{ V}$$

$$\text{Impedance of each branch, } Z_{ph} = \frac{V_{ph}}{I_{ph}} = \frac{400}{10} = 40\ \Omega$$

$$\text{Power factor, } \cos\phi = \frac{R_{ph}}{Z_{ph}}$$

Resistance of each branch, $R_{ph} = Z_{ph} \cos\phi = 40 \times 0{\cdot}866 =$ **34.64 Ω** *(Ans.)*

$$\text{Inductive reactance, } X_{ph} = \sqrt{Z_{ph}^2 - R_{ph}^2} = \sqrt{(40)^2 - (34.64)^2} = 20\ \Omega$$

$$\text{Inductance, } L = \frac{X_{ph}}{2\pi f} = \frac{20}{2\pi \times 50} = \mathbf{63.66\ mH}\ (Ans.)$$

Power drawn , $P = 3V_{ph}\, I_{ph} \cos\phi = 3 \times 400 \times 10 \times 0{\cdot}866 =$ **10392 W** *(Ans.*

$$\text{Power factor of the load, } \cos\phi = \frac{R}{Z} = \frac{15}{19{\cdot}58} = 0{\cdot}7661 \text{ lagging}$$

Power absorbed in each phase, $P_{ph} = V_{ph}\, I_{ph} \cos\phi$

$= 200 \times 10{\cdot}215 \times 7661 =$ **1565 W** *(Ans.)*

Total power absorbed, $P = \sqrt{3}\, V_L\, I_L \cos\phi = \sqrt{3} \times 200 \times 17{\cdot}69 \times 0{\cdot}7661 =$ **4695 W** *(Ans.)*

PRACTICE EXERCISE

1. *Three 100 Ω resistors are connected first in star and then in delta across 415 V, 3-phase supply. Calculate the line and phase currents in each case and also the power taken from the source.*

   (*Ans.* 1.155 A; 800 W; 2 A; 3·464 A; 2400 W)

2. *Three similar coils each having a resistance of 8 ohm and an inductance of* 0·0191 H *in series in each phase is connected across a* 400 *V, three phase,* 50 *Hz supply. Calculate the line current, power input, kVA and kVAR taken by the load.*

   (*Ans.* 23·1 A; 0·6 lagging; 9·6 kW; 16; 12.8)

3. *A 3-phase, 400 V supply is connected to a 3-phase star connected balanced load. The line current is 20 A and the power consumed by the load is 12 kW. Calculate the impedance of the load, phase current and power factor.*

   (*Ans.* 20 A; 0·866; 10 Ω; 0.0184 H)

4. *A 3-phase balanced delta connected load is connected to a 3-phase, 400 V, 50 Hz supply. It draws a line current of* 34·64 *A at* 0·8 *power factor lagging. Determine resistance and inductance of each branch. Also determine the power drawn by each phase.*

   (*Ans.* 34.64 Ω; 63.66 mH; 10392 W; 1565 W; 4695 W)

Example 5. *A 3-f, 3-wire, Y-connected system has 150 V between phase to phase. Each phase has* $Z = 5 \angle -30°$. *Find current in each phase and total power drawn from the mains. Draw the phasor diagram.*

7. *Three similar coils each having a resistance of 15 ohm and an inductance of 0·04 henry are connected in star to a 3-phase 50 Hz supply, 200 volts between lines. Calculate the line current. If they are now connected in delta, calculate the phase current, line current and the total power absorbed in each phase.*

   (*Ans:* 5·9 A; 17·69 A; 1565 W; 4695 W)

SUMMARY

1. *Alternating voltage*: A voltage that changes its polarity and magnitude at regular inervals of time is called an *alternating voltage*.
2. *Sinusoidal alternating quantity*: An alternating quantity that varies according to sine of angle θ is known as sinusoidal alternating quantity.
3. *Equation of an alternating quantity*:
   $$e = E_m \sin \theta = E_m \sin \omega t; \; i = I_m \sin \theta = I_m \sin \omega t$$
4. *Wave form*: The shape of the curve obtained by plotting the instantaneous values along *y*-axis and time or angle $\theta (= \omega t)$ along *x*-axis is called wave form or wave shape.
5. *Cycle*: When an alternating quantity goes through a complete set of +ve and –ve value or goes through 360 electrical degrees, it is said to have completed a cycle.
6. *Alternation*: One half of cycle is called an alternation.
7. *Time period* (*T*): Time taken to complete one cycle in second is called time period.
8. *Frequency* (*f*): The number of cycles made per second by an alternating quantity is called its frequency. It is measured in c/s or Hz (Hertz).
9. *Relations*: $f = 1/T$ and $\omega = 2\pi f$
10. Instantaneous value (v or i): The value of an alternating quantity at any instant is called its instantaneous value.
11. *Amplitude* (V_m or I_m): The maximum value obtained by an alternating quantity during a cycle is called its amplitude or maximum value or peak value or crest value.
12. *Average value* (V_{av} or I_{av}): The arithmetic average of all the instantaneous values considered of an alternating quantity over one cycle is called its *average value*.
   $$I_{av} = \frac{i_1 + i_2 + i_3 + ... + i_n}{n}$$
   For sinusoidal current, $I_{av} = 2\, I_m/\pi = 0{\cdot}637\, I_m$.
13. *Effective or r.m.s. value* ($V_{r.m.s.}$ or $I_{r.m.s.}$): That steady current which when flows through a resistor of known resistance for a given time produces the same amount of heat as produced by the alternating current when flows through the same resistor for the same time is called effective or r.m.s. value of the alternating current.
   $$I_{r.m.s.} = \sqrt{\frac{i_1^2 + i_2^2 + i_3^2 + ... + i_n^2}{n}}$$
   For sinusoidal current, $I_{r.m.s.} = I_m/\sqrt{2} = 0{\cdot}707\, I_m$.
14. *Form factor*: $I_{r.m.s.}/I_{av}$; For sinusoidal quantities, its value is 1.11.
15. *Peak factor*: $I_m/I_{r.m.s.}$; For sinusoidal quantities, its value is 1.414.
16. *Phase*: The phase of an alternating quantity at an instant is defined as the fractional part of a cycle through which the quantity has advanced from a selected origin. It has less importance, in practice.

17. *Phase difference*: The angular displacement between the maximum positive values of two alternating quantities having the same frequency is called the phase difference between them.

Mathematically ; If $i_1 = I_{m1} \sin \omega t$ the $i_2 = I_{m2} \sin (\omega t \pm \phi)$

18. *Leading quantity*: An alternating quantity that attains its + ve maximum value prior to the other is called leading quantity.

19. *Lagging quantity*: An alternating quantity that attains its +ve maximum value after the other is called lagging quantity.

20. *A.C. Circuit:* The path for the flow of alternating current is called a.c. circuit.

21. *Pure resistive circuit*: Circuit contains pure resistance.

$v = V_m \sin \omega t$; $i = I_m \sin \omega t$; $I_m = V_m/R$; $P = VI$; $p.f. = 1$;

Current is in phase with voltage vector.

22. *Pure inductive circuit*: Circuit contains pure inductance.

$v = V_m \sin \omega t$; $i = I_m \sin (\omega t - \pi/2)$; $I_m = V_m/X_L$; $X_L = 2 \pi f L$; $P = 0$; $p.f. = 0$ lag; Current lags behind voltage vector of 90° (*i.e.* $\pi/2$)

23. *Pure capacitive circuit*: Circuit contains pure capacitance.

$v = V_m \sin \omega t$; $i = I_m \sin (\omega t + \pi/2)$; $I_m = V_m/X_C$; $X_C = 1/2 \pi f C$; $P = 0$; $p.f. = 0$ lead; Current leads the voltage vector by 90° (*i.e.* $\pi/2$).

24. *R—L series circuit*: Circuit contains resistance and inductance in series. It is generally called inductive circuit. $v = V_m \sin \omega t$; $i = I_m \sin (\omega t - \phi)$; $\phi = \tan^{-1} X_L/R$; $I_m = V_m/Z$; $Z = \sqrt{R^2 + X_L^2}$; $P = VI \cos \phi = I^2 R$; $p.f. = \cos \phi = R/Z$ lag;

25. *Power triangle*: The three components of a right angled power triangle are:

(*i*) *True power:* It is the base or horizontal component of the power triangle which represents the actual or true power consumed in an a.c. circuit.

$P = VI \cos \phi$; unit of true power is watt (W), kilowatt (kW) or megawatts (MW)

(*ii*) *Reactive power:* It is the perpendicular or vertical component of the power triangle which represents reactive power of the circuit. This is the power which reacts in circuit, in fact, it is that power which is supplied by the source in a quarter cycle and the same is fed back in the next quarter cycle.

$P_r = VI \sin \phi$; units of reactive power are volt-ampere-reactive (VAR), kVAR or MVAR.

(*iii*) *Apparent power*: It is the hypotenuse of the power triangle. It is the power which looks to be consumed in the circuit but actually it is not so.

$P_a = VI$; units of apparent power are volt-ampere (VA), kVA or MVA.

26. *Power factor and its importance*: p.f. $= \cos \phi = R/Z =$ true power/apparent power

Ill-effects of poor p.f. (*i*) greater conductor size, (*ii*) poor efficiency, (*iii*) poor voltage regulation and (*iv*) larger *kVA* rating of equipment.

27. *R—C series circuit*: Circuit contains resistance and capacitance in series. It is generally called capacitive circuit. $v = V_m \sin \omega t$; $i = I_m \sin (\omega t + \phi)$; $\phi = \tan^{-1} X_C/R$; $I_m = V_m/Z$; $Z = \sqrt{R^2 + X_C^2}$; $P = VI \cos \phi = I^2 R$; P.f. $= \cos \phi = R/Z$ leading;

28. *R—L—C series circuit*: Circuit contains resistance, inductance and capacitance in series. This circuit behaves as under:

(*i*) When $X_L > X_C$..........inductive circuit..........p.f. lagging ....0 to 1

(*ii*) When $X_L < X_C$..........capacitive circuit........p.f. leading ....0 to 1

(*iii*) When $X_L = X_C$..........resistive circuit.........p.f. unity....1 only.

29. *Series resonance*: An $R - L - C$ circuit is said to be in series resonance, when

$$X_L = X_C;\ Z_r = \sqrt{R^2 + (X_L - X_C)^2} = R$$

Resonance frequency, $f_r = 1/2\,\pi\,\sqrt{LC}$

Current is maximum *i.e.* $I_r = V/Z_r = V/R$

$$\text{Q-factor} = 2\,\pi\,f_r\,L/R = \frac{1}{R}\sqrt{\frac{L}{C}}$$

30. *A.C. Parallel Circuit*: An a.c. circuit in which number of branches (each branch contains number of component in series) are connected in parallel is called an a.c. parallel circuit.

31. *Method of solving parallel a.c. circuits:* The following methods are used:

(*i*) Phasor or vector method (*ii*) Admittance method (*iii*) Symbolic or j-method

Phasor or vector method: The magnitude and direction of the currents flowing through various branches is determined are represented as phasors. The phasors are resolved to determine the resultant.

32. Parallel resonance: A parallel circuit, containing inductor and capacitor in parallel, is said to be in parallel resonance when circuit current is in phase with the applied voltage, *i.e.*

$$I_C = I_L \sin\phi;\ \text{Resonance frequency, } f_r = \frac{1}{2\,\pi}\sqrt{\frac{1}{LC} - \frac{R^2}{L^2}}$$

$$\text{If } R \text{ is very small and neglected, } f_r = \frac{1}{2\,\pi\sqrt{LC}}$$

$$\text{Current is minimum, } i.e.,\ I_r = V/Z_r;\ Q\text{-factor} = 2\,\pi\,f_r\,L/R = \frac{1}{R}\sqrt{\frac{L}{C}}$$

33. *Polyphase system*: An a.c. system having a group of (two or more than two) equal voltages of same frequency arranged to have equal phase difference between them (adjacent e.mfs.) is called *polyphase system.*

34. *Advantage of 3-phase system:* Constant power; higher rating (1.5 times), economy in power transmission; superiority of 3-phase induction motor etc.

35. *Generation of 3-phase e.m.fs.*: $e_{a1a2} = E_m \sin\omega t$; $e_{b1b2} = E_m \sin(\omega t - 120°)$; $e_{c1c2} = E_m \sin(\omega t - 240°)$

36. *Interconnection of three-phases*: (*i*) Star-connection: $E_L = \sqrt{3}\,E_{ph}$; $I_L = I_{ph}$

(*ii*) Delta-connection: $E_L = E_{ph}$; $I_L = \sqrt{3}\,I_{ph}$

Power, $P = \sqrt{3}\,V_L\,I_L \cos\phi = 3\,V_{ph}\,I_{ph} \cos\phi$

SHORT ANSWER QUESTIONS

1. What do you understand by alternating current ?
2. What is difference between ac and dc ?
3. What is meant by instantaneous value ?
4. What is meant by peak value of an alternating quantity ?
5. What are scalar and vector quantities ?
6. What do you mean by phase ?
7. What is an in-phase condition ?
8. What is an out of phase condition ?
9. List the various factors governing the emf generation ?
10. Define average value of an alternating quantity.
11. What is *RMS* value of an alternating quantity ?
12. Why are alternating voltages and currents expressed in *RMS* values instead of average values ?
13. Define form factor.
14. Define peak factor of an alternating quantity.
15. What is significance of form factor ?
16. What is the significance of peak factor ?
17. Differentiate between form factor and peak factor.
18. How will you calculate average value for symmetrical waves ?
19. What is a phasor ?
20. What is the significance of the phasor representation of an alternating quantity ?
21. What is meant by phase difference ?
22. What are leading and lagging quantities ?
23. What is an *A.C.* circuit ?
24. What are the differences between *DC* circuit and *AC* circuit
25. What is the phase difference between voltage and current in a resistor ?
26. What is the average power in a resistor ?
27. What is the phase difference between current and voltage in an inductor (pure inductance ?
28. Define inductive reactance.
29. What is the value of average power in an inductor ?
30. What is the phase difference between voltage and current in a capacitor ?
31. Define capacitive reactance.
32. What is the value of average power in a capacitor.
33. What is a series *RL* circuit ?
34. What is the impedance of *RL* series circuit ?
35. What is apparent power of an *AC* circuit ?
36. What is real power of an A.C. circuit ?
37. What is power factor ?

38. What is reactive power of an *a.c.* circuit ?
39. What is active component of an *a.c.* current ?
40. What is a reactive component of an *a.c.* current ?
41. What do you understand by *RC* series circuit ?
42. What is the impedance of *RC* series circuit ?
43. What is meant by *RLC* series circuit ?
44. What is the impedance of *RLC* series circuit ?
45. What is parallel *ac* circuit ?
46. What is mean by resonance circuit ?
47. What happens when *RLC* series circuit is at resonance ?
48. How does series resonance result in voltage amplification ?
49. The *Q* of a coil is 20 what does it mean ?
50. What are important points to be noted for a parallel circuit when it is at resonance ?
51. At parallel resonance why impedance of the circuit is pure resistance ?
52. Three-phase system is preferred over single-phase system, why ?
53. The knowledge of phase sequence is very important in industries and power system, why ?
54. In star connections the voltage between the two line terminals is not double to that of phase voltage but it is $\sqrt{3}$ times, why ?
55. What do you understand by the following terms:
(*i*) Three phase balanced supply, (*ii*) Three phase balanced load, (*iii*) Three phase unbalanced supply, (*iv*) Three phase unbalanced load, (*v*) Single phasing.
56. Write down the advantages of 3-phase system over single phase.
57. What are the two ways of connecting a 3-phase system ? Draw their phasor diagrams and write down the relationship between phase and line voltages and currents for these systems.
58. Show that the total power in a 3-phase balanced load is $P = \sqrt{3}\ VI \cos\phi$ where V is the rms line voltage, I is the rms line current and ϕ is the phase angle between phase voltage and phase current.
59. Compare 3-phase star and delta-connected systems

NUMERICAL FOR PRACTICE

1. A current is given by $i = 45.34 \sin 377\ t$, find (*i*) maximum value of current; (*ii*) r.m.s. value of current; (*iii*) frequency of current; (*iv*) radians through which its phasor has advanced after 0·01 second; (*v*) number of degrees in (*iv*) and (*vi*) value of current at instant mentioned in part (*iv*). (**Ans.** *45.34 A; 32 A; 60 Hz; 3·77 rad.; 216°; –26.6 A*)
2. An alternating voltage is given as $u = 200 \sin 314\ t$ determine its (*i*) maximum value; (*ii*) effective value; (*iv*) form factor; (*v*) value of voltage after 0·0025 second taking reckoning time from the instant when voltage is zero and becoming positive; (*vi*) time after which voltage attains 200 V for the first time.
(**Ans.** *200 V; 141.42 V; 50 Hz; 1·11; 141.42 V; 1·67 ms*)
3. A sinusoidal voltage of 50 Hz has a maximum value of $100\sqrt{2}$ V. At what time measured from a positive maximum value will the instantaneous voltage be 70.7 V ?(**Ans.** *3·33 ms*)

4. Calculate the r.m.s. value of a periodic voltage having the following values for equal intervals changing suddenly from one value to the next: 0, 10, 30, 60, 30, 10, 0, – 10, – 30, – 60, – 30, – 10 etc. (**Ans.** *30·55 V*)

5. For a full wave rectified alternating current, determine (*i*) average value; (*ii*) r.m.s. value; (*iii*) form factor and (*iv*) peak factor. (**Ans.** $0{\cdot}637\ I_m$; $0{\cdot}707\ I_m$; *1·11; 1·4142*)

6. Represent the following alternating currents as phasors on the same axes and determine phase angle of all other currents with I_1: $i_1 = 15 \sin(\omega t + \pi/3)$; $i_2 = 16 \cos(\omega t - \pi/6)$; $i_3 = 20 \cos(\omega t - \pi/3)$; $i_4 = 10 \sin(\omega t - \pi/6)$. (**Ans.** 60°; 30°; 90°)

7. An e.m.f. $e_1 = 50 \sin \omega t$ and the other $e_2 = 30 \sin(\omega t - \pi/6)$ act together in the same circuit. Find the resultant e.m.f. (**Ans.** $77{\cdot}45 \sin(\omega t + 11{\cdot}17°)$)

8. A 230 V water heater operating for 3 hours daily on a.c. mains is found to give 250 k calorie per hour, the efficiency of heater system being 90%. Determine (*i*) the r.m.s. and the maximum value of current; (*ii*) resistance of the heater; (*iii*) also write down the equations for voltage and current when the supply frequency is 50 Hz.
(**Ans.** *1·41 A; 1·993 A; 163·1 Ω;* $v = 325{\cdot}27 \sin 314\,t$; $i = 1{\cdot}993 \sin 314\,t$)

9. An inductive coil has negligible resistance and inductance of 0·1 henry. It is connected across a 220 V, 50 Hz supply. Find the current and power. Also write down the expression for instantaneous value of applied voltage and current.
(**Ans.** *7 A; zero;* $v = 311 \sin 314\,t$, $i = 9{\cdot}9 \sin(314\,t - \pi/2)$)

10. A capacitor has a capacitance of 30 microfarad. Find its capacitive reactance for frequency of 25 and 50 Hz. Find in each case the current if the supply voltage is 400 V.
(**Ans.** *212·2 Ω; 106·1 Ω; 2·073 A; 4·146 A*)

11. A coil having a resistance of 7 Ω and an inductance of 31·8 m H is connected to 230 V, 50 Hz supply. Calculate (*i*) the circuit current, (*ii*) phase angle, (*iii*) power factor and (*iv*) power consumed. (**Ans.** *18·85 A; 55° lag; 0·573 lag; 2484·24 W*)

12. A voltage $e = 200 \sin 100\,\pi\,t$ is applied to a coil having $R = 200$ ohm and $L = 638$ millihenry. Find the expression for the current and also determine the power taken by the coil. (**Ans.** $0{\cdot}706 \sin(314\,t - 45{\cdot}06°)$; *50 W*)

13. A 100 V, 60 W lamp is to be operated at 220 V, 50 Hz mains. What (*i*) pure resistance (*ii*) pure inductance or (*iii*) pure capacitance should be, placed in series with the lamp, which enable it to run without being damaged ? Which method would be preferred and why ?
(**Ans.** *200 Ω; 1·038 H; 9·75 μF*)

14. Two coils are connected in series having resistance and inductive reactance 5 and 6 ohm, 3 and 7 ohm respectively. A sinusoidal voltage of 200 V, 50 Hz is applied across the combination. Calculate (*i*) current, power factor and power absorbed in the whole circuit; (*ii*) voltage drop across each coil; (*iii*) power factor and power absorbed in each coil.
(**Ans.** *13·1 A, 0·5241 lag, 1373W; 102·3 V, 99·77 V; 0·64 lag, 858 W, 0·394 lag, 515 W*)

15. A voltage of 125 V at 50 Hz is applied across a non-inductive resistor connected in series with a condenser. The current in the circuit is 2·2 A. The power loss in the resistor is 96·8 W and that is the condenser is negligible. Calculate the resistance and the capacitance.
(**Ans.** *20 Ω; 59·85 μ F*)

16. An a.c. circuit having a resistance of 10 ohm, an inductance of 0·2 henry and a capacitance of 100 μF in series, is connected across a single-phase 110V, 50 Hz supply. Calculate the following for this circuit. (*i*) Resultant reactance; (*ii*) Impedance; (*iii*) Current; (*iv*) Voltages across *R*, *L* and *C*; (*v*) Phase difference between the current and supply voltage; (*vi*) Draw the phasor diagram of the circuit.

(**Ans.** *31 ohm; 32·573 ohm; 3·377 A, 33·77V, 212·18 V, 107·49 V; 72·12°*)

17. A coil of resistance 100 Ω and inductance 100 μ H is connected in series with a 100 p F capacitor. The circuit is connected to a 50 V variable frequency supply. Calculate (*i*) the resonant frequency; (*ii*) current at resonance; (*iii*) voltage across *L* and *C* at resonance and (*iv*) *Q*-factor of the circuit. (**Ans.** *1·59 × 10⁶ Hz; 0·5 A; 500 V; 500 V; 10*)

18. A capacitor of capacitance *C* farads is connected in series with a coil having 75 Ω resistance and 12 henry inductance. Calculate the value of *C* when the circuit is connected across 220 V, 60 Hz supply. (**Ans.** *0·587 μ F*)

19. Determine the parameter of an *R* – *L* – *C* series circuit that will resonate at 10000 Hz, has a bandwidth of 1000 Hz and draws 15·3 W from a 200 V generator operating at the resonant frequency of the circuit. (**Ans.** *2·614 ohm, 416 m H, 609 p F*)

20. A coil of resistance 15 ohm and inductance 0·05 henry is connected in parallel with a non-inductive resistor of 20 ohm. Find (*i*) the current in each branch of the circuit, (*ii*) the total current supplied, and (*iii*) the phase angle of the combination, when a voltage of 230 volt at 50 hertz is applied across the circuit. Draw the relevant phasor diagram.

(**Ans.** *10·59 A , 11·5 A; 20·31 A; 22·14° lagging*)

21. Two coils are connected in parallel across a 200 V, 50 Hz supply. At the supply frequency, their impedances are 6 and 10 ohm respectively and their resistances are 2 and 3 ohm respectively, calculate: (*i*) The current in each coil; (*ii*) the total current; (*iii*) The total power. (**Ans.** *33·33 A; 20 A; 53·3 A; 3421·6 W*)

22. A single phase motor draws 10 A at 230 V at a power factor of 0·8 lagging. Calculate the value of the capacitor which when connected across the terminals of the motor will bring the line current in phase with the voltage. Frequency of supply may be assumed as 50 Hz.

(**Ans.** *166 μ F*)

23. A parallel circuit consists of a 2·5 μ F capacitor and a coil whose resistance and inductance are 15 Ω and 260 m H respectively. Determine (*i*) the resonant fiequency; (*ii*) *Q* of the coil and (*iii*) dynamic impedance of the circuit. (**Ans.** *197 Hz, 21·45, 6933·33 Ω*)

24. A coil has a resistance of 500 Ω and inductance of 350 μ H. Find the capacitance of a capacitor which when connected in parallel with the coil, will produce resonance with a supply frequency of 10^6 Hz. If a second capacitor of capacitance 30 p F is connected in parallel with the first capacitor, find the frequency at which resonance will occur.

(**Ans.** *68·8 p F, 0·825 M Hz*)

25. Three similar coils each having a resistance of 15 ohm and an inductance of 0·04 henry are connected in star to a 3-phase, 50 Hz supply, 200 volt between lines. Calculate the line current. If they are now connected in delta, calculate the phase current, line current and the total power absorbed in each phase.

(**Ans.** *5·9 A, 10·215 A, 17·69 A, 1565 W, 4695 W*)

26. Three identical impedances are connected in delta to a 3-phase, 400 V, 50 Hz supply. The line current is 34·65 A and total power taken from supply is 14.14 kW. Calculate the value of resistance and inductance of each phase. (**Ans.** *11·78 Ω, 51·45 mH*)

27. Three inductive coils each of inductances 50 mH are connected in star to a 3-phase, 200 V, 50 Hz system. Calculate the inductance of each coil, which when connected in delta to the same supply, will taken the same line current. **(Ans.** *150 mH)*

28. Three 100 ohm non-inductive resistances are connected in (*i*) star (*ii*) delta across a 400 V, 50 Hz supply. Calculate the power taken from the supply in each case. In the event of one of the resistance getting opened what would be the value of the total power taken from the supply in each case. **(Ans.** *1600 W, 4800 W; 800 W, 3200 W)*

29. Show that the power consumption, when three identical impedances are connected in delta across a balanced 3-phase supply is three times that when the same impedances are connected in star across the same 3-phase supply.

30. A 400 V, 3-phase system is connected to a balanced star load with the load impedance in each phase as 40 $\angle 60°$ ohm. Find (*i*) the line and phase currents and (*ii*) draw phasor diagram showing line voltages, line currents and phase currents.**(Ans.** *5·775 A; 5·775 A)*

MULTI-CHOICE QUESTIONS

1. *The r.m.s. value of an alternating current is given by steady current which when flowing through a given circuit for a given time produces*
(*a*) more heat than produced by a.c. when flowing through the same circuit
(*b*) the same amount of heat as produced by a.c. when flowing through the same circuit.
(*c*) less heat than produced by a.c. when flowing through the same circuit.
(*d*) 144 joule.

2. *The average value of sinusoidal current is given by the relation*
(*a*) $I_m/\sqrt{2}$ (*b*) 0·707 I_m (*c*) $2\,I_m/\pi$ (*d*) none of above.

3. *In case of an unsymmetrical alternating quantity, the average value must always be taken over*
(*a*) the half cycle (*b*) the quarter cycle
(*c*) the whole cycle (*d*) any fraction of the cycle.

4. *The r.m.s. value of sinusoidal alternating current is given by the relation*
(*a*) $I_m/\sqrt{2}$. (*b*) 0·637 I_m (*c*) $2\,I_m/\pi$ (*d*) $I_m/2$

5. *The amplitude factor of sinusoidal current is*
(*a*) 1·11 (*b*) 1·414 (*c*) 0.707 (*d*) 0·637

6. *The r.m.s. value is a fraction of maximum value of sinusoidal current, the value of the fraction is*
(*a*) 1/1·414 (*b*) 0·637 (*c*) $\sqrt{2}$ (*d*) none of above.

7. *The r.m.s. value of an a.c. signal is 10 volt. The peak to peak value will be*
(*a*) 6·37 V (*b*)14·14 V (*c*) 141 V (*d*) 28·28 V

8. *A wave completes one cycle in 10μ seconds, its frequency will be*
(*a*) 0.1 M Hz (*b*) 100,000 Hz (*c*) 100 k Hz (*d*) All of these.

9. *An alternating quantity which attains its positive maximum value prior to the other is called the*
(*a*) in phase quantity (*b*) lagging quantity (*c*) leading quantity (*d*) none of above.

10. *If the effective value of the sinusoidal voltage is 222 volt. Its average value will be*

(*a*) 200 V (*b*) 246.42 V (*c*) 313.02 V (*d*) none of above

11. *The inductive reactance is measured in ohm because it*

(*a*) Opposes the alternating current

(*b*) Helps the alternating current

(*c*) is the product of frequency and inductance

(*d*) has a back e.m.f.

12. *In an R-L series circuit, if R = X_2, the phase difference between ac voltage and current will be*

(*a*) 0° (*b*) 30° (*c*) 45° (*d*) 90°

13. *If an a.c. voltage is applied across a capacitor, the alternating current flows through the circuit because*

(*a*) of high peak value of voltage. (*b*) capacitor offers low opposition

(*c*) electrons can pass through capacitor (*d*) capacitor charges and discharges according to the supply frequency.

14. *As frequencies increases, the value of capacitive reactance*

(*a*) increases (*b*) decreases

(*c*) remains the same (*d*) First decreases then increases

15. *In a pure inductive a.c. circuit:*

(*a*) voltage leads the current vector by 90° (*b*) current lags the voltage vector by 90°

(*c*) current is in phase with voltage vector. (*d*) both *a* and *b*

16. *In an a.c. circuit, the true expression for apparent power is*

(*a*) $V_{av} I_{av}$ (*b*) $VI \cos \phi$ (*c*) $V_{r.m.s.} I_{r.m.s.}$ (*d*) $VI \sin \phi$

17. *If the frequency of power supply offered to a pure capacitive circuit is doubled,*

(*a*) the current will also be doubled

(*b*) the current will reduce to half

(*c*) the opposition to current decreases to half

(*d*) both *a* and *b*

18. *The power factor of a series resonance circuit is*

(*a*) zero (*b*) 0·5 lagging (*c*) unity (*d*) 0·5 leading

19. *The current in a parallel resonance circuit will be*

(a) zero (*b*) infinity (*c*) very small (*d*) very large

20. *An R—L—C series circuit contains resistance 25 Ω, inductance 8 m H and capacitance of 80 μ F, the p.f. of the circuit will be*

(*a*) unity (*b*) zero (*c*) lagging (*d*) leading

21. *For the same rating, the size of 3-phase machine to that of 1-phase machine will be*

(*a*) more (*b*) less (*c*) same (*d*) none of the above.

22. *To transmit the same amount of power over a particular distance at a given voltage, the amount of conducting material required in a 3-phase system to that of 1-phase system will be*
(*a*) 1·5 times (*b*) 0·5 times (*c*) $\sqrt{3}$ times (*d*) 0·75 times

23. *The phase sequence of a 3-phase system is RYB. The same phase sequence can be represented as*
(*a*) *RBY* (*b*) *BYR* (*c*) *YBR* (*d*) *YRB*

24. *In a balanced 3-phase, star-connected system, the phase difference between phase voltages and their respective line voltages are*
(*a*) 30° (*b*) 60° (*c*) 120° (*d*) 45°

25. *In a balanced 3-phase, star-connected system, the relation between phase voltage (V_{ph}) and line voltage (V_L) is*
(*a*) $V_{ph} = \sqrt{3}\ V_L$ (*b*) $V_{ph} = 0{\cdot}577\ V_L$ (*c*) $V_{ph} = V_L\sqrt{2}$ (*d*) none of above

26. *A 3-phase load is said to be a balanced load, if all the three phases have the same*
(*a*) impedance (*b*) power factor (*c*) both *a* and *b* (*d*) none of above

27. *Three 100 ohm resistors are connected in star across a 400 V, 3-phase supply, if one of the resistors is disconnected, then the line current will be*
(*a*) 2 A (*b*) 4 A (*c*) $4/\sqrt{3}\ A$ (*d*) $4\sqrt{3}\ A$

28. *Three 50 ohm resistors are connected in delta across a 400 V, 3-phase supply, if one of the resistor is disconnected, then the phase current will be*
(*a*) $8/\sqrt{3}\ A$ (*b*) $8\times\sqrt{3}\ A$ (*c*) 8 A (*d*) $4\sqrt{3}\ A$

29. *In a 3-ohase, balanced load, the power consumed is given by the relation*
(*a*) $\sqrt{3}\ V_L\ I_L \cos\phi$ (*b*) $3\ V_{ph}\ I_{ph} \cos\phi$ (*c*) both *a* and *b* (*d*) none of the above

30. *When three 10 ohm resistors are connected in star across a 400 V, 3-phase supply, each resistor must have a power rating.*
(*a*) 5290 W (*b*) 2300 W (*c*) 4000 W (*d*) 4600 W

31. *Three delta connected resistors absorb 30 kW when connected to a 400 V, 3-phase supply. When they are connected in star across the same supply the power absorbed will be*
(*a*) 60 kW (*b*) 90 kW (*c*) 20 kW (*d*) 10 kW

32. *Three identical resistances, each of 15 Ω, are connected in star across a 400 V, 3-phase supply. The value of resistance the each phase of the equivalent delta-connected load would be*
(*a*) 5 Ω (*b*) 45 Ω (*c*) 30 Ω (*d*) 7.5 Ω

TEST QUESTIONS

1. Explain how a sinusoidal e.m.f. is generated.
2. Derive an expression for the instantaneous value of alternating voltage varying sinusoidally.
3. Explain the terms amplitude, cycle, time period, frequency and phase difference as applied to alternating wave forms.
4. Define r.m.s. value and average value of an alternating quantity.
5. Show that the r.m.s. value of a sinusoidal a.c. voltage of amplitude V_m is $V_m/\sqrt{2}$.

6. Derive the expression for average and r.m.s. value of a sinusoidally varying quantity.
7. Show that the form factor of the sinusoidal wave form is 1.11.
8. Derive the values of form factor and peak factor of a sinusoidally varying quantity.
9. Explain how sinusoidal quantities can be represented by vectors.
10. Explain the component method for addition of alternating quantities.
11. Distinguish between
    (*i*) Time period and frequency (*ii*) Cycle and alternation
    (*iii*) Peak factor and form factor (*iv*) Average and *rms* value of an alternating current.
12. Explain why the phasor of current across an inductor lags its voltage by 90° and the phasor of current across a capacitor leads its voltage by 90°.
13. Determine phase angle relationship between alternating voltage and current in a purely inductive and a purely capacitance circuit under steady state condition.
14. Explain with mathematical expression that power consumed in a pure inductance in zero.
15. Explain with mathematical expression that power consumed in a pure capacitance is zero.
16. Show that power consumed in a pure resistive circuit is not constant but it is fluctuating.
17. Develop the expression for the power consumed in an ac single-phase *R-L* serirs circuit.
18. What do you mean by 'watt-less current'.
19. What do you understand by real power, reactive power and apparent power ?
20. What is power factor ? Discuss its practical importance.
21. Explain how capacitor is used to improve the pf.
22. What is power factor in *R—L—C* series circuit; when $X_L > X_C$, $X_L < X_C$, $X_L = X_C$?
23. Define *Q*-factor for the series resonant circuit and express it in terms of the circuit parameters.
24. If the voltage of *R*-phase is represented by $v_R = V_m \sin \omega t$, write down similar expression for the voltage of *Y*-phase and *B*-phase. The phase sequence being *RBY*.
25. What are the advantages of 3-phase system over 1-phase system ?
26. What do you understand by 3-phase power supply ? Describe star and delta-connections by showing line voltages and phase voltages, line currents and phase currents.
27. Derive the relation between phase and line voltages and currents for a balanced 3-phase star-connected system.

    Or

    For star connected system in a 3-phase circuit prove that $V_L = \sqrt{3}\ V_{ph}$ and $I_L = I_{ph}$.

    Or

    Derive the relationship between the line and phase voltages of an alternator.
28. Write down the relationship between line voltage and line current with phase voltage and phase current in star-connected and delta-connected circuits.
29. Derive expression for power in 3-phase star and delta connections.
30. Obtain the line and phase relationships for voltages and currents for a delta-connected 3-phase power system. Also obtain the expression of total power and per phase power.

31. Develop an expression for the total power in a balanced 3-phase load.

32. Describe the basic features of a balanced 3-phase system.

33. Show that power consumption, when three identical impedances are connected in delta across a balanced 3-phase supply, is three times that when the same impedances are connected in star across the same 3-phase supply.

ANSWER TO MULTI-CHOICE QUESTIONS

Multi-choice questions:

1. (*b*)	2. (*c*)	3. (*c*)	4. (*a*)	5. (*b*)
6. (*a*)	7. (*d*)	8. (*d*)	9. (*c*)	10. (*a*)
11. (*a*)	12. (*c*)	13. (*d*)	14. (*b*)	15. (*d*)
16. (*c*)	17. (*d*)	18. (*c*)	19. (*c*)	20. (*d*)
21. (*b*)	22. (*d*)	23. (*c*)	24. (*a*)	25. (*b*)
26. (*c*)	27. (*a*)	28. (*c*)	29. (*c*)	30. (*a*)
31. (*d*)	32. (*b*)			

✪✪✪✪✪

TRANSFORMERS

LEARNING OBJECTIVES

- What are diamagnetic, paramagnetic and ferromagnetic materials?
- What is the significance of *B-H* curve?
- What is a transformer and what is its necessity?
- What is an ideal transformer?
- How an ideal transformer is different to an actual transformer?
- What is the effect of leakage fluxes in a transformer?
- Why a transformer behaves as an inductive circuit?
- What is the meaning of voltage regulation of a transformer?
- What are the major losses which occur in a transformer?
- Why a transformer has highest efficiency among all electrical machines?
- What is an auto-transformer?
- Why 3-phase transformers are preferred over 1-phase transformers?
- What are the different accessories attached with power transformers for their satisfactory opperations?

INTRODUCTION

The operation of all the electrical machines such as transformers, *d.c.* machines, synchronous machines, induction motors etc. rely upon their magnetic circuits. The closed path followed by the magnetic lines of force is called a *magnetic circuit*. The operation of all the electrical devices and machines depend upon the magnetism produced by their magnetic circuits. Therefore, to obtain the required characteristics of these devices, their magnetic circuits have to be designed carefully.

Transformer is a major electrical machine or device which is used in the power system. At generating stations, it is used to step-up the voltage for economical reasons whereas, it is used to step-down the voltage at receiving stations for safety reasons (for utilisation of electrical energy voltage is reduced to 220 volt)

In this chapter, we shall focus our attention on magnetic materials, transformer and its working.

3.1 MAGNET

A **magnet** is a material or object that exhibits a magnetic field around it (see Fig.3.1) and has the ability to attract iron (ferromagnetic) pieces. Although, magnetic field is invisible but it is responsible for the most notable property of a magnet. A magnet has a north and south pole at opposite ends.

A magnet contains electrons that have both uneven orbits and uneven spins. Those magnetic atoms are aligned in nice straight rows inside each domain. And those domains are also lined up all in the same direction. If a piece of metal satisfies ALL these conditions only then it become a magnet.

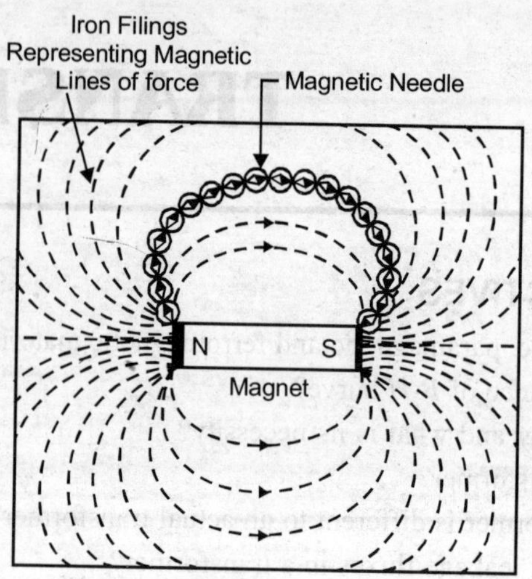

Fig.3.1: Magnet bar with its field

3.2 TYPES OF MAGNETIC MATERIALS

All materials can be classified in terms of their magnetic behavior. The two most common types of magnetism are diamagnetism and paramagnetism, which account for the magnetic properties of most of the periodic table of elements at room temperature. These elements are usually referred to as nonmagnetic, whereas those which are referred to as magnetic are actually classified as ferromagnetic.

When a material is placed within a magnetic field, the magnetic forces of the material's electrons will be affected. This effect is known as Faraday's Law of Magnetic Induction. However, materials can react quite differently to the presence of an external magnetic field. This reaction is dependent on a number of factors, such as the atomic and molecular structure of the material, and the net magnetic field associated with the atoms. The magnetic moments associated with atoms have three origins. These are the electron motion, the change in motion caused by an external magnetic field, and the spin of the electrons.

In most atoms, electrons occur in pairs. Electrons in a pair spin in opposite directions. So, when electrons are paired together, their opposite spins cause their magnetic fields to cancel each other. Therefore, no net magnetic field exists. Alternately, materials with some unpaired electrons will have a net magnetic field and will react more to an external field. Most materials can be classified as diamagnetic, paramagnetic or ferromagnetic.

Diamagnetic materials: The material which have a weak, negative susceptibility to magnetic fields are called diamagnetic materials.

These materials are slightly repelled by a magnetic field and they do not retain the magnetic properties when the external field is removed. In diamagnetic materials all the electron are paired so there is no permanent net magnetic moment per atom. Diamagnetic properties arise from the realignment of the electron paths under the influence of an external magnetic field. Most elements in the periodic table, including copper, silver, and gold, are diamagnetic.

Paramagnetic materials: The material which have a small, positive susceptibility to magnetic fields are called paramagnetic materials.

These materials are slightly attracted by a magnetic field and they do not retain the magnetic properties when the external field is removed. Paramagnetic properties are due to the presence of some unpaired electrons, and from the realignment of the electron paths caused by the external magnetic field. Paramagnetic materials include magnesium, molybdenum, lithium, and tantalum.

Ferromagnetic materials: The material which have a large, positive susceptibility to an external magnetic field are called ferromagnetic materials.

These materials exhibit a strong attraction to magnetic fields and are able to retain their magnetic properties after the external field has been removed. Ferromagnetic materials have some unpaired electrons so their atoms have a net magnetic moment. They get their strong magnetic properties due to the presence of magnetic domains. In these domains, large numbers of atom's moments (1012 to 1015) are aligned parallel so that the magnetic force within the domain is strong. When a ferromagnetic material is in the un-magnetized state, the domains are nearly randomly organized and the net magnetic field for the part as a whole is zero. When a magnetizing force is applied, the domains become aligned to produce a strong magnetic field within the part. Iron, nickel, and cobalt are examples of ferromagnetic materials.

3.3 ELECTROMAGNETISM

The phenomenon by which magnetism is produced by the effect of electric current is called **electromagnetism.**

Important Terms

While studying magnetic circuits, generally, we come across the following terms:

1. **Magnetic field:** The region around a magnet where its poles exhibit a force of attraction or repulsion is called *magnetic field.*
2. **Magnetic flux (ϕ):** The amount of magnetic lines of force set-up in a magnetic circuit is called *magnetic flux*. Its unit is weber (Wb). It is analogous to *electric current I* in electric circuit.
3. *The* **magnetic flux density** *at a point is the flux per unit area at right angles to the flux at that point.*

   It is, generally, represented by letter '*B*'. Its unit is Wb/m^2 or Tesla, *i.e.*

$$B = \frac{\phi}{A}\ \text{Wb/m}^2 \quad \text{or} \quad \text{T} \ (1\ \text{Wb/m}^2 = 1 \times 10^4\ \text{Wb/cm}^2)$$

4. **Permeability:** *The ability of a material to conduct magnetic lines of force through it is called the* **permeability** *of that material.*

   It is generally represented by μ (*mu*, a Greek letter). The greater the permeability of a material, the greater is its conductivity for the magnetic lines of force and *vice-versa*. The permeability of air or vacuum is the poorest and is represented as μ_0 (where $\mu_0 = 4\,\pi \times 10^{-7}$ H/m).

Relative permeability: The absolute (or actual) permeability μ of a magnetic material is much greater than absolute permeability of air μ_0. The relative permeability of a magnetic material is given in comparison with air or vacuum.

Hence, the ratio of the permeability of material μ to the permeability of air or vacuum μ_0 is called the relative permeability μ_r of the material.

i.e.
$$\mu_r = \frac{\mu}{\mu_0} \quad \text{or} \quad \mu = \mu_0\,\mu_r$$

Obviously, the relative permeability of air would be $\mu_0/\mu_0 = 1$. The value of relative permeability of all the non-magnetic materials is also 1. However, its value is as high as 8000 for solf iron, whereas, its value for mumetal (iron 22% and nickel 78%) is as high as 1 20 000.

5. **Magnetic field intensity:** The force acting on a unit north pole (1 Wb) when placed at a point in the magnetic field is called the magnetic intensity of the field at that point. It is denoted by *H*.

In magnetic circuits, it is defined as mmf per unit length of the magnetic path. It is denoted by *H*, mathematically,

$$H = \frac{\text{m.m.f.}}{\text{length of magnetic path}} = \frac{NI}{l} \text{ AT / m}$$

6. **Magnetomotive force (m.m.f.):** The magnetic pressure which sets-up or tends to set-up magnetic flux in a magnetic circuit is called *magnetomotive force.* As per work law it may be defined as under:

The work done in moving a unit magnetic pole (1 Wb) once round the magnetic circuit is called *magnetomotive force.* In general

$$m.m.f. = NI \text{ ampere-turns (or AT)}$$

It is analogous to *e.m.f.* in an electric circuit.

7. **Reluctance (S):** The opposition offered to the magnetic flux by a magnetic circuit is called its *reluctance.*

It depends upon length (l), area of cross-section (a) and permeability ($\mu = \mu_0 \mu_r$) of the material that makes up the magnetic circuit. It is measured in AT/Wb.

$$\text{Reluctance, } S = \frac{l}{a\,\mu_0\,\mu_r}$$

It is analogous to *resistance* in an electric circuit.

8. **Permeance.** It is a measure of the ease with which flux can be set-up in the material. It is just reciprocal of reluctance of the material and is measured in *Wb/AT* or *henry.*

$$\text{Permeance} = \frac{1}{\text{reluctance}} = \frac{a\,\mu_0\,\mu_r}{l} \text{ Wb/AT or H}$$

It is analogous to *conductance* in an electric circuit.

9. **Reluctivity:** It is specific reluctance and analogous to *resistivity* in electric circuit.

3.4 MAGNETISATION OR B – H CHARACTERISTICS

The graph plotted between flux density *B* and magnetising force *H* of a material is called the *magnetisation or B – H Curve* of that material.

The general shape of the *B – H* Curve of a magnetic material is shown in fig.3.2 The shape of the curve is non-linear This indicates that the relative permeability ($\mu_r = B/\mu_0 H$) of a magnetic material is not constant but it varies. The value of μ_r largely depends upon the value of flux density.

The first part of magnetisation curve is *OA*. In *OA* flux density rises with the rise in magnetising force *H*.

The second part of the magnetisation curve is *AB*. In *AB* we see that the curve becomes steep and straight *i.e.* after point *A*, the rise in flux density (*B*) is quite high even when the magnetising force *H* increases slightly.

The third portion of the magnetisation curve *is B* onwards. The curve becomes almost a horizontal straight line. This straight lining of magnetisation curve is called magnetic saturation and point *B* is called point of saturation. After *B* the rise in flux density *B* is negligible. This shows that with the rise in magnetising force, flux density increases but if the magnetising force goes on increasing, then a stage will arrive when change in flux density (*B*) will be negligible. This condition is called magnetic saturation and after magnetic saturation, the flux density remains constant.

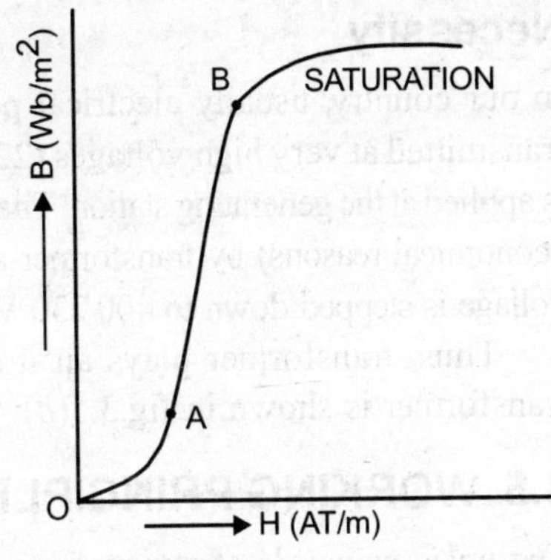

Fig.3.2: B-H curve

The *B – H* curves of some of the common magnetic materials are shown in fig.3.3. The *B – H* curve for a non-magnetic material is shown in fig.3.4. It is a straight line curve since $B = \mu_0 H$ or $B \propto H$ as the value of μ_0 is constant.

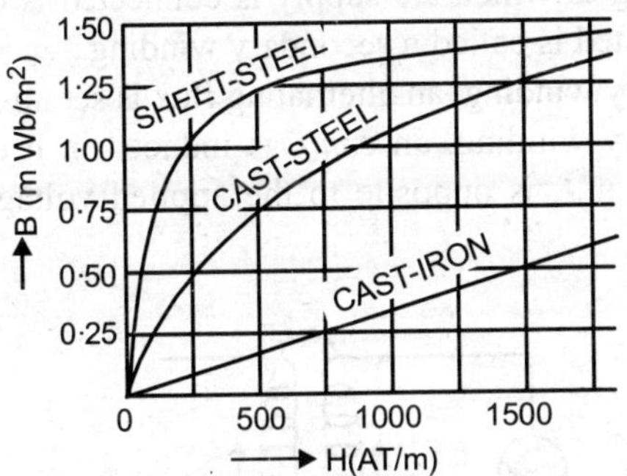

Fig.3.3: B-H curve of various magnetic materials

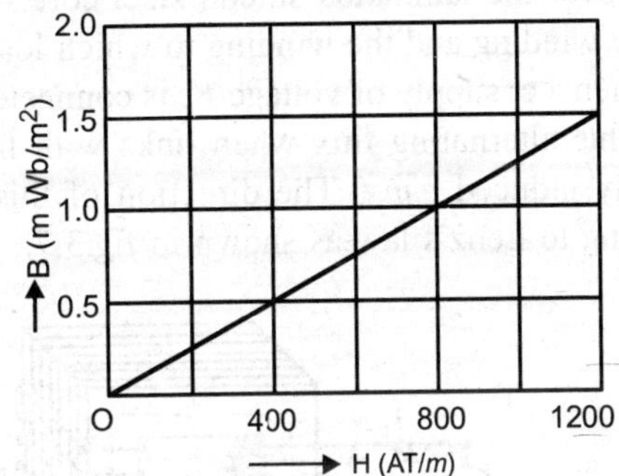

Fig.3.4: B-H Curve for non-magnetic materials

3.5 TRANSFORMER

A *transformer* is a static device which transfers a.c. electrical power from one circuit to the other at the same frequency but the voltage level is usually changed as shown in fig.3.5(*a*). When the voltage is raised on the output side ($V_2 > V_1$), the transformer is called a step up transformer, whereas, the transformer in which the voltages is lowered on the output side ($V_2 < V_1$) is called a step down transformer.

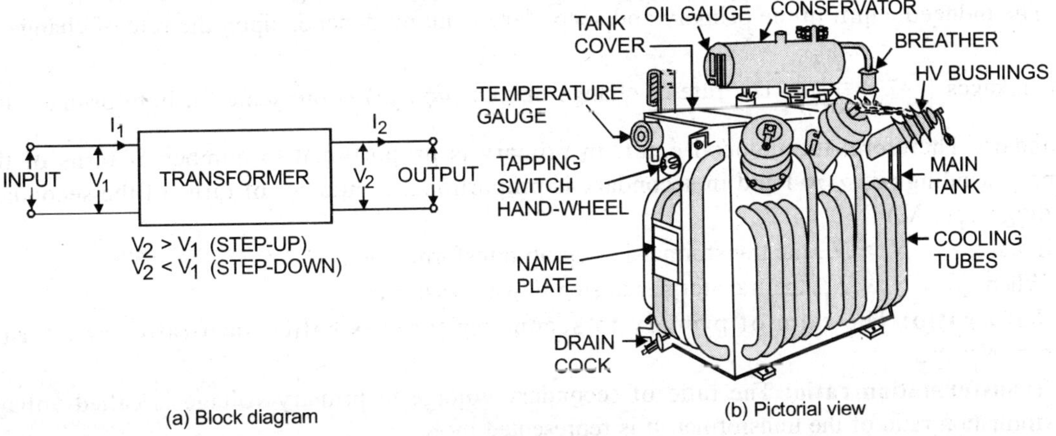

Fig.3.5: Transformer

Necessity

In our country, usually electrical power is generated at 11 kV. For economical reasons, a.c. power is transmitted at very high voltages (220 kV or 400 kV) over long distances, therefore, a step up transformer is applied at the generating station. Then to feed different areas, voltages are stepped down to different levels (for economical reasons) by transformer at various substations. Ultimately for utilisation of electrical power, the voltage is stepped down to 400/230 V for safety reasons.

Thus, transformer plays an important role in the power system. The pictorial view of a power transformer is shown in fig.3.5(*b*). The important accessories are labeled on it.

3.5 WORKING PRINCIPLE OF A TRANSFORMER

The basic principle of a transformer is ***electromagnetic induction.***

A simple form of a transformer is shown in fig.3.6. it essentially consists of two separate windings placed over the laminated silicon steel core. The winding to which a.c supply is connected is called primary winding and the winding to which load is connected is called a secondary winding.

When a.c. supply of voltage V_1 is connected to primary winding, an alternating flux is set up in the core., This alternating flux when links with the secondary winding, an *e.m.f.* is induced in it called mutually induced *e.m.f.* The direction of this induced e.m.f. is opposite to the applied voltage V_1, according to Lenz's law as shown in fig.3.7.

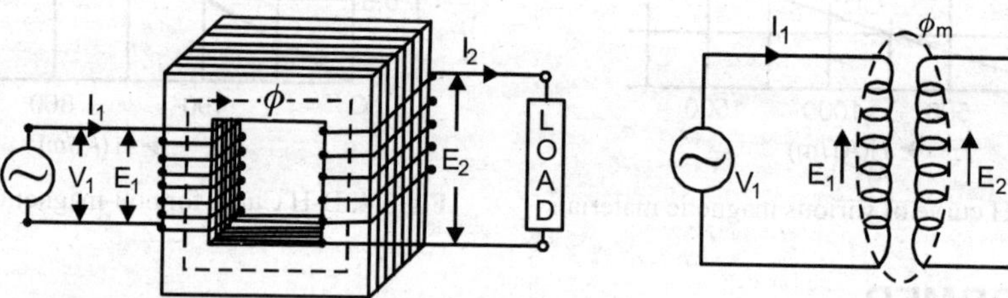

Fig.3.6: Transformer core and windings **Fig.3.7:** Coupling of primary and secondary

The same alternating flux also links with the primary winding and produces self induced e.m.f. E_1 This induced e.m.f E_1 also acts in opposite direction to the applied voltage V_1 according to Lenz's law.

Although, there is no electrical connection between primary and secondary winding but electrical power is transferred from primary circuit to the secondary circuit through mutual flux.

The induced e.m.f. in the primary and secondary winding depends upon the rate of change of flux linkages $\left(i.e.\ N\frac{d\phi}{dt}\right)$. The rate of change of flux ($d\phi / dt$) is the same for both primary and secondary. Therefore, the induced e.m.f. in primary is proportional to number of turns of the primary winding ($E_1 \propto N_1$) and in secondary is proportional to number of turns of the secondary winding ($E_2 \propto N_2$).

$\therefore$ In case, $N_2 > N_1$, the transformer is step up transformer and

When $N_2 < N_1$, the transformer is step down transformer.

Turn ratio: The ratio of primary to secondary turns is called turn ratio, *i.e.*, turn ratio = N_1 / N_2.

Transformation ratio: The ratio of secondary voltage to primary voltage is called voltage transformation ratio of the transformer. It is represented by K.

$$K = \frac{E_2}{E_1} = \frac{N_2}{N_1} \text{ (since } E_2 \propto N_2 \text{ and } E_1 \propto N_1)$$

3.7 TRANSFORMER CONSTRUCTION

The main elements of a transformer are; two coils and a laminated steel core. The two coils are insulated from each other as well as from the steel core. The core of the transformer is constructed from laminations of sheet or silicon steel assembled to provide a continuous magnetic path. At usual flux densities the silicon steel material has low hysteresis losses. The core is laminated to minimise the eddy current loss. The laminations are insulated from each other by a light coating of varnish or by an oxide layer. The thickness of laminations varies from 0.35 mm to 0.5 mm for a frequency of 50 Hz.

According to the core construction and the manner in which the primary and secondary are placed around it, the transformers are named as (*i*) core type transformers (*ii*) shell type transformers.

Core-type transformers

In a simple core type transformer the magnetic core is built up of laminations to form a rectangular frame. The laminations are cut in the form of L-shape strips as shown in fig.3.8. In order to avoid high reluctance at the joints where laminations are butted against each other the alternate layers are stacked differently to eliminate continuous joint as shown fig.3.9.

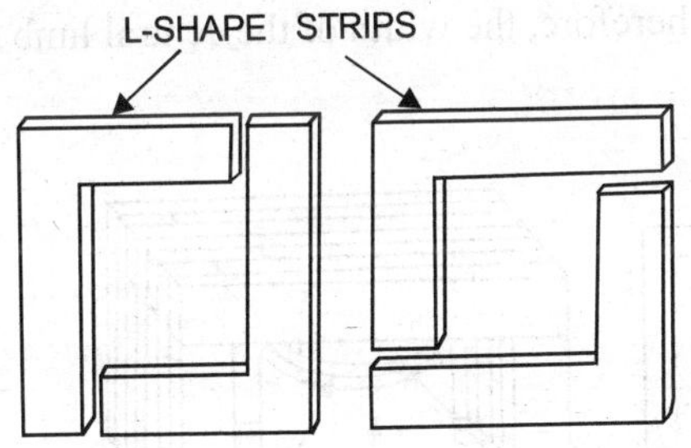

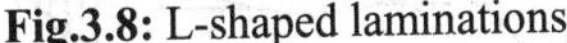

Fig.3.8: L-shaped laminations

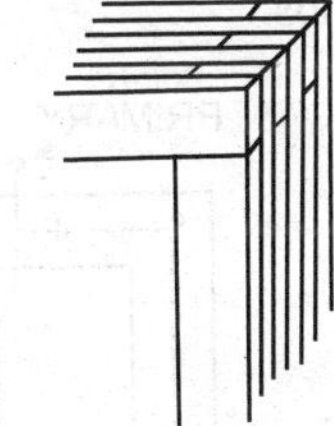

Fig.3.9: Staggering of laminations

In actual transformer construction, the primary and secondary windings are interleaved to reduce the leakage flux. Half of each winding is placed side by side or concentrically on either limb or leg of the core as shown in fig.3.10(a). However, for simplicity, the two windings are shown in fig.3.10(b) located on separate limbs of the core.

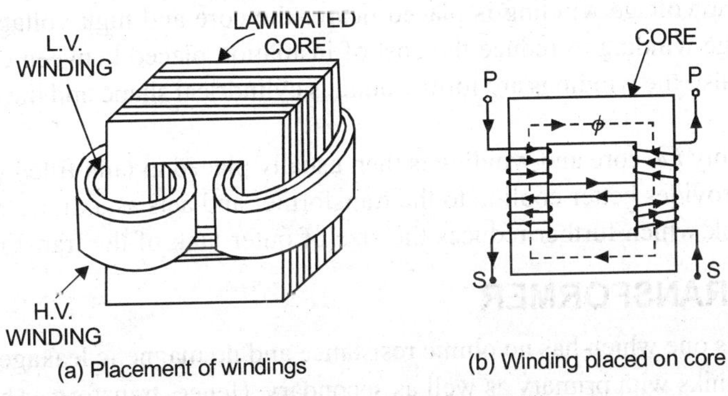

(a) Placement of windings

(b) Winding placed on core

Fig.3.10: Transformer core with windings

While placing these windings, an insulation layer (bakelite former) is provided between core and lower winding and between the two windings. To reduce the insulation, low voltage winding is always placed nearer the core as shown in fig.3.10(*a*). The windings used are form wound (usually cylindrical in shape) and the laminations are inserted later on.

Shell-type transformers

In case of shell type transformer, each lamination is cut in the form of long strips of E's and I's as shown in fig.3.11. In order to avoid high reluctance at the joints where the laminations are butted against each other, the alternate layers are stacked differently to eliminate continuous joints.

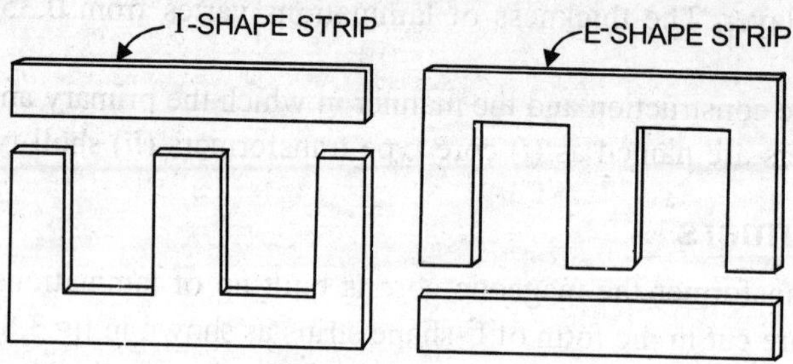

Fig.3.11: E&I laminations

In a shell-type transformer, the core has three limbs. The central limb carries whole of the flux, where as the side limbs carry half of the flux. Therefore, the width of the central limb is about double to that of the outer limbs.

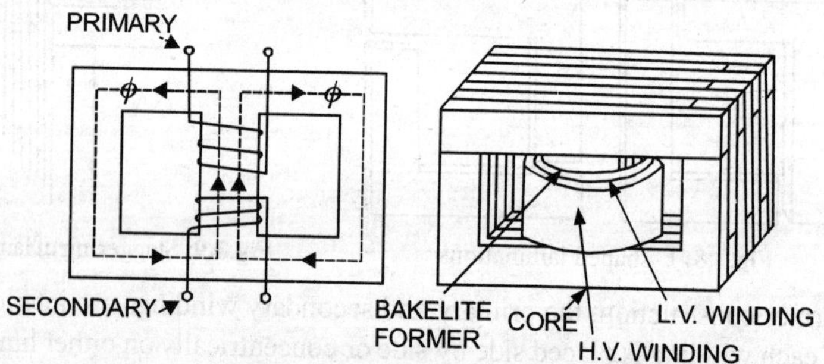

Fig.3.12: Winding placed on core

Both the primary and secondary windings are placed on the central limb side by side or concentrically (see fig.3.12). The low voltage winding is placed nearer the core and high voltage winding is placed outside the low voltage winding to reduce the cost of insulation placed between core and low voltage winding. In this case also the windings are form wound is cylindrical shape and the core laminations are inserted later on.

The whole assembly *i.e.* core and winding is then usually placed in tank filled with transformer oil. The transformer oil provides better cooling to the transformer and acts as a dielectric medium between winding and outer tank which further reduces the size of outer tank of the transformer.

3.8 AN IDEAL TRANSFORMER

An ideal transformer is one which has no ohmic resistance and no magnetic leakage flux *i.e.* all the flux produced in the core links with primary as well as secondary. Hence, transformer has no copper losses and core losses. It means an ideal transformer consists of two purely inductive coils wound on a loss free core. Although in actual practice, it is impossible to realize such a transformer, yet for convenience, it is better to start with an ideal transformer and then extend it to an actual transformer.

In an ideal transformer there is no power loss, therefore, output must be equal to input.

i.e. $E_2 I_2 \cos\phi = E_1 I_1 \cos\phi$ or $E_2 I_2 = E_1 I_2$ or $\dfrac{E_2}{E_1} = \dfrac{I_1}{I_2}$

since $E_2 \propto N_2$; $E_1 \propto N_1$ and $E_1 \cong V_1$; $E_2 \cong V_2$

$$\frac{V_2}{V_1} = \frac{E_2}{E_1} = \frac{N_2}{N_1} = \frac{I_1}{I_2} = K \text{ (transformation ratio)}$$

Hence, primary and secondary currents are inversely proportional to their respective turns. *The ratio of secondary turns to primary turns is called* **transformation ratio** *of the transformer and is represented by K.*

Behaviour and phasor diagram

Consider an ideal transformer whose secondary is open as shown in fig.3.13(a). When its primary is connected to sinusoidal alternating voltage V_1, a current I_m flows through it. Since the primary coil is pure inductive, the current I_m lags behind the applied voltage V_1 by 90°. This current sets up alternating flux (or mutual flux f_m) in the core and magnetises it. Hence it is called magnetising current. Flux is in phase with I_m as shown in the phasor diagram in fig.3.13(*b*). The alternating flux links with both primary and secondary windings. When it links with primary, it produces self induced *e.m.f.* E_1 in opposite direction to that of applied voltage V_1. When it links with secondary winding, it produces mutually induced *e.m.f.* E_2 in opposite direction to that of applied voltage. Both the *e.m.f.s.* E_1 and E_2 are shown in phasor diagram in fig.3.13(*b*).

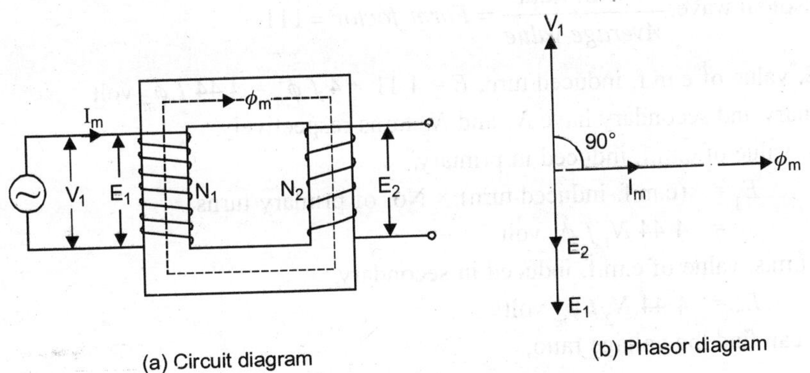

Fig.3.13: Ideal transformer

3.9 TRANSFORMER ON D.C.

A transformer can not work on d.c supply. If a rated d.c voltage is applied across the primary, a flux of constant magnitude will be set up in the core. Hence, there will not be any self induced e.m.f. (which is only possible with the rate of change of flux linkages) in the primary winding to oppose the applied voltage. As the resistance of the primary winding is very low, the primary current will be quite high as given by the ohm's law.

$$\text{Primary current} = \frac{\textit{d.c. applied voltage}}{\textit{resistance of primary winding}}$$

This current is much more than the rated full load current of primary winding. Thus it will produce lot of heat ($I^2 R$) loss and burns the insulation of the primary winding, consequently the transformer will be damaged. That is why, d.c is never applied to a transformer.

3.10 E.M.F EQUATION

When sinusoidal voltage is applied to the primary winding of a transformer, a sinusoidal flux, as shown in fig.3.14. is set up in the iron core which links with primary and secondary winding.

Let, ϕ_m = Maximum value of flux in Wb;

f = supply frequency in Hz (or c/s);

N_1 = No. of turns in primary;

N_2 = No. of turns in secondary.

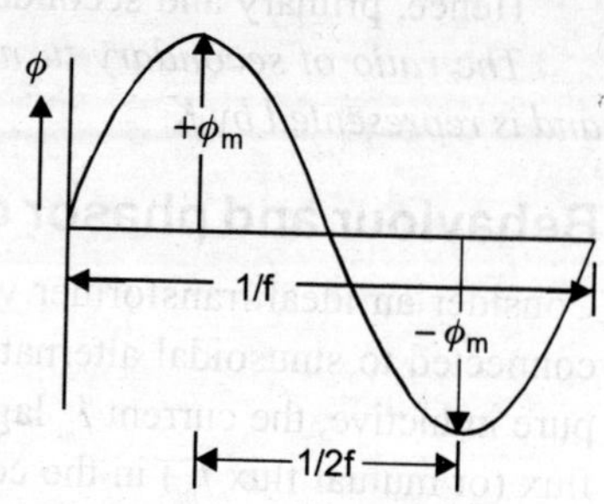

Fig.3.14: Sinusoidal flux

As shown in fig.10, flux changes from $+\phi_m$ to $-\phi_m$ in half a cycle *i.e.* $\frac{1}{2f}$ second,

Average rate of change of flux

$$= \frac{\phi_m - (-\phi_m)}{1/2f} = 4f\,\phi_m \text{ Wb/s}$$

Now, the rate of change of flux per turn is the average induced e.m.f. per turn in volt.

∴ Average e.m.f. induced per turn = $4 f \phi_m$ volt

For a sinusoidal wave, $\frac{\textit{R.M.S. value}}{\textit{Average value}} = \textit{Form factor} = 1{\cdot}11$

∴ R.M.S. value of e.m.f. induced/turn, $E = 1.11 \times 4 f \phi_m = 4.44 f \phi_m$ volt

Since primary and secondary have N_1 and N_2 turns respectively.

∴ R.M.S. value of e.m.f. induced in primary,

E_1 = (e.m.f. induced/turn) × No. of primary turns

$= 4{\cdot}44\, N_1 f \phi_m$ volt ...(*i*)

Similarly, r.m.s. value of e.m.f. induced in secondary,

$E_2 = 4{\cdot}44\, N_2 f \phi_m$ volt ...(*ii*)

Again, we can find the voltage ratio,

$$\frac{E_2}{E_1} = \frac{4{\cdot}44 N_2\, f\phi_m}{4{\cdot}44 N_1\, f\phi_m} \;\; or \;\; \frac{E_2}{E_1} = \frac{N_2}{N_1} = K \;\; \text{(transformation ratio)}$$

Equation (*i*) and (*ii*) can be written in the form of maximum flux density B_m using relation,

$f_m = B_m \times A_i$ (where A_i is iron area)

∴ $E_1 = 4{\cdot}44\, N_1 f B_m A_i$ volts and $E_2 = 4{\cdot}44\, N_2 f B_m A_i$ volt.

Example 3.1: *The primary winding of a 50 Hz single phase transformer has 500 turns and is fed from 6500 V supply. The secondary winding has 25 turns. Find peak value of flux in the core and secondary voltage.*

Solution:

Here, $f = 50$ Hz; $N_1 = 500$; $E_1 = 6500$ V; $N_2 = 25$

Now, $E_1 = 4{\cdot}44\, N_1 f \phi_m$

Peak value of flux, $\phi_m = \dfrac{E_1}{4{\cdot}44\, N_1 f}$

$$= \frac{6500}{4{\cdot}44 \times 500 \times 50} = \mathbf{0{\cdot}05856\ Wb = 58.56\ mWb}\ (Ans.)$$

Secondary voltage, $E_2 = \dfrac{N_2}{N_1} \times E_1 = \dfrac{25}{500} \times 6500 = \mathbf{325\ V}\ (Ans.)$

Example 3.2: *A 3300/300 V, 50 Hz, single-phase transformer is built on a core having an effective cross-sectional area of 130 cm² and 80 turns on the low voltage winding. Calculate (a) the value of the maximum flux density (b) number of turns on the high voltage winding.*

Solution: Here, $E_1 = 3300$ V; $E_2 = 300$ V; $f = 50$ Hz; $N_2 = 80$;

$A = 130\ \text{cm}^2 = 130 \times 10^{-4}\ \text{m}^2$

We know, $E_2 = 4{\cdot}44\, N_2 f A B_m$

Maximum value of flux density, $B_m = \dfrac{E_2}{4{\cdot}44 f N A}$

$$= \frac{300}{4.44 \times 50 \times 80 \times 130 \times 10^{-4}} = \mathbf{1.3\ T}\ (Ans.)$$

(*b*) Number of turns on high-voltage winding, $N_1 = \dfrac{E_1}{E_2} \times N_2 \quad \left(\text{since } \dfrac{N_2}{N_1} = \dfrac{E_2}{E_1}\right)$

$$= \frac{3300}{300} \times 80 = \mathbf{880}\ (Ans.)$$

Example 3.3: *A 30 kVA transformer has 600 turns on the primary and 50 turns on the secondary winding. The primary is connected to 3300 V, 50 Hz mains, calculate (i) primary and secondary currents at full load; (ii) The secondary e.m.f. and (iii) The maximum flux in the core. Neglect magnetic leakage, resistance of the winding and the primary no load current in relation to the full load current.*

Solution: (*i*) At full load, $I_1 = \dfrac{30 \times 10^3}{3300} = \mathbf{9.1\ A}\ (Ans.)$

Now $\dfrac{I_1}{I_2} = \dfrac{E_2}{E_1} = \dfrac{N_2}{N_1}$

secondary current, $I_2 = \dfrac{N_1}{N_2} \times I_1 = \dfrac{600}{50} \times 9.1 = \mathbf{109{\cdot}1\,A}\ (Ans.)$

(*ii*) Secondary e.m.f., $E_2 = \dfrac{N_2}{N_1} \times E_1 = \dfrac{50}{600} \times 3300 = \mathbf{275\ V}\ (Ans.)$

(*iii*) Using relation, $E_1 = 4{\cdot}44\, N_1 f \phi_m$

$$3300 = 4{\cdot}44 \times 600 \times 50 \times \phi_m$$

or
$$\phi_m = \frac{3300}{4{\cdot}44 \times 600 \times 50} = \textbf{24.77 mWb} \ (Ans)$$

Example 3.4: *A single-phase, 50 Hz transformer has 35 primary and 350 secondary turns. The net cross sectional area of core is 300 cm². If the primary winding is connected to a 230 V, 50 Hz supply, calculate (i) Peak value of flux density in the core; (ii) Voltage induced in the secondary winding. Neglect losses, what is the primary current when the secondary current is 120 amperes ?*

Solution: $E_1 = 4{\cdot}44\, N_1 f B_m A_i$

where, $E_1 = 230$ V; $N_1 = 35$; $f = 50$ Hz and $A_i = 300 \times 10^{-4}$ m²

or $230 = 4{\cdot}44 \times 35 \times 50 \times 300 \times 10^{-4} \times B_m$

$\therefore$ Max flux density, $B_m = $ **0·987 Tesla (*or* Wb/m²)** (*Ans.*)

Using relation,
$$K = \frac{N_2}{N_1} = \frac{E_2}{E_1} = \frac{I_1}{I_2}$$

$$E_2 = \frac{N_2}{N_1} \times E_1 = \frac{350}{35} \times 230 = \textbf{2300 V} \ (Ans)$$

$$I_1 = \frac{N_2}{N_1} \times I_2 = \frac{350}{35} \times 100 = \textbf{1000 A} \ (Ans)$$

Example 3.5: *A 200 kVA, 3300/240 volts, 50 Hz. Single –phase transformer has 100 turns on the secondary winding. Assuming an ideal transformer, calculate (i) primary and secondary current on full load; (ii) the maximum value of flux; (iii) the number of primary turns.*

Solution:
$$I_1 = \frac{200 \times 1000}{3300} = \textbf{60·6 A} \ (Ans.);$$

$$I_2 = \frac{200 \times 1000}{240} = \textbf{833·3 A} \ (Ans.)$$

$$E_2 = 4{\cdot}44\, N_2 f \phi_m;$$

$$240 = 4{\cdot}44 \times 100 \times 50 \times \phi_m$$

$$\therefore \phi_m = \textbf{10.81 m Wb} \ (Ans.)$$

Now, $\dfrac{N_1}{N_2} = \dfrac{E_1}{E_2}$ or $\dfrac{N_1}{100} = \dfrac{3300}{240}$ *or* $N_1 = $ **1375 turns** (***Ans***)

PRACTICE EXERCISE

Q.1 *The primary winding of a 50 Hz single phase transformer has 480 turns and is fed from 6400 V supply. The secondary winding has 20 turns. Find peak value of flux in the core and secondary voltage.* (***Ans.*** 0.06Wb, 266.67 V)

Q.2 *A 3300/250 V, 50 Hz, single-phase transformer is built on a core having an effective cross-sectional area of 125 cm² and 71 turns on the low voltage winding. Calculate (a) the value of the maximum flux density (b) number of turns on the high voltage winding.* (***Ans.*** 1·269 T, 937)

Q.3 *A 25 kVA transformer has 500 turns on the primary and 40 turns on the secondary winding. The primary is connected to 3000 V, 50 Hz mains, calculate (i) primary and secondary currents at full load; (ii) The secondary e.m.f. and (iii) The maximum flux in the core. Neglect magnetic leakage, resistance of the winding and the primary no load current in relation to the full load current.*

*(**Ans.** 8·33A, 104.15 A, 240V, 0.027 Wb)*

Q.4 *A single-phase, 50 Hz transformer has 30 primary and 350 secondary turns. The net cross sectional area of core is 250 cm². If the primary winding is connected to a 230 V, 50 Hz supply, calculate (i) Peak value of flux density in the core; (ii) Voltage induced in the secondary winding. Neglect losses, what is the primary current when the secondary current is 100 amperes ?*

*(**Ans.** 1·38 Tesla or Wb/m², 2683.33 V, 1166.67 A)*

Q.5 *A 200 kVA, 3300/240 volts, 50 Hz. Single –phase transformer has 80 turns on the secondary winding. Assuming an ideal transformer, calculate (i) primary and secondary current on full load; (ii) the maximum value of flux; (iii) the number of primary turns.*

*(**Ans.** 60.6A, 833.3A, 13.51mWb, 1100 turns)*

3.11 PRACTICAL (ACTUAL) TRANSFORMER

An actual transformer has primary and secondary winding resistance. These windings also have inductances due to leakage fluxes which develop inductive reactance. These effects cause voltage drop in a transformer. Also there is copper loss in primary and secondary winding due to their resistances. Since alternating magnetic flux is set-up in core that causes iron or magnetic (hysteresis and eddy current) losses in it.

All these facts are further explained in articles to follow

3.12 RESISTANCE OF TRANSFORMER WINDINGS

Actual transformers have some primary and secondary winding resistance represented by R_1 and R_2 respectively. These resistances are shown external to the windings in fig.3.15.

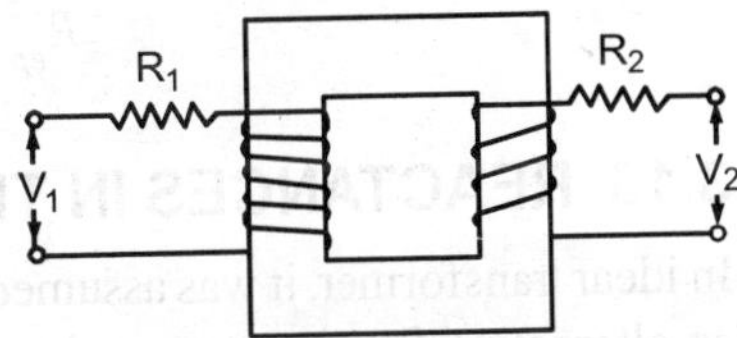

Fig.3.15: Resistance of primary and secondary winding

Equivalent resistance

To make the calculations easy the resistance of the two windings can be transferred to either side. The resistance is transferred from one side to the other in such a manner that percentage voltage drop remains the same when represented on either side.

Let the primary resistance R_1 be transferred to the secondary side and the new value of this resistance be R_1' called equivalent resistance of primary referred to secondary side as shown in fig.3.16(*a*). I_1 and I_2 be the full load primary and secondary currents respectively.

$$\text{Then } \frac{I_2 R_1'}{V_2} \times 100 = \frac{I_1 R_1}{V_1} \times 100 \text{ (\% voltage drops)}$$

$$\text{or } R_1' = \frac{I_1}{I_2} \times \frac{V_2}{V_1} \times K^2 R_1$$

∴ Total equivalent resistance referred to secondary.

$$R_{es} = R_2 + R_1' = R_2 + K^2 R_1$$

Now consider resistance R_2, when it is transferred to primary, let its new value be R_2' called equivalent resistance of secondary referred to primary as shown in fig.3.16(*c*).

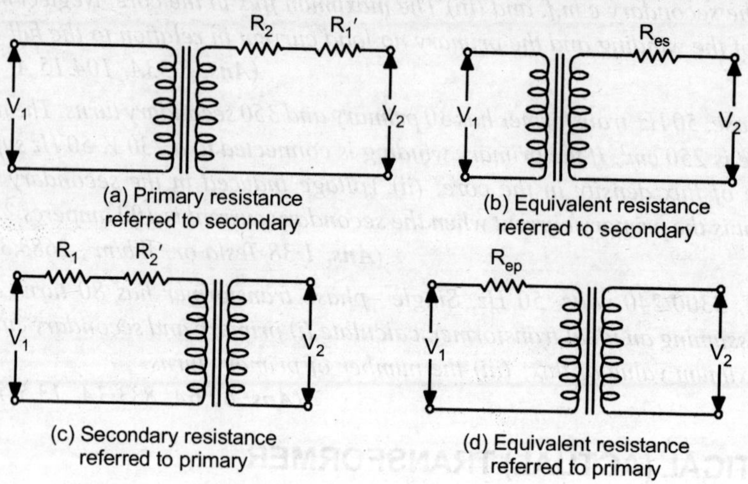

Fig.3.16: Transfer of resistance

Then $\dfrac{I_1 R_2'}{V_1} \times 100 = \dfrac{I_2 R_2}{V_2} \times 100$ or $R_2' = \dfrac{I_2}{I_1} \times \dfrac{V_1}{V_2} \times R_2 = \dfrac{R_2}{K^2}$

∴ Total equivalent resistance referred to primary,

$$R_{ep} = R_1 + R_2¢ = R_1 + \frac{R_2}{K^2}$$

3.13 REACTANCES IN TRANSFORMER WINDINGS

In idear transformer, it was assumed that when a.c. supply is given to the primary winding of a transformer, an alternating flux is set up in the core and whole of this flux links with both the primary and secondary windings. However, in an actual transformer, both the windings produce some flux that links only with the winding that produces it.

The flux that links with both the windings of the transformer is called **mutual flux** *and the flux which links only with one winding of the transformer and not to the other is called* **leakage flux.**

The primary ampere turns produce some flux ϕ_{l_1} which is set up in air and links only with primary winding, as shown in fig.3.17(a), is called *primary leakage flux.*

Similarly, secondary ampere turns produce some flux ϕ_{l_2} which is set up in air and links only with secondary winding is called *secondary leakage flux.*

The primary leakage flux ϕ_{l_1} is proportional to the primary current I_1 and secondary leakage flux ϕ_{l_2} is proportional to secondary current I_2. The primary leakage flux f_{l_1} produces self inductance L_1 (= $N_1 \phi_{l_1} / I_1$) which in turn produces leakage reactance $X_1 (= 2\pi f L_1)$. Similarly, secondary leakage flux ϕ_{l_2} produces leakage reactance X_2 (= $2\pi f L_2$). The leakage reactance (inductive) have been shown external to the windings in fig.3.17(*b*).

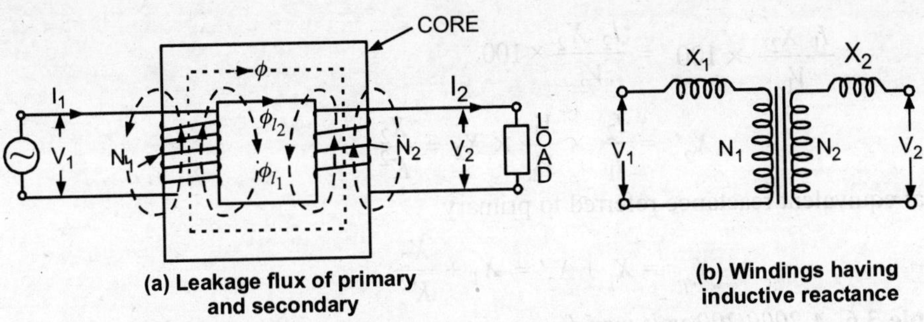

(a) Leakage flux of primary and secondary

(b) Windings having inductive reactance

Fig.3.17: Inductive reactance of primary and secondary

Equivalent reactance

To make the calculations easy the reactances of the two winding can be transferred to any one side. The reactance from one side to the other is transferred in such a manner that percentage voltage drop remains the same when represented on either side.

Let the primary reactance X_1 be transferred to the secondary and the new value of this reactance is X_1' called equivalent reactance of primary referred to secondary, as shown in fig.3.18(*a*).

Then $$\frac{I_2 X_1'}{V_2} \times 100 = \frac{I_1 X_1}{V_1} \times 100 \quad \text{(% voltage drops)}$$

or $$X_1' = \frac{I_1}{I_2} \times \frac{V_2}{V_1} \times X_1 = K^2 X_1$$

∴ Total equivalent reactance referred to secondary.

$$X_{es} = X_2 + X_1' = X_2 + K^2 X_1$$

Now, let us consider secondary reactance X_2 when it is transferred to primary side its new value is X_2' called equivalent reactance of secondary referred to primary, as shown in fig.3.18(*c*).

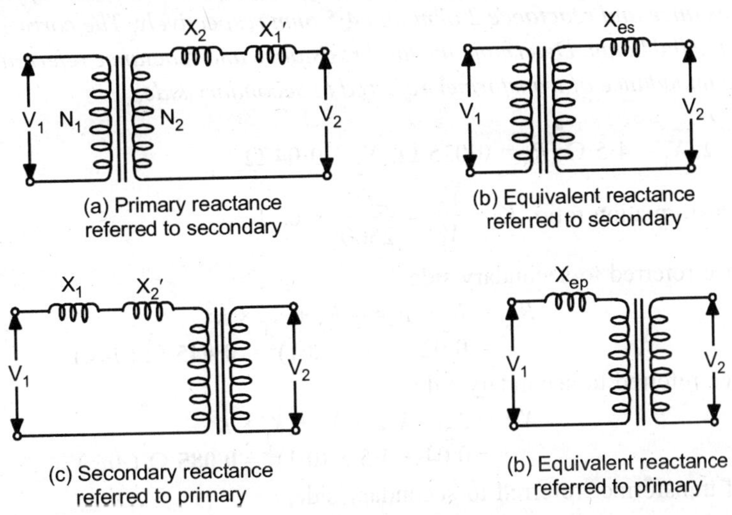

(a) Primary reactance referred to secondary

(b) Equivalent reactance referred to secondary

(c) Secondary reactance referred to primary

(b) Equivalent reactance referred to primary

Fig.3.18: Transfer to reactance

Then $\quad \dfrac{I_1 X_2'}{V_1} \times 100 = \dfrac{I_2 X_2}{V_2} \times 100$

or $\quad X_2' = \dfrac{I_2}{I_1} \times \dfrac{V_1}{V_2} \times X_2 = \dfrac{X_2}{K^2}$

$\therefore$ Total equivalent reactance referred to primary.

$$X_{ep} = X_1 + X_2' = X_1 + \frac{X_2}{K^2}$$

Example 3.6. *A 2000/200 volt transformer has a primary resistance 2.5 ohm and reactance 4.5 ohm, the secondary resistance 0.025 ohm and reactance 0·045 ohm. Determine total resistance and reactance referred to primary side.*

Solution: Primary resistance, $R_1 = 2·5$ ohm

Primary reactance, $X_1 = 4·5$ ohm

Secondary resistance, $R_2 = 0·025$ ohm

Secondary reactance, $X_2 = 0·045$ ohm

Transformation ratio, $K = \dfrac{V_2}{V_1} = \dfrac{200}{2000} = 0.1$

Total resistance referred to primary side,

$$R_{ep} = R_1 + \frac{R_2}{K^2} = 2.5 + \frac{0.025}{(0.1)^2} = \mathbf{5\ \Omega}\ (\textit{Ans.})$$

Total reactance referred to primary side,

$$X_{ep} = X_1 + \frac{X_2}{K^2} = 4.5 + \frac{0.045}{(0.1)^2} = \mathbf{9\ \Omega}\ (\textit{Ans.})$$

Example 3.7. *A single phase transformer having voltage ratio 2500/250V (primary to econdary) has a primary resistance and reactance 2 ohm and 4·5 ohm respectively. The corresponding secondary values are 0·025 and 0·04 ohm. Determine the total resistance and reactance referred to secondary side. Also calculate the impedance of transformer referred to secondary side.*

Solution:

Here, $R_1 = 2\ \Omega$; $X_1 = 4·5\ \Omega$; $R_2 = 0·025\ \Omega$; $X_2 = 0·04\ \Omega$

Transformation ratio, $K = \dfrac{V_2}{V_1} = \dfrac{250}{2500} = 0·1$

Total resistance referred to secondary side,

$$R_{es} = R_2 + R_1\text{¢} = R_2 + R_1 \times K^2$$
$$= 0·025 + 2 \times (0·1)^2 = \mathbf{0·045}\ \Omega\ (\textit{Ans.})$$

Total reactance referred to secondary side,

$$X_{es} = X_2 + X_1\text{¢} = X_2 + X_1 \times K^2$$
$$= 0·04 + 4·5 \times (0·1)^2 = \mathbf{0·085}\ \Omega\ (\textit{Ans.})$$

Impedance of transformer referred to secondary side,

$$Z_{es} = \sqrt{(R_{es})^2 + (X_{es})^2} = \sqrt{(0.045)^2 + (0.085)^2}$$
$$= \mathbf{0·096}\ \Omega\ (\textit{Ans.})$$

PRACTICE EXERCISE

Q.1 *A 2000/200 volt transformer has a primary resistance 2.3 ohm and reactance 4.2 ohm, the secondary resistance 0.025 ohm and reactance 0·04 ohm. Determine total resistance and reactance referred to primary side.* (**Ans.** *4·8 Ω, 8·2 Ω*)

Q.2 *A single phase transformer having voltage ratio 2500/250V (primary to secondary) has a primary resistance and reactance 1·8 ohm and 4·2 ohm respectively. The corresponding secondary values are 0·02 and 0·045 ohm. Determine the total resistance and reactance referred to secondary side. Also calculate the impedance of transformer referred to secondary side.* (**Ans.** *0·038 Ω .0·087 Ω, 0·095 Ω.*)

3.14 CIRCUIT DIAGRAM OF AN ACTUAL TRANSFORMER

An actual transformer has (*i*) Primary and secondary resistances R_1 and R_2, (*ii*) primary and secondary leakage reactances X_1 and X_2 and (*iii*) iron and copper losses. The equivalent circuit of an actual transformer is shown in fig.3.19.

Primary impedance,

$$\overline{Z_1} = R_1 + j X_1$$

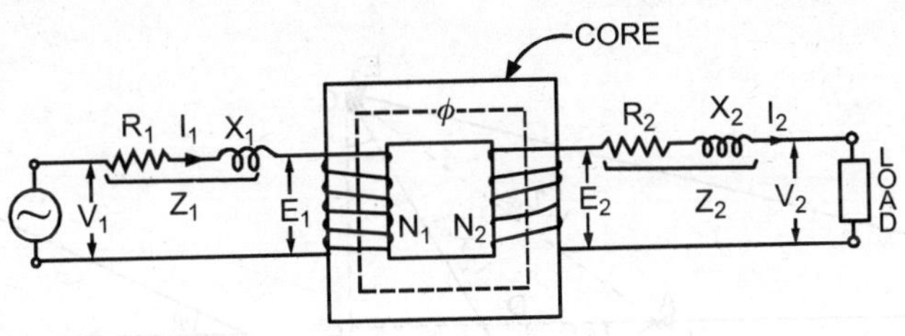

Fig.3.19: Complete circuit diagram of an actual transformer

Supply voltage is V_1. The resistance and leakage reactance of primary winding are responsible for some voltage drop in primary winding.

$$\therefore \quad \overline{V_1} = \overline{E_1} + \overline{I_1}\,(R_1 + j\,X_1) = \overline{E_1} + \overline{I_1}\,\overline{Z_1}$$

Secondary impedance, $\overline{Z_2} = R_2 + j X_2$

Similarly, the resistance and leakage reactance of secondary winding are responsible for some voltage drop in secondary winding. Hence,

$$\overline{V_2} = \overline{E_2} - \overline{I_2}\,(R_2 + j\,X_2) = \overline{E_2} - \overline{I_2}\,\overline{Z_2}$$

3.15 SIMPLIFIED EQUIVALENT CIRCUIT

The simplified equivalent circuit of a transformer is drawn by representing all the parameters of the transformer either on the secondary or on the primary side. The no load current I_0 is neglected as its value is very small as compared to full load current, therefore, I_1¢ $= I_1$

(*i*) Equivalent circuit when all the quantities are referred to secondary.

The primary resistance when referred to secondary side, its valuc is $R_1' - K^2 R_1$ and the total or equivalent resistance of transformer referred to secondary, $R_{es} = R_2 + R_1'$. Similarly, the primary reactance when referred to secondary side, its value is $X_1' = K^2 X_1$ and the total or equivalent reactance of transformer referred to secondary, $X_{es} = X_2 + X_1'$. All the quantities when referred to the secondary side are shown in fig.3.20.

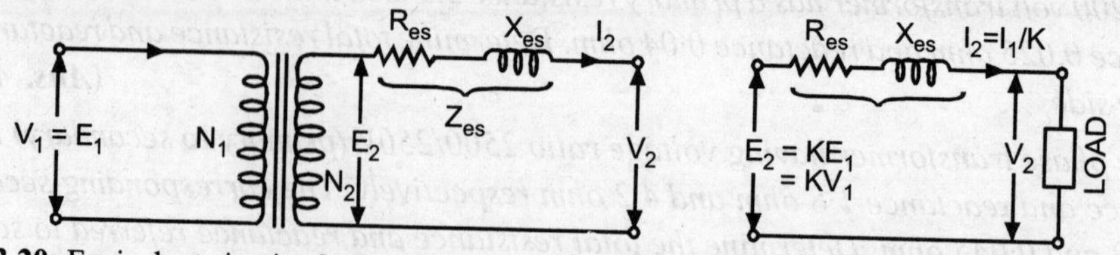

Fig.3.20: Equivalent circuit of a transformer when all the quantities are referred to secondary side

Total or equivalent impedance referred to secondary side,

$$Z_{es} = R_{es} + j\,X_{es}$$

There is some voltage drop in resistance and reactance of transformer referred to secondary. Hence,

$$\overline{V_2} = \overline{E_2} - \overline{I_2}\,(R_{es} + j\,X_{es}) = \overline{E_2} - \overline{I_2}\,\overline{Z_{es}}$$

3.16 EXPRESSION FOR NO LOAD SECONDARY VOLTAGE

For a loaded transformer when all the quantities are referred to secondary side, its phasor diagram can be drawn as shown in fig.3.21.

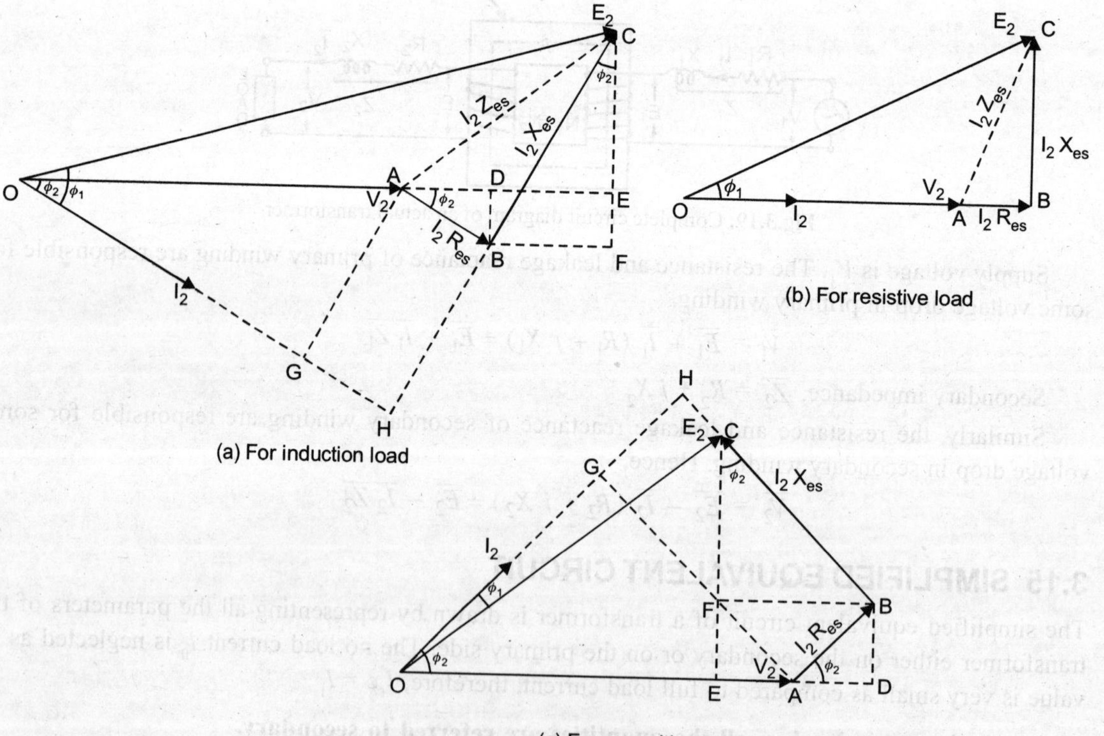

(a) For induction load

(b) For resistive load

(c) For capacitive load

Fig.3.21: Phasor diagrams

Complete the phasor diagram as shown in fig.3.21. From the phasor diagram we can derive the approximate as well as exact expressions for no load secondary voltage.

***(i)* Approximate expression**

(a) *for lagging p.f.* (inductive load),

Consider right angle triangle *OEC* [see fig 3.21(*a*)].

$$OC \cong OE = OA + AD + DE = OA + AD + BF$$

or $$E_2 = V_2 + I_2 R_{es} \cos\phi_2 + I_2 X_{es} \sin\phi_2$$

(b) *for unity p.f.* (resistive load),

Consider rt. ∠ triangle OBC [see fig.3.21(*b*)]

$$OC \cong OB = OA + AB; \; E_2 = V_2 + I_2 R_{es}$$

(c) *for leading p.f.* (capacitive load),

Consider rt. ∠ triangle *OEC* [see fig.3.21(c)]

$$OC \cong OE = OA + AD - DE = OA + AD - BF$$

or $$E_2 = V_2 + I_2 R_{es} \cos\phi_2 - I_2 X_{es} \sin\phi_2$$

***(ii)* Exact expression:**

(a) *for a lagging p.f.* (inductive load),

Consider rt. ∠ triangle *OHC* [see fig.3.21(*a*)]

$$OC = \sqrt{(OH)^2 + (HC)^2} = \sqrt{(OG+GH)^2 + (HB+BC)^2}$$

$$= \sqrt{(OG+AB)^2 + (GA+BC)^2}$$

or $$E_2 = \sqrt{(V_2 \cos\phi_2 + I_2 R_{es})^2 + (V_2 \sin\phi_2 + I_2 X_{es})^2}$$

$$\text{Primary p.f., } \cos\phi_1 = \frac{OH}{OC} = \frac{OG+GH}{OC} = \frac{OG+AB}{OC} = \frac{V_2\cos\phi_2 + I_2 R_{es}}{E_2}$$

(b) *for unity p.f.,* (resistive load)

Consider rt. ∠ triangle *OBC* [see fig.3.21(*b*)]

$$OC = \sqrt{(OB)^2 + (BC)^2}$$

or $$OC = \sqrt{(OA+AB)^2 + (BC)^2}$$

or $$E_2 = \sqrt{(V_2 + I_2 R_{es})^2 + (I_2 X_{es})^2}$$

$$\text{Primary p.f., } \cos\phi_1 = \frac{OB}{OC} = \frac{OA+AB}{OC} = \frac{V_2 + I_2 R_{es}}{E_2}$$

(c) *for leading p.f.* (capacitive load),

Consider rt. ∠ triangle *OHC* [see fig.3.21(*c*)]

$$OC = \sqrt{(OH)^2 + (HC)^2} = \sqrt{(OG+GH)^2 + (HB-BC)^2}$$

$$= \sqrt{(OG+AB)^2 + (GA-BC)^2}$$

or $$E_2 = \sqrt{(V_2 \cos\phi_2 + I_2 R_{es})^2 + (V_2 \sin\phi_2 - I_2 X_{es})^2}$$

$$\text{Primary p.f., } \cos \phi_1 = \frac{HC}{OC} = \frac{OG+GH}{OC} = \frac{OG+AB}{OC} = \frac{V_2 \cos\phi_2 + I_2R_{es}}{E_2}$$

3.17 VOLTAGE REGULATION

When a transformer is loaded, with a constant supply voltage, the terminal voltage changes due to voltage drop in the internal parameters of the transformer *i.e.* primary and secondary resistances and inductive reactances. The internal voltage drop also depends upon the load and its power factor. The algebraic difference between the no-load and full-load terminal voltage is measured in terms of voltage regulation.

At a constant supply voltage, the change in secondary terminal voltage from no-load to full-load with respect to no load voltage is called ***voltage regulation*** *of the transformer.*

Let, E_2 = Secondary terminal voltage at no-load.
V_2 = Secondary terminal voltage at full-load.

$$\text{Then, voltage regulation} = \frac{E_2 - V_2}{E_2} \text{ (per unit)}$$

$$\text{In the form of percentage,} \quad \% \text{ Reg} = \frac{E_2 - V_2}{E_2} \times 100$$

When all the quantities are referred to the primary side of the transformer;

$$\% \text{ Reg} = \frac{V_1 - E_1}{V_1} \times 100$$

3.18 APPROXIMATE EXPRESSION FOR VOLTAGE REGULATION

The approximate expression for the no-load secondary voltage is derived in article 15.

(*i*) *For inductive load*: $E_2 = V_2 + I_2 R_{es} \cos \phi_2 + I_2 X_{es} \sin \phi_2$

or $E_2 - V_2 = I_2 R_{es} \cos \phi_2 + I_2 X_{es} \sin \phi_2$

or $$\frac{E_2 - V_2}{E_2} \times 100 = \frac{I_2 R_{es}}{E_2} \times 100 \cos \phi_2 + \frac{I_2 - X_{es}}{E_2} \times 100 \sin \phi_2$$

where, $\frac{I_2 X_{es}}{E_2} \times 100$ = percentage resistance drop and

$\frac{I_2 X_{es}}{E_2} \times 100$ = percentage reactance drop

∴ % Reg = % resistance drop × cos ϕ_2 + % resistance drop

Similarly

(*ii*) *For resistive load*: % Reg = % resistance drop

(*iii*) *For capacitive load*:

∴ % Reg = % resistance drop × cos ϕ_2 – % reactance drop × sin ϕ_2

Example 3.8: *A single phase transformer with a ratio 1: 2 has primary resistance of 0.25 ohm and reactance of 0.5 ohm and the corresponding values for the secondary are 0.8 ohm and 1.8 ohm respectively. Determine the no load secondary terminal voltage of the transformer if it is delivering 10 A and 400 V at 0.8 p.f. lagging.*

Solution: Primary resistance, $R_1 = 0.25\ \Omega$

Primary reactance, $X_1 = 0.5\ \Omega$

Secondary resistance, $R_2 = 0.8\ \Omega$

Secondary reactance, $X_2 = 1.8\ \Omega$

Secondary current, $I_2 = 10\text{ A}$

Load p.f., $\cos\phi_2 = 0.8$ *lagging*

Secondary terminal voltage at load, $V_2 = 400\text{ V}$

Transformation ratio, $K = \frac{2}{1} = 2$

Primary resistance referred to secondary $R_1' = K^2 R_1 = 2 \times 2 \times 0.25 = 1.0\ \Omega$

Primary reactance referred to secondary $X_1' = K^2 X_1 = 2 \times 2 \times 0.5 = 2.0\ \Omega$

Equivalent resistance referred to secondary $R_{es} = R_2 + R_1' = 0.8 + 1.0 = 1.8\ \Omega$

Equivalent reactance referred to secondary $X_{es} = X_2 + X_1' = 1.8 + 2.0 = 3.8\ \Omega$

Load p.f., $\cos\phi_2 = 0.8 \quad \therefore \sin\phi_2 = \sin\cos^{-1} 0.8 = 0.6$

Secondary terminal voltage at no load, $E_2 = \sqrt{(V_2\cos\phi_2 + I_2R_{es})^2 + (V_2\sin\phi_2 + I_2X_{es})^2}$

$= \sqrt{(400\times0.8+10\times1.8)^2 + (400\times0.6+10\times3.8)^2}$

$= \sqrt{(338)^2 + (278)^2}$ = **437.64 V** *(Ans.)*

Example 3.9: *A single phase transformer with a ratio 5: 1 has primary resistance of 0·4 ohm and reactance of 1·2 ohm and the secondary resistance of 0·01 and reactance of 0·04 ohm. Determine the percentage regulation when delivering 125 A at 600 V at (i) 0·8 p.f. lagging (ii) 0·8 p.f. leading.*

Solution: Data given, $R_1 = 0{\cdot}4\ \Omega$; $X_1 = 1{\cdot}2\ \Omega$; $R_2 = 0{\cdot}01\ \Omega$;

$X_2 = 0{\cdot}04\ \Omega$; $I_2 = 125$ A and $V_2 = 600$ V

Transformation ratio, $K = 1/5 = 0{\cdot}2$

Now, $R_1' = K^2 R_1 = (0{\cdot}2)^2 \times 0{\cdot}4 = 0{\cdot}016\ \Omega$;

$R_{es} = R_2 + R_1' = 0{\cdot}01 + 0{\cdot}016 = 0{\cdot}026\ \Omega$

$X_1' = K^2 X_1 = (0{\cdot}2)^2 \times 1{\cdot}2 = 0{\cdot}048\ \Omega$

$X_{es} = X_2 + X_1' = 0{\cdot}04 + 0{\cdot}048 = 0{\cdot}088\ \Omega$

(*i*) For p.f. $\cos\phi_2 = 0{\cdot}8$ lag; $\sin\phi_2 = \sin\cos^{-1} 0{\cdot}8 = 0{\cdot}6$

Secondary induced voltage,

$E_2 = V_2 + I_2 R_{es} \cos\phi_2 + I_2 X_{es} \sin\phi_2$

$= 600 + 125 \times 0{\cdot}026 \times 0{\cdot}8 + 125 \times 0{\cdot}088 \times 0{\cdot}6$

$= 600 + 2{\cdot}6 + 6{\cdot}6 = 609{\cdot}2$ V

$\%\text{ Reg} = \frac{E_2 - V_2}{E_2} \times 100 = \frac{609{\cdot}2 - 600}{609{\cdot}2} \times 100$ = **1·51 %** *(Ans.)*

(*ii*) For p.f., $\cos\phi_2 = 0{\cdot}8$ leading; $\sin\phi_2 = 0{\cdot}6$

Secondary induced voltage,

$$E_2 = V_2 + I_2 R_{es} \cos \phi_2 - I_2 X_{es} \sin \phi_2$$
$$= 600 + 125 \times 0{\cdot}06 \times 0{\cdot}8 - 125 \times 0{\cdot}088 \times 0{\cdot}6$$
$$= 600 + 2{\cdot}6 - 6{\cdot}6 = 596 \text{ V}$$
$$\% \text{ Reg} = \frac{E_2 - V_2}{E_2} \times 100 = \frac{596 - 600}{596} \times 100 = -\mathbf{0{\cdot}67\ \%} \ (Ans.)$$

Example 3.10: *Calculate the regulation of transformer in which ohmic loss is 1% of the output and reactance drop 5% of the voltage when the power factor is (i) 0·8 lagging (ii) 0·8 leading (iii) unity.*

Solution: Ohmic loss or resistance drop = 1%; Reactance drop = 5%

(*i*) When p.f., $\cos \phi_2 = 0{\cdot}8$ lagging; $\sin \phi_2 = \sin \cos^{-1} 0{\cdot}8 = 0{\cdot}6$

$$\% \text{ Reg} = \% \text{ resistance drop} \times \cos \phi_2 + \% \text{ reactance drop} \times \sin \phi_2$$
$$= 1 \times 0{\cdot}8 + 5 \times 0{\cdot}6 = \mathbf{3{\cdot}8\%} \ (Ans.)$$

(*ii*) When p.f., $\cos \phi_2 = 0{\cdot}8$ leading; $\sin \phi_2 = \sin \cos^{-1} 0{\cdot}8 = 0{\cdot}6$

$$\% \text{ Reg} = \% \text{ resistance drop} \times \cos \phi_2 - \% \text{ reactance drop} \times \sin \phi_2$$
$$= 1 \times 0{\cdot}8 - 5 \times 0{\cdot}6 = -\mathbf{2{\cdot}2\ \%} \ (Ans.)$$

(*iii*) When p.f. is unity % Reg = % resistance drop = **1%** (*Ans.*)

PRACTICE EXERCISE

Q.1 *A 10 kVA, 1-phase transformer with 2000/400 V at no load, has resistance and leakage reactance of primary winding 5·5 W and 12 W respectively, the corresponding values of secondary winding being 0·5 W and 0·45 W. Determine the value of secondary voltage at full load, 0·8 pf lagging, when the primary applied voltage is 2000 V.* (**Ans.** 377.65V)

Q.2 *A 10 kVA, 2000/400 V, single phase transformer has resistance and leakage reactcnce as follows:*

Primary winding: Resistance = 5.5 ohms, Reactance = 12 ohms

Secondary winding: Resistance = 0.2 ohm, Reactance = 0.45 ohm

Determine the value of the secondary voltage at full load, 0.8 p.f. leading, when the primary supply voltage is 2000 V. (**Ans.** 405.55V)

Q.3 *The primary and secondary windings of a 40 kVA, 6600/250 V single phase transformer have resistance of 10 ohm and 0.02 ohm respectively. The total leakage reactance is 35 ohm as referred to primary winding. Find full load regulation of at a p.f. 0.8 lagging.* (**Ans.** 3.56%)

3.18 LOSSES IN A TRANSFORMER

The losses which occur in an actual transformer are:

(*i*) Core or iron losses (*ii*) Copper losses

(*i*) **Core or iron losses:** When ac supply is given to the primary winding of a transformer an alternating flux is set up in the core, therefore, hysteresis and eddy current losses occur in the magnetic core.

(*a*) *Hysteresis loss:* When the magnetic material is subjected to reversal of flux, power is required for the continuous reversal of molecular magnets. This power is dissipated in the form of heat and is known as hysteresis loss ($P_h = K_h V f B_m^{1.6}$). This loss can be minimized by using silicon steel material for the construction of core.

(*b*) *Eddy current loss:* Since flux in the core of a transformer is alternating, it links with the magnetic material of the core itself also. This induces an e.m.f. in the core and circulates eddy currents.

Power is required to maintain these eddy currents. This power is dissipated in the form of heat and is known as eddy current loss ($P_e = K_e V f^2 t^2 B_m^2$). This loss can be minimised by making the core of thin laminations.

It is already seen in article–8 that the flux set up in the core of the transformer remains constant from no load to full load. Hence, iron loss are independent of the load and are known as constant losses.

(ii) **Copper losses:** Copper losses occur in both the primary and secondary windings due to their ohmic resistance. If I_1, I_2 are the primary and secondary currents and R_1, R_2 are the primary and secondary resistances respectively.

Then, total copper losses = $I_1^2 R_1 + I_2^2 R_2 = I_1^2 R_{ep} = I_2^2 R_{es}$

The currents in the primary and secondary winding vary according to the load, therefore, these losses vary according to the load and are known as variable losses.

Example 3.11. *A 50 kVA, 2000/200 V, single phase, 50 Hz transformer has primary resistance of 3·5 ohm and reactance of 4·5 ohm. The secondary resistance and reactance are 0·015 ohm and 0·02 ohm respectively. Find (i) equivalent resistance, reactance and impedance referred to the primary side (ii) Total copper losses in the transformer.*

Solution: Transformer rating = 50 kVA, 2000/200 V, 1-ϕ, 50 Hz

Primary resistance, $R_1 = 3 \cdot 5\ \Omega$

Primary reactance, $X_1 = 4 \cdot 5\ \Omega$

Secondary resistance, $R_2 = 0 \cdot 015\ \Omega$

Secondary reactance, $X_2 = 0 \cdot 02\ \Omega$

(*i*) Transformation ratio, $K = \dfrac{V_2}{V_1} = \dfrac{200}{2000} = 0 \cdot 1$

Equivalent resistance referred to primary side,

$$R_{ep} = R_1 + R_2'$$

$$= R_1 + \frac{R_2}{K^2} = 3 \cdot 5 + \frac{0 \cdot 015}{(0 \cdot 1)^2} = 3 \cdot 5 + 1 \cdot 5 = \mathbf{5\ \Omega}\ (\textit{Ans.})$$

Equivalent reactance referred to primary side,

$$X_{ep} = X_1 + X_2'$$

$$= X_1 + \frac{X_2}{K^2} = 4 \cdot 5 + \frac{0 \cdot 02}{(0 \cdot 1)^2} = 4 \cdot 5 + 2 \cdot 0 = \mathbf{6 \cdot 5\ \Omega}\ (\textit{Ans.})$$

Equivalent impedance referred to primary side,

$$Z_{ep} = \sqrt{R_{ep}^2 + X_{ep}^2} = \sqrt{(15)^2 + (6 \cdot 5)^2} = \mathbf{8 \cdot 2\ \Omega}\ (\textit{Ans.})$$

(*ii*) Primary rated current, $I_1 = \dfrac{50 \times 1000}{2000} = 25$ A

Full-load copper losses in transformer,

$$P_c = I_1^2 R_{ep} = (25)^2 \times 5 = \mathbf{3125\ W}\ (\textit{Ans.})$$

3.19 EFFICIENCY OF A TRANSFORMER

The efficiency of a transformer is defined as the ratio of output to the input power, the two being measured in same units (either in watts or in kW).

Transformer efficiency, $\eta = \dfrac{\text{output power}}{\text{input power}} = \dfrac{\text{output power}}{\text{output power + losses}}$

or $\eta = \dfrac{\text{output power}}{\text{output power + iron losses + coper losses}} = \dfrac{V_2 I_2 \cos\phi_2}{V_2 I_2 \cos\phi_2 + P_i + P_c}$

where, V_2 = Secondary terminal voltage

I_2 = Full load secondary current

$\cos\phi_2$ = p.f. of the load

P_i = Iron losses = Hysteresis losses + eddy current losses (constant losses)

P_c = Full load copper losses = $I^2{}_2 R_{es}$ (variable losses)

If x is the fraction of the full load, the efficiency of the transformer at this fraction is given by the relation;

$$\eta_x = \frac{x \times \text{output at full load}}{x \times \text{output at full load} + P_i + x^2 P_c} = \frac{xV_2 I_2 \cos\phi_2}{x V_2 I_2 \cos\phi_2 + P_i + x^2 I_2^2 R_{es}}$$

The copper losses vary as the square of the fraction of the load.

3.20 CONDITION FOR MAXIMUM EFFICIENCY

The efficiency of a transformer at a given load and p.f. is expressed by the relation

$$\eta = \frac{V_2 I_2 \cos\phi_2}{V_2 I_2 \cos\phi_2 + P_i + I_2^2 R_{es}} = \frac{V_2 \cos\phi_2}{V_2 \cos\phi_2 + P_i / I_2 + I_2 R_{es}}$$

The terminal voltage V_2 is approximately constant. Thus for a given p.f., efficiency depends upon the load current I_2. In expression (*i*), the numerator is constant and the efficiency will be maximum if denominator is minimum. Thus the maximum condition is obtained by differentiating the quantity in the denominator w.r.t. the variables I_2 and equating that to zero *i.e.*

$$\frac{d}{d I_2}\left(V_2 \cos\phi_2 + \frac{P_i}{I_2} + I_2 R_{es} \right) = 0 \text{ or } 0 - \frac{P_i}{I_2^2} + R_{es} = 0$$

or $$I_2^2 R_{es} = P_i$$

i.e. **Copper losses = Iron losses**

Thus, the efficiency of a transformer will be maximum when copper (or variable) losses are equal to iron (or constant) losses.

$$\therefore \quad \eta_{max} = \frac{V_2 I_2 \cos\phi_2}{V_2 I_2 \cos\phi_2 + 2 P_i} \quad [\text{since } P_c = P_i]$$

From equation (*ii*), the value of output current I_2 at which the efficiency of the transformer will be maximum is given by;

$$I_2 = \sqrt{\frac{P_i}{R_{es}}}$$

If x is the fraction of full load kVA at which the efficiency of the transformer is maximum.

Then, copper losses = $x^2 P_c$ (where P_c is the full load Cu losses)

Lron losses = P_i

For maximum efficiency, $x^2 P_c = P_i; x = \sqrt{\frac{P_i}{P_c}}$

∴ Output kVA corresponding to maximum efficiency

$$= x \times \text{full load kVA} = \text{full load kVA} \times \sqrt{\frac{P_i}{P_c}}$$

$$= \text{full load kVA} \times \sqrt{\frac{\textit{iron losses}}{\textit{copper losses at full load}}}$$

Example 3.12: *In a 25 kVA, 2000/200 V power transformer the iron and full load copper losses are 350 W and 400 W respectively. Calculate the efficiency at unity power factor at (i) full load and (ii) half load.*

Solution: $\eta_x = \frac{x\ kVA \times 1000 \times \cos\phi}{x\ kVA \times 1000 \times \cos\phi + P_i + x^2 P_c}$

where, $\cos f = 1$: $P_i = 350$ W; $P_c = 400$ W

(*i*) At full-load $x = 1$

$$\therefore \quad \eta = \frac{1 \times 25 \times 1000 \times 1}{1 \times 25 \times 1000 \times 1 + 350 + 1 \times 1 \times 400} \times 100 = \mathbf{97{\cdot}087\ \%}\ (Ans)$$

(*ii*) At half-load; $x = 0{\cdot}5$

$$\therefore \quad \eta = \frac{0{\cdot}5 \times 25 \times 1000 \times 1}{0{\cdot}5 \times 25 \times 1000 \times 1 + 350 + (0{\cdot}5)^2 \times 400} \times 100 = \mathbf{96{\cdot}525\ \%}\ (Ans)$$

Example 3.13: *At full load current, the iron and copper losses in a 100 kVA transformer are each equal to 2·5 kW. Find the efficiency of the transformer at a load of 75 kVA and 0·8 power factor lagging.*

Solution: Efficiency of a transformer at any fraction x of full load,

$$\eta_x = \frac{x\ kVA \times \cos\phi}{x\ kVA \times \cos\phi + P_i + x^2 P_c} \times 100$$

Where, Transformer rated capacity = 100 kVA

Operating load = 75 kVA

Fraction of load, $x = \frac{75}{100} = 0 \cdot 75$

Iron losses, $P_i = 2{\cdot}5\ k\Omega$

Full load copper losses, $P_c = 2{\cdot}5\ k\Omega$

Power factor of load, $\cos\phi = 0{\cdot}8$ lagging

$$\therefore \quad \eta_x = \frac{0 \cdot 75 \times 100 \times 0 \cdot 8}{0 \cdot 75 \times 100 \times 0 \cdot 8 + 2 \cdot 5 + (0 \cdot 75)^2 \times 2 \cdot 5} \times 100$$

$= \mathbf{93{\cdot}88\%}$ (*Ans.*)

Example 3.14: *The efficiency of a 1000 KVA, 110/220 V, 50 Hz single phase transformer is 98·5% at half load at 0·8 power factor leading and 98·8% at full load unit power factor. Find (i) Iron loss (ii) Full load copper loss.*

Solution. Efficiency at any fraction x of the load;

$$\eta_x = \frac{x\,kVA \times \cos\phi}{x\,kVA \times \cos\phi + P_i + x^2\,P_c} \times 100$$

where, P_c is full-load copper loss and P_i is the iron loss.

(*i*) When, $x = 0{\cdot}5$; $\cos\phi = 0{\cdot}8$; $\eta_x = 98{\cdot}5\%$

$$\therefore \quad 98{\cdot}5 = \frac{0\cdot5 \times 1000 \times 0\cdot8}{0\cdot5 \times 1000 \times 0\cdot8 + P_i + (0\cdot5)^2\,P_c} \times 100$$

$$\text{or} \quad 400 + Pi + 0{\cdot}25\,P_c = \frac{400 \times 100}{98\cdot5} \quad \text{or } P_i + 0{\cdot}25\,P_c = 6{\cdot}1 \text{ kW} \qquad \ldots(i)$$

(*ii*) When, $x = 1$; $\cos\phi = 1$; $\eta_x = 98{\cdot}8\%$

$$\therefore \quad 98{\cdot}8 = \frac{1 \times 1000 \times 1}{1 \times 1000 \times 1 + Pi + (1)^2\,Pc} \times 100$$

$$\text{or} \quad 1000 + P_i + P_c = \frac{1000 \times 100}{98\cdot8}$$

$$\text{or} \quad P_i + P_c = 12{\cdot}15 \text{ kW} \qquad \ldots(ii)$$

Subtracting equation (*i*) from (*ii*), we get,

$$0{\cdot}75\,P_c = 6{\cdot}05$$

$$\text{or} \quad P_c = \mathbf{8{\cdot}07\ kW}\ (Ans.)$$

$$\text{and} \quad P_i = 12{\cdot}15 - 8{\cdot}07 \text{ kW} = \mathbf{4{\cdot}08\ kW}\ (Ans.)$$

Example 3.15: *In a 25 kVA, 2000 V/200 V transformer the iron and copper losses are 200 W and 400 W respectively. Calculate the efficiency at half load and 0·8 power factor lagging. Determine also the maximum efficiency and the corresponding load.*

Solution: Transformer rating = 25 kVA; Iron losses, $P_i = 200$ W;

Full load copper losses, $P_c = 400$ W

$$\text{Fraction of load, } x = \frac{1}{2} = 0{\cdot}5; \text{ Power factor, } \cos\phi = 0{\cdot}8$$

$$\text{Efficiency,} \quad \eta_x = \frac{x \times \text{kVA} \times 1000 \cos\phi}{x \times \text{kVA} \times 1000 \cos\phi + P_i + x^2\,P_C} \times 100\%$$

$$= \frac{(0\cdot5) \times 25 \times 10^3 \times 0\cdot8}{0\cdot5 \times 25 \times 10^3 \times 0\cdot8 + 200 + (0\cdot5)^2 \times 400} \times 100\%$$

$$= \mathbf{97{\cdot}08\ \%}\ (Ans.)$$

For maximum efficiency $\quad P_i = x^2\,P_c$

Fraction of load at which efficiency is maximum,

$$x = \sqrt{\frac{P_i}{P_c}} = \sqrt{\frac{200}{400}} = 0\cdot707$$

$$\text{Maximum efficiency,} \quad \eta_{max} = \frac{x \times \text{kVA} \times 1000 \times \cos\phi}{x \times \text{kVA} \times 1000 \times \cos\phi + 2P_i}$$

$$= \frac{0 \cdot 707 \times 25 \times 10^3 \times 0 \cdot 8}{0 \cdot 707 \times 25 \times 10^3 \times 0 \cdot 8 + 2 \times 200} = \mathbf{97 \cdot 25\%} \ (Ans.)$$

Load at max. efficiency $= x$ rated kVA $= 0 \cdot 707 \times 25 =$ **17·675 kVA** (*Ans.*)

PRACTICE EXERCISE

Q.1 *The primary and secondary windings of a 500 kVA transformer have resistance of 0·42 ohm and 0·0011 ohm respectively. The primary and secondary voltages are 600 V and 400 V respectively. The iron loss is 2·9 kW. Calculate the efficiency at half full load at a power factor of 0·8 lagging.* **(*Ans.* 98.07%)**

Q.2 *A 220/400 V, 10 kVA, 50Hz, single-phase transformer has copper loss of 120 W at full load. If it has an efficiency of 98% at full load, unity power factor, determine the iron losses. What would be the efficiency of the transformer at half full–load at 0.8 p.f. lagging.* **(*Ans.* 84.08W;97.23%)**

Q.3 *A 40 kVA transformer has a core loss of 400 watt and full-load copper loss of 800 watt. If the power factor of the load is 0·9 lagging, calculate:*

(i) the full-load efficiency, and

(ii) percentage of the full-load at which the maximum efficiency occurs. **(*Ans.* 96.77%; 70.7%)**

3.21 AUTO-TRANSFORMER

An auto-transformer is a transformer with only one winding wound on a laminated core. A part of this winding is common to both primary and secondary sides. On load, a part of the load current is obtained direct from the supply and remaining part is obtained by transformer action. In an ordinary transformer the primary and secondary windings are electrically insulated from each other but connected magnetically [see fig.3.22(a)], whereas, in an auto-transformer the primary and secondary windings are connected magnetically and electrically. Fig.3.22(b). shows the primary winding *AB* from which a tapping at *C* is taken, such that *CB* acts as a secondary winding. The supply voltage is applied across *AB* and the load is connected across *CB*. The tapping may be fixed or variable.

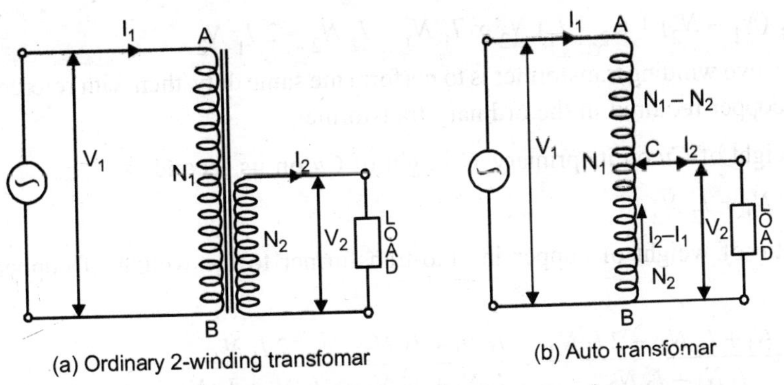

(a) Ordinary 2-winding transfomar (b) Auto transfomar

Fig.3.22: Ordinary two winding and auto-transformer

When a.c. voltage V_1 is applied across *AB*, an alternating flux is set up in the core, it induces an e.m.f. E_1 in the winding *AB*. The part of this e.m.f. is taken in the secondary circuit.

Let V_1 = primary applied voltage; V_2 = secondary voltage across the load;

I_1 = primary current; I_2 = load current.
N_1 = number of turns between A and B and
N_2 = number of turns between C and B

Neglecting no-load current, leakage reactance and losses, $V_1 = E_1$ and $V_2 = E_2$

$$\therefore \text{ Transformation ratio, } K = \frac{V_2}{V_1} = \frac{N_2}{N_1} = \frac{I_1}{I_2}$$

As the secondary ampere turns are opposite to primary ampere turns so the current I_2 is in phase opposition to I_1. The secondary voltage is less than the primary, therefore, current I_2 is more than I_1. The resultant current flowing through section BC is $(I_2 - I_1)$.

Now, ampere turns due to section BC = current × turns

$$= (I_2 - I_1)\,N_2 = \left(\frac{I_1}{K} - I_1\right) \times N_1 K = I_1\,N_1\,(1 - K) \qquad \ldots(i)$$

Ampere turns due to section $AC = I_1\,(N_1 - N_2)$

$$= I_1\,N_1\left(1 - \frac{N_2}{N_1}\right) = I_1\,N_1\,(1 - K) \qquad \ldots(ii)$$

Equations (*i*) and (*ii*) show that the ampere turns due to section BC and AC balance each other which is the characteristic of transformer action.

Saving of copper:

Volume, and hence weight of copper, is proportional to the length and area of X-section of the conductor. The length of conductor is proportional to number of turns whereas area of X-section is proportional to the current flowing through it. Hence the weight of copper is proportional to the product of current and number of turns.

Now, with reference to fig.3.22(b), weight of copper required in an auto-transformer.

W_a = weight of Cu in section AC + weight + Cu in section CB

$$\therefore \quad W_a \propto I_1\,(N_1 - N_2) + (I_2 - I_1)\,N_2 \propto I_1\,N_1 + I_2\,N_2 - 2\,I_1\,N_2$$

If an ordinary two winding transformer is to perform the same duty, then with reference to fig.3.22(a). Total weight of copper required in the ordinary transformer.

W_0 = weight of Cu on its primary + weight of Cu on its secondary.

$$\therefore \quad W_0 \propto I_1\,N_1 + I_2\,N_2$$

Now, the ratio of weight of copper in auto-transformer to the weight of copper in an ordinary transformer,

$$\frac{W_a}{W_0} = \frac{I_1 N_1 + I_2 N_2 - 2 I_1 N_2}{I_1 N_1 + I_2 N_2} = \frac{I_1 N_1 + I_2 N_2}{I_1 N_1 + I_2 N_2} - \frac{2 I_1 N_2}{I_1 N_1 + I_2 N_2}$$

$$= 1 - \frac{2 I_1 N_2 / I_1 N_1}{I_1 N_1 / I_1 N_1 + I_2 N_2 / I_1 N_1} = 1 - K$$

or $\quad W_a = (1 - K)\,W_0$

Saving of copper affected by using an auto-transformer

= wt. of cu required in an ordinary transformer – wt. of copper required in an auto-transformer.

$$= W_0 - W_a = W_0 - (1 - K)\, W_0 = KW_0$$

$\therefore$ Saving = $K \times$ Wt. of copper required for two winding transformer

Hence, saving in copper increases as the transformation ratio approaches to unity, therefore, auto transformers are used when K in nearly equal to unity.

Uses of auto-transformer: (*i*) It is used as an *auto-transformer starter* to give upto 50 to 60% of full voltage to the stator of a squirrel cage induction motor during starting.

(*ii*) It is used to give a small boost to a distribution cable to correct the voltage drop.

(*iii*) It is used as a regulating transformer (the top point C is variable in this case).

Disadvantages: Although auto-transformers have less cost, better regulation and low losses as compared to the ordinary two winding transformer of same rating. But still they are not widely used due to one major disadvantage that the secondary winding is not insulated from the primary. If an auto transformer is used to supply low voltage from a high voltage and there is a break in the secondary winding, the full *primary voltage comes across the secondary terminals which may be dangerous to the operator and equipment (load).* Therefore, it is advisable not to use an auto transformer for interconnecting high voltage and low voltage system. Their use is only limited to the places where slight variation of output voltage from the input voltage is required.

Example 3.16: *The primary and secondary voltages of an auto transformer are 500 V and 400 V respectively. Show with the aid of diagram the current distribution in the winding when the secondary current is 100 A. calculate the economy of copper in this particular case.*

Solution: Transformation ratio, $K = \dfrac{V_2}{V_1} = \dfrac{400}{500} = 0{\cdot}8$

Using the relation, $\dfrac{I_1}{I_2} = K$

Primary current, $I_1 = KI_2 = 0{\cdot}8 \times 100 =$ **80 A** (*Ans*)

Current in section BC = $I_2 - I_1 = 100 - 80 =$ **20 A** (*Ans*)

For distribution of current in different sections of the winding please see fig.31 (*b*).

Economy or saving of copper = $KW_0 = 0.8\, W_0$

Percentage saving = $0{\cdot}8 \times 100 =$ **80%** (*Ans*)

3.22 THREE-PHASE TRANSFORMERS

Three phase system is invariably adopted for generation, transmission and distribution of electrical power due to economical reasons. Usually, power is generated at the generating stations at 11 kV (or 33 kV), whereas, it is transmitted at 400 kV, 220 kV, 132 kV or 66 kV due to economical reasons. At the receiving stations, the voltage level is decreased and power is transmitted through shorter distances. While delivering power to the consumers, the voltage level is decreased to as low as 400V (line value) for safety reasons.

To handle 3-phase electrical power and to increase the voltage level at the generating stations and to decrease the voltage level at the receiving stations, 3-phase step-up and step down transformers are employed.

3.23 MERITS OF THREE-PHASE TRANSFORMERS

The voltage level in 3-phase system at the generating stations and at the receiving stations can be changed by employing three single-phase transformers (inter connecting them in star or delta) or by employing one three phase transformer. Generally, one three phase transformer is preferred over single phase transformers because of the following reasons.

(*i*) It requires smaller quantity of iron and copper.

(*ii*) It has smaller size and can be accommodated in smaller tank and hence needs smaller quantity of oil for cooling.

(*iii*) It has less weight and occupies less space.

(*iv*) It needs fewer number of bushings.

(*v*) Over and above, it costs nearly 15% lesser than a bank of three single phase transformers of equal rating.

(*vi*) It operates at slightly better efficiency and regulation.

Hence, three phase transformers are invariably employed in the power system to step-up and step–down the voltages. Although, these transformers suffer from the following disadvantage.

(*i*) It is more difficult and costly to repair 3-phase transformers.

(*ii*) It is difficult to transport single large unit of 3-phase transformer than to transport three single phase transformers individually.

3.24 CONSTRUCTION OF THREE-PHASE TRANSFORMERS

Form construction point of view, 3-phase transformers are also classified as

(*i*) Core type transformers (*ii*) Shell type transformers

(*i*) **Core type transformers:** In core type 3-phase transformers, the core has three limbs of equal area of cross-section. Three limbs are joined by two horizontal (top and bottom) members called yokes. The area of cross- section of all the limbs and yokes is the same since at every instant magnitude of flux set-up in each part is the same. The core consists of laminations of silicon steel material having oxide film coating on both the sides for insulation. The laminations are usually of *E* and *I* shape and are staggered alternately.

The complete section of a 3-phase core type transformer with its plan is shown in fig.3.23. This type of transformers are usually wound with circular cylindrical coils. The low voltage (*LV*) winding is wound nearer the core and high voltage (*HV*) winding is wound over low voltage winding as shown in fig.3.23. Insulation is always provided between the core and low voltage winding and between low voltage winding and high voltage winding.

(*ii*) **Shell type transformers:** In shell type transformers, the core construction is such that the windings are embedded in the core instead of surrounding the iron as shown in fig.3.24. The area of cross section of the central limbs is double to that of the side limbs and horizontal members.

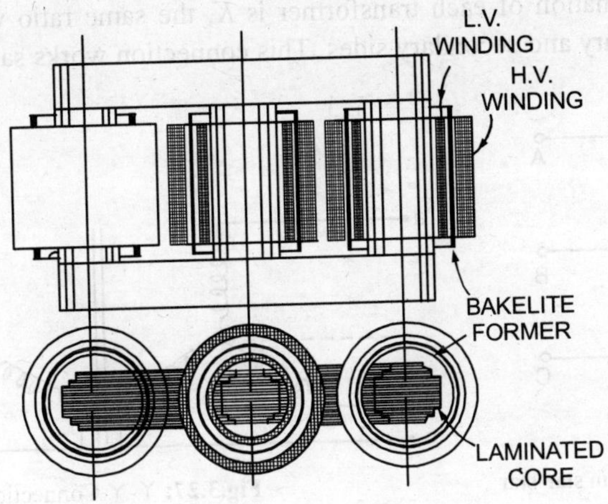

Fig.3.23: Three-phase transformer (core-type)

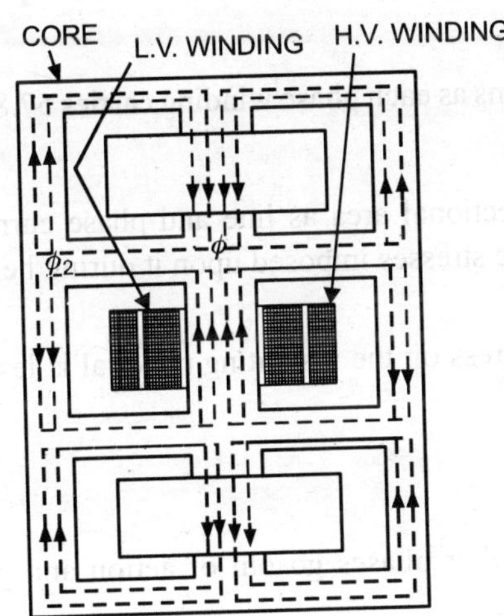

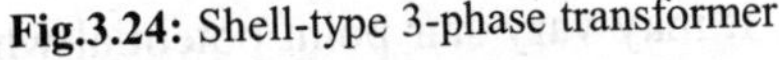

Fig.3.24: Shell-type 3-phase transformer

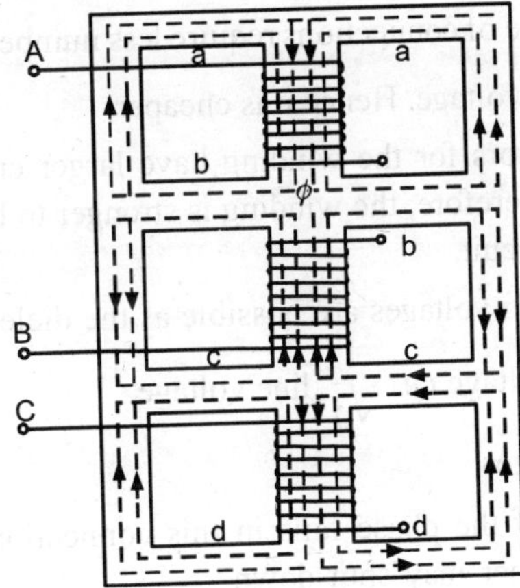

Fig.3.25: flux distribution in the core of a 3-phase shell-type transformer

The low voltage and high voltage windings of the three-phases are wound on the central limbs. These windings are placed vertically in the three portions as shown in fig.3.25.

3.25 CONNECTIONS OF THREE-PHASE TRANSFORMERS

The three primary and three secondary windings of a three-phase transformers may be connected in star od delta, accordingly the transformers are named as:

(*i*) Star-star connected 3-phase transformers

(*ii*) Delta-Delta connected 3-phase transformers

(*iii*) Delta-Star connected 3-phase transformers

(*iv*) Star-Delta connected 3-phase transformers

(*i*) **Star-Star connected 3-phase transformers:** Bunches of three windings of 3-phase transformer connected in Star (Y) on both primary and secondary sides are shown in fig.3.26 and fig.3.27. If the ratio of transformation of each transformer is K, the same ratio will exist between the line voltages on the primary and secondary sides. This connection works satisfactorily only when load is balanced.

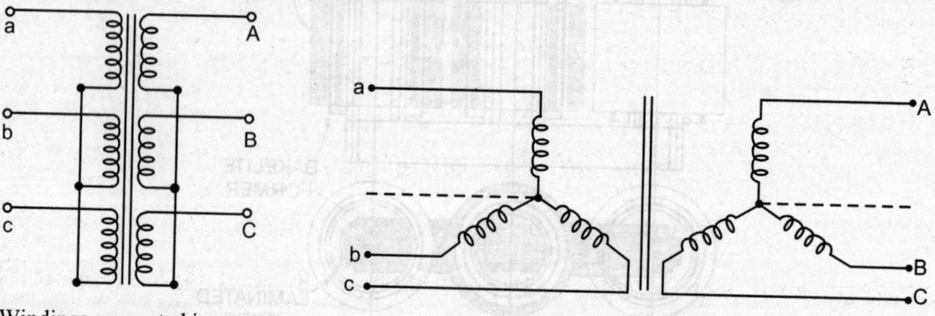

Fig.3.26: Windings connected in star-star

Fig.3.27: Y-Y Connections

The advantages and disadvantages of these connections are

Advantages

(*i*) This type of connections require less number of turns as each phase winding carries 57.8% $\left(\frac{E_L}{\sqrt{3}}\right)$ of the line voltage. Hence it is cheaper.

(*ii*) Conductors for the winding have larger cross- sectional area as line and phase currents are the same, therefore, the winding is stronger to bear the stresses imposed upon it during heavy loads or short circuit.

(*iii*) Very high voltages are possible as the dielectric stress on the insulating material is less due to the lesser voltage *i.e.* $\frac{1}{\sqrt{3}}$ line voltage.

Disadvantages

(*i*) If one of the phase fails in this connection the other phases go out of action and possibly the transformer may shut down.

(*ii*) In case, there is unbalancing of the secondary phases, it may vary star point potential which may assume any value with respect to earth if it is not earthed. The shifting of neutral point can be prevented by connecting the star point of primary to the star point of the alternator.

(*iii*) Even if the star point of the primary in earthed, still if there is a third harmonic present in the alternator voltage, it causes interference to the telephone lines run on the same routes. The triple frequency currents can be eliminated by means of tertiary winding on each transformer.

(*iv*) The regulation of the phases will be poor if the star points of both the primary and secondary are not earthed.

(ii) Delta-Delta connected 3-phase transformers: Bunches of these windings of a 3-phase transformer are connected in delta on both the primary and secondary sides are shown in fig.3.28. and fig.3.29. The line voltages are now simply the voltages across the individual transformer windings and there is no visible neutral.

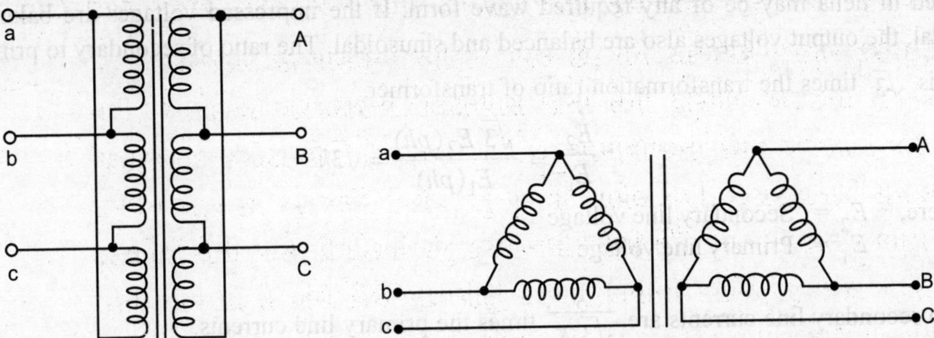

Fig.3.28: Windings connected in delta-delta **Fig.3.29:** D -D Connections

In this case, it is necessary that the transformation ratio of all the three sets of windings be same, otherwise circulating currents round the windings will be set up even at no load. The voltages on the primary and secondary sides of a given line will be more or less in the same phase, only small difference may be found due to the reactance of the winding. These connections have the following advantages:

1. In this case, there is no distortion of the flux, since the third harmonic component of the exciting current (magnetising current) finds a return path in each of the primary windings.
2. No difficulty is experienced from unbalanced loading as in the Y/Y connection. The three phase voltages remain practically equal regardless of the degree of load unbalance.
3. The cross sectional area of the wire is reduced as the phase current is $\frac{1}{\sqrt{3}}$ times the line current, and hence it is comparatively cheaper as the copper used is less.
4. These connections are more suitable for large transformers working on low voltages and high currents.
5. If one of the three windings becomes disabled for any reason, the delta-delta connections continue to operate uninterrupted in open delta or *V-V* connection. Mostly for this advantage the delta connections are preferred.

It has only one disadvantage that no neutral is available. The neutral can be created by star connecting three equal resistances, inductances and capacitances to the three line conductors.

(*iii*) Delta-Star connected 3-phase transformers. In these connections, primary is connected in delta and secondary in star as shown in fig.3.30. and fig.3.31. The primary input voltage then must be equal the primary voltage of the individual transformer winding and the exciting current drawn by the bank

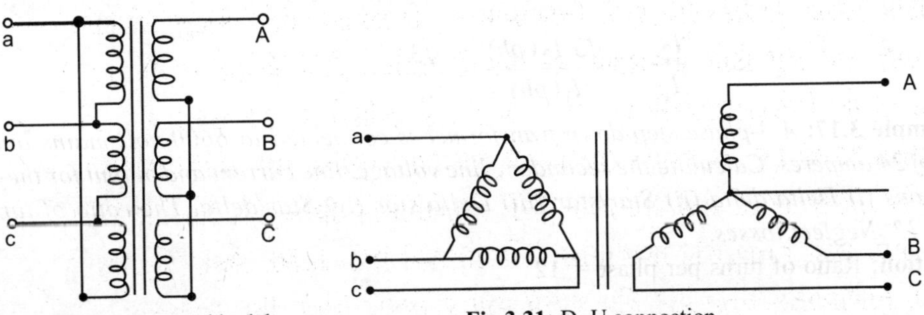

Fig.3.30: Windings connected in delta-star **Fig.3.31:** D- U connection

connected in delta may be of any required wave form. If the impressed voltages are balanced and sinusoidal, the output voltages also are balanced and sinusoidal. The ratio of secondary to primary line voltage is $\sqrt{3}$ times the transformation ratio of transformer.

i.e. $$\frac{E_2}{E_1} = \frac{\sqrt{3}\,E_2(ph)}{E_1(ph)} = \sqrt{3}k$$

Where, E_2 = Secondary line voltage
E_1 = Primary line voltage

Also secondary line currents are $\frac{1}{\sqrt{3}\,K}$ times the primary line currents.

$$\frac{I_2}{I_1} = \frac{I_2(ph)}{\sqrt{3}\,I_1(ph)} \quad \text{or} \quad \frac{I_2}{I_1} = \frac{1}{\sqrt{3}\,K}$$

Out of all the arrangements of connections, this type is one commonly used in distribution system, because it can be used to supply both 3 phase power equipment and single phase lighting circuits.

The advantages of delta-star connections over the delta-delta connections is that for high voltages less insulation is required.

(iv) Star-Delta connected 3-phase transformers. When the primaries are connected in star and secondaries in delta as shown in fig.3.32. and 3.33, exactly the opposite of that which is described above in delta-star would hold. Such a combination then would be suitable for stepping up the voltages at the generating stations.

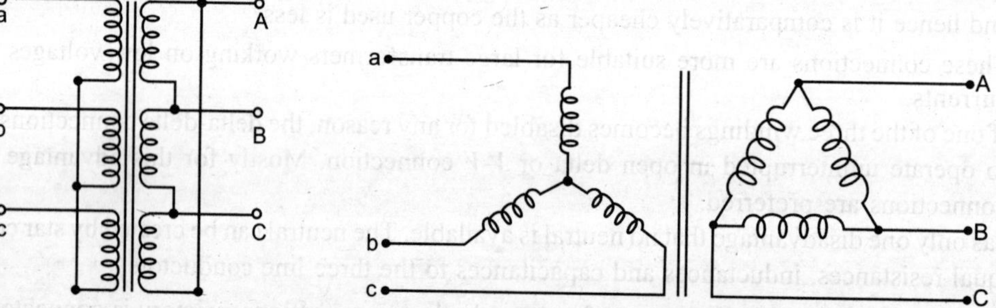

Fig.3.32: Windings connected in star-delta

Fig.3.33: U-D connections

In these connections: $$\frac{E_2}{E_1} = \frac{E_2(ph)}{\sqrt{3}\,E_1(ph)} = \frac{1}{\sqrt{3}}K$$

and $$\frac{I_2}{I_1} = \frac{\sqrt{3}\,I_2(ph)}{I_1(ph)} = \frac{\sqrt{3}}{K}$$

Example 3.17: *A 3-phase step down transformer is connected to 6600 volt mains and takes a current of 24 amperes. Calculate the secondary line voltage, line current and output for the following connections (i) Delta/delta (ii) Star/star (iii) Delta/star (iv) Star/delta. The ratio of turns of per phase is 12. Neglect losses.*

Solution: Ratio of turns per phase = 12

Transformation ratio, $K = \dfrac{1}{12}$

(i) **Delta-Delta connections:**

In Delta connections, line voltage = phase voltage.

Primary line voltage = 6600

Primary phase voltage = 6600 V

$$\text{Secondary phase voltage} = \frac{6600}{12} = 550 \text{ V}$$

Secondary line voltage = **550 V** (*Ans*)

$$\text{Line current} = \sqrt{3} \times \text{phase current}$$

$$\therefore \quad \text{Primary phase current} = \frac{24}{\sqrt{3}}$$

$$\text{Secondary phase current} = \frac{24}{\sqrt{3}} \times 12$$

$$\text{Secondary line current} = \sqrt{3} \times \frac{24}{\sqrt{3}} \times 12 = \mathbf{288\ A}\ (Ans)$$

$$\text{Output} = \sqrt{3}\, V_L\, I_L = \frac{\sqrt{3} \times 550 \times 288}{1000} \text{ kVA} = \mathbf{274.36\ kVA}\ (Ans)$$

(ii) **Star-Star Connections:**

In star connections, line voltage = $\sqrt{3}$ × phase voltage

Also, line current = phase current.

$$\text{Primary phase voltage} = \frac{6600}{\sqrt{3}}$$

$$\text{Secondary phase voltage} = \frac{6600}{\sqrt{3} \times 12}$$

$$\text{Secondary line voltage} = \sqrt{3} \times \frac{6600}{\sqrt{3} \times 12} = \mathbf{550\ V}\ (Ans)$$

Primary line current = 24 A

Secondary line current = 24 ×12 = **288 A** (*Ans*)

$$\text{Output in kVA} = \frac{\sqrt{3} \times 550 \times 288}{1000} = \mathbf{274.36\ kVA}\ (Ans)$$

(iii) **Delta/Star Connections:**

Primary phase voltage = 6600 volts

$$\text{Secondary phase voltage} = \frac{6600}{12} = 550 \text{ V}$$

$$\text{Secondary line voltage} = \sqrt{3} \times 550 = \mathbf{952 \cdot 63\ V}\ (Ans)$$

$$\text{Primary phase current} = \frac{24}{\sqrt{3}}$$

$$\text{Secondary phase current} = \frac{24}{\sqrt{3}} \times 12 = 116 \cdot 27\ A$$

$$\text{Secondary line current} = \mathbf{116.27\ A}\ (Ans)$$

$$\text{Output in kVA} = \frac{\sqrt{3} \times 952 \cdot 63 \times 166 \cdot 27}{1000} = \mathbf{274 \cdot 36\ kVA}\ (Ans)$$

(*iv*) **Star/Delta connections:**

$$\text{Primary phase voltage} = \frac{6600}{\sqrt{3}}$$

$$\text{Secondary phase voltage} = \frac{6600}{\sqrt{3}} \times \frac{1}{12} = \mathbf{317 \cdot 54\ V}\ (Ans)$$

$$\text{Secondary line voltage} = \mathbf{317 \cdot 54\ V}\ (Ans)$$

$$\text{Primary line current} = \text{primary phase current} = \mathbf{24\ A}\ (Ans)$$

$$\text{Secondary phase current} = 24 \times 12 = 288\ A$$

$$\text{Secondary line current} = \sqrt{3} \times 288 = \mathbf{498 \cdot 83\ A}\ (Ans)$$

$$\text{Output in kVA} = \frac{\sqrt{3} \times 317 \cdot 54 \times 498 \cdot 83}{1000} = \mathbf{274.36\ kVA}\ (Ans)$$

3.26 POWER TRANSFORMER AND ITS AUXILIARIES

The transformers used in the power system for transfer of electric power or energy from one circuit to the other are called power transformers. The rating of a transformer includes voltage, frequency and kVA. The kVA rating is the kVA output that a transformer can deliver at the rated voltage and frequency under general service conditions without exceeding the standard limit of temperature rise (usually 45° to 60°C) The power transformer has the following important parts :

1. **Magnetic circuit:** The magnetic circuit comprises of transformer core. The transformer core may be core type or shell type in construction. The power transformers used in the power system are mostly three phase transformers. In a core type 3-phase transformer core has three limbs of equal area of cross-section.
2. **Electrical circuit:** In three phase transformers there are three primary (*H.V.*) windings and three secondary (*L.V.*) windings. Whole of the *L.V.* winding is wound over one limb next to the core, then whole of the *H.V.* winding is wound over the *L.V.* winding. In between the *L.V.* winding and *H.V.* winding and between core and *L.V.* winding insulation is provided.
3. **Transformer oil:** Transformer oil is a mineral oil obtained by fractional distillation of crude petroleum. The oil is used only in the oil cooled transformers. The oil not only carries the heat produced due to losses in the transformer, by convection from the windings and core to the transformer tank, but also has even more important function of insulation.

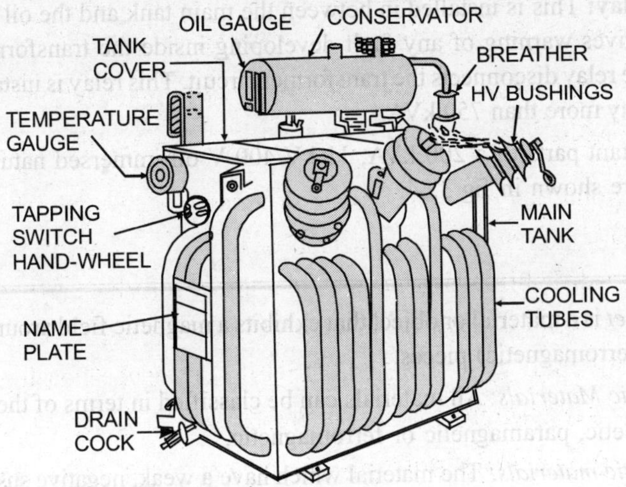

Fig.3.34: Distribution transformer

When transformer delivers power, heat is produced due to the iron and copper losses in the transformer. This heat must be dissipated effectively otherwise the temperature of the winding will increase. The raise in temperature further increases the losses. Thus, the efficiency of the transformer will decrease. As there is no rotating part in the transformer, it is difficult to cool down the transformer as compared to rotating machines. Various methods are adopted to cool down the transformers of different rating. The common methods are air natural cooling, oil immersed natural cooling, oil immersed forced oil circulation natural cooling, oil immersed forced oil circulation with air blast cooling, oil immersed forced oil circulation with water cooling etc.

Generally, for cooling of distribution transformers, oil immersed natural cooling method is adopted. Cooling tubes or small cooling radiators are used with the main tanks, as shown in fig.3.34, to increase the surface area for the dissipation of heat.

4. **Tank Cover:** A number of parts are arranged on the tank cover of which most important are:

   (i) **Bushing:** The internal winding of the transformer are connected to the lines through copper rods or bars which are insulated from the tank cover, these are known as bushings. Upto 33 kV ordinary porcelain bushing can be used. Above this voltage oil filled bushings or condenser bushing are employed.

   (ii) **Oil conservator tank:** Oil conservator is also known as an oil expansion chamber. It is a small cylindrical air tight and oil tight vessel. The oil conservator is connected with a tube to the main transformer tank at the tank cover. This tank is partially filled with oil. The expansion and contraction of oil, changes the oil level in the conservator.

   (iii) **Breather:** The transformer oil should not be allowed to come in contact with atmospheric air, since a small amount of moisture causes a great decrease in the dielectric strength of transformer oil. All the tank fittings are made air tight. When oil level in the oil conservator changes, air moves in and out of the conservator. This action is known as breathing.

   The breathed air is made to pass through an apparatus called breather to abstract moisture. Breather, contains *Silica-gel* or some other drying agent such as *calcium chloride*. This ensures that only dry air enters the transformer tank.

(iv) **Buchholz Relay:** This is installed in between the main tank and the oil conservator. It is a gas relay which gives warning of any fault developing inside the transformer, and if the fault is dangerous, the relay disconnects the transformer circuit. This relay is installed in the transformer having capacity more than 750 kVA.

All the important parts of a 200 kVA, 11 kV/400 V oil immersed natural cooled distribution transformer are shown in fig 3.34.

SUMMARY

1. *Magnet:* A *magnet* is a material or object that exhibits a magnetic field around it and has the ability to attract iron (ferromagnetic) pieces.
2. *Types of Magnetic Materials*: All materials can be classified in terms of their magnetic behaviour such as diamagnetic, paramagnetic or ferromagnetic.
   (i) *Diamagnetic materials:* The material which have a weak, negative susceptibility to magnetic fields are called *diamagnetic materials*.
   (ii) *Paramagnetic materials:* The material which have a small, positive susceptibility to magnetic fields are called *paramagnetic materials*.
   (iii) *Ferromagnetic materials:* The material which have a large, positive susceptibility to an external magnetic field are called *ferromagnetic materials*.
3. *Important terms*: The important terms related to magnetism are:
   (i) *Magnetic field*: The region around a magnet where its poles exhibit a force is called *magnetic field*.
   (ii) *Magnetic flux*: The magnitude of magnetic lines of force set-up in a magnetic circuit as called *magnetic flux*.
   (iii) *Mangetic flux density*: The flux per unit area at right angles to the flux at a point is called *magnetic flux density* at that point.
   (iv) *Permeability*: The ability of a material to conduct magnetic lines of force through it is called the *permeability* of that material.
   (v) *m.m.f.:* The work done in moving a unit magnetic pole once round the magnetic circuit is called m.m.f., m.m.f. = NI AT.
   (vi) *Reluctance (S)*: The opposition offered to the magnetic flux by a magnetic circuit is called its reluctance. $S = \dfrac{l}{a\,\mu_0\,\mu_r}$ AT/Wb
   (vii) *Magnetic flux (f)*: The amount of magnetic lines of force set-up in a magnetic circuit is called magnetic flux.
   (viii) *Relation between m.m.f., reluctance and flux.*

$$f = \frac{\text{m.m.f.}}{\text{reluctance}} \quad \text{or } f = \frac{NI}{l}\, a\,\mu_0\,\mu_r.$$

   (ix) *Permeance*: It is reciprocal to reluctance.
4. *B—H Curve*; The graph between B and H of a material is called $B - H$ curve of that material. It shows the properties of the material.

5. *Transformer*: A static device which transfers a.c electrical power from one circuit to the other at same frequency but usually at different voltage level is called a transformer. It works on the basic principle of electromagnetic induction (mutually induced e.m.f).

6. *Ideal transformer*: Losses are neglected; $E_1 I_1 \cos f = E_2 I_2 \cos f$

7. *Transformer on d.c.*: It is never applied on d.c. otherwise it will burn since there is no counter of self induced e.m.f. and it draws heavy current on d.c.

8. *E.M.F. equation:* $E_2 = 4.44\, N_2 f \phi_m = 4.44\, N_2 f B_m\, A_i$

9. *Transformer winding resistance:* $R_1; R_2; R_1' = R_1 \times K^2; R_{es} = R_2 + R_1'$;
$R_2' = R_2/K^2; R_{ep} = R_1 + R_2'$

10. *Leakage flux*: A part of the flux produced by a winding which is not linking with the other is called leakage flux. It develops leakage reactance X_1 and X_2, $X_1' = X_1 K^2; X_{es} = X_2 + X_1'; X_2' = X_2/K^2$; $X_{ep} = X_1 + X_2'$

11. *Voltage regulation*: At constant supply voltage, the change in secondary terminal voltage from no-load to full load is called voltage regulation. It is generally taken as percentage of no-load voltage.

$$\% \text{ Reg} = \frac{E_2 - V_2}{E_2} \times 100$$

12. *Losses in a transformer*: 1. Iron loss (P_i), 2. Cu. loss (P_c at full-load)

13. *Efficiency of a transformer*: $\eta = \dfrac{V_2 I_2 \cos\phi_2}{V_2 I_2 \cos\phi_2 + P_i + P_c}$; $\eta_x = \dfrac{x\, V_2\, I_2 \cos\phi_2}{x\, V_2\, I_2 \cos\phi_2 + P_i + x^2\, P_c}$

14. *Condition for max. efficiency*: Cu. loss = Iron loss; $I_2 = \sqrt{P_i / R_{es}}$

15. *Auto-transformer*: A transformer, having only one winding, a part of which acts as primary and the other as secondary is called an auto transformer. Saving in copper = K × Wt. of Cu. required for two winding transformer.

16. *Three-phase transformer*: A static device that transfers 3-phase power from one circuit to the other keeping frequency to be the same but (usually) changing the voltage level is called a *3-phase transformer*.

17. *Merits of 3-phase transformers*: They require smaller quantity of iron and copper, have smaller size, occupy less space, light in weight, better efficiency and regulation, smaller cost etc.

18. *Construction of 3-phase transformers*: They are either core type or shell type.

19. *Connections of three-phase transformers*: According to the interconnection of primaries and secondaries, these transformers may be classified as (*i*) Star-star connected (*ii*) Delta-delta connected (*iii*) Delta-star connected (*iv*) star-delta connected.

20. *Distribution/power transformer and its auxiliaries*: A transformer employed in the power system to step-up or step-down the voltage as per the need is called *power transformer*.

The main auxiliaries of these transformers are (*i*) conservator tank (*ii*) cooling tubes (*iii*) breather, (*iv*) Buchholz relay (*v*) temperature gauge etc.

NUMERICAL FOR PRACTICE

1. A sinusoidal flux 0.2 Wb (max.) links with 55 turns of a transformer secondary coil. Calculate the r.m.s. value to the induced e.m.f. in the secondary. The supply frequency is 50 Hz.

(***Ans.*** 244.2 V)

2. The design requirements of a 6600/400 V, 50 Hz, single phase, core type transformer are: e.m.f. per turn 15 V; maximum flux density 1.5 Tesla (i.e.Wb/m^2). Find a suitable number of primary and secondary turns and the net cross sectional area of core. (***Ans.*** *440; 26·67; 450 cm^2*)

3. A single phase transformer has 400 primary and 1000 secondary turns. The net cross sectional area of the core is 60 cm^2. If the primary winding is connected to a 50Hz supply at 500 V, calculate the value of maximum flux density in the core and the e.m.f. induced in secondary winding. Draw the vector diagram representing the condition. (***Ans.*** *0·938 Tesla, 1250 V*)

4. A 33 kVA, 2200/220 V, 50 Hz single phase transformer has the following parameters: Primary winding (hv side); resistance $r_1 = 2{\cdot}4\ \Omega$, leakage reactance $x_1 = 6{\cdot}0\ \Omega$. Secondary winding (*lv* side): resistance $r_2 = 0{\cdot}03\ \Omega$, leakage reactance $x_2 = 0{\cdot}07\ \Omega$. Find the primary resistance and leakage reactance referred to secondary.

(***Ans.*** *0·024* W *and* 0·06 W)

5. A single-phase transformer with a ratio of 4: 1 has primary resistance and reactance of 0·5 ohm and 1·5 ohm and the corresponding values for the secondary are 0·034 ohm and 0·1 ohm respectively. Determine the percentage regulation when delivering 120 A at 600 V at p.f. (*i*) 0·707 lagging (*ii*) 0·8 leading. (***Ans.*** *3·53%; – 3·26%*)

6. Calculate the value of voltage regulation at 0·8 p.f. lagging for a transformer with resistance drop 2% and reactance drop 4% of the voltage. (***Ans.*** *4·0%*)

7. A 230/460 V, single-phase transformer has a primary resistance of 0·2 ohm and reactance 0·5 ohm. The corresponding values for the secondary are 0·75 Ω and 1·8W Ω respectively. Find the secondary terminal voltage when supplying 10 A at 0·8 p.f. lagging. (***Ans.*** *424·8 V*)

8. The primary and secondary windings of a 500 kVA transformer have resistance of 0·42 ohm and 0·0011 ohm respectively. The primary and secondary voltages are 6600 V and 400 V respectively. The iron loss is 2·9 kW. Calculate the efficiency at half full load at a power factor of 0·8 lagging. (***Ans.*** *98·07%*)

9. A 50 kVA transformer has an efficiency of 98% at full-load at 0.8 p.f and an efficiency of 96.9% at $\frac{1}{4}$ full load, 0.8 p.f. Determine the iron loss and full load copper loss.

(***Ans.*** *287 W; 529 W*)

10. A 440/110 V transformer has an effective primary resistance of 0.3 ohm and a secondary resistance of 0.02 ohm. If iron loss on normal input voltage is 150 W, calculate the secondary current at which maximum efficiency will occur. What is the value of this maximum efficiency for unity power factor load ? (***Ans.*** *62·22 A; 94·8 %*)

11. In a 25 kVA, 1100/400 V, single phase transformer, the iron and copper loss at full load are 350 and 400 watt respectively. Calculate the efficiency on unity power factor at half load. Determine the load on maximum efficiency. (***Ans.*** *96·52%; 23·85 kW*)

12. A 2000/200 V transformer has primary resistance and reactance of 2 ohm and 4 ohm respectively. The corresponding secondary values are 0·025 ohm and 0·04 ohm. Determine: (*i*) Equivalent resistance and reactance of primary referred to secondary; (*ii*) Total resistance and reactance referred to secondary; (*iii*) Equivalent resistance and reactance of secondary referred to primary; (*iv*) Total resistance and reactance referred to primary.

(***Ans.*** *0·02 Ω; 0·04 Ω; 0·045 Ω; 0·08 Ω; 2·5 Ω; 4 Ω; 4·5 Ω; 8 Ω*)

MULTI-CHOICE QUESTIONS

1. *during unsaturated portion of a B-H curve*

(*a*) B < H (*b*) B > H (*c*) B ∝ H (*d*) B = O

2. *Transformer core is laminated*

(*a*) because it is difficult to fabricate solid core

(*b*) because laminated core provides high flux density

(*c*) to reduce eddy current losses.

(*d*) to avoid hysteresis losses.

3. *A transformer with output of 250 kVA at 3000 V, has 600 turns on its primary and 60 turns on secondary winding. What will be the transformation ratio of the transformer ?*

(*a*) 10 (*b*) 0·1 (*c*) 100 (*d*) 0·01

4. *If R_1 is the resistance of primary winding of the transformer and K is transformer ratio (N_2/N_1) then the equivalent primary resistance referred to secondary will be*

(*a*) KR_1^2 (*b*) KR_1 (*c*) K^2R_1 (*d*) R_1/K^2

5. *The condition for maximum efficiency of the transformer is that*

(*a*) copper losses are half of the iron losses

(*b*) copper losses are square of the iron losses

(*c*) copper losses are equal to the iron losses

(*d*) copper losses are zero

6. The induced e.m.f. in the transformer secondary will depend on

(*a*) frequency of the supply only.

(*b*) Number of turns in secondary only.

(*c*) Frequency and flux in core.

(*d*) Frequency, number of secondary turns and flux in the core.

7. A transformer is never connected in the d.c. line because

(*a*) there is no need to step up or step down the d.c. voltage

(*b*) Faraday's law is not valid as the rate of changed of flux is zero.

(*c*) Losses in the d.c. circuit are high.

(*d*) It is not economical.

8. The eddy current loss in the transformer occurs in the
(*a*) primary winding (*b*) core (*c*) secondary winding (*d*) none of the above.
9. Which of the following electrical machines has the highest efficiency?
(*a*) d.c. generator (*b*) a.c. generator (*c*) transformer (*d*) induction motor.
10. The condition for maximum efficiency of the transformer is that
(*a*) copper losses are half of the iron losses (*b*) copper losses are square of the iron losses
(*c*) copper losses are equal to the iron losses (*d*) copper losses are zero
11. If the iron losses and full load copper losses are given then the load at which the efficiency of a transformer is maximum, is given by
(*a*) full load × $\frac{\text{iron loss}}{\text{f.l.cu loss}}$ (*b*) full load × $\sqrt{\frac{\text{iron loss}}{\text{f.l.cu loss}}}$
(*c*) full load × $\left(\frac{\text{iron loss}}{\text{f.i. cu. loss}}\right)$ (*d*) full load × $\sqrt{\frac{\text{f.I.cu. loss}}{\text{iron loss}}}$
12. The transformation ratio of a 3-phase transformer is given by the relation.
(*a*) $\frac{E_{2(L)}}{E_{1(L)}}$ (*b*) $\frac{E_{2(ph)}}{E_{1(ph)}}$ (*c*) $\frac{E_{2(L)}}{E_{1(ph)}}$ (*d*) $\frac{E_{2(ph)}}{E_{1(L)}}$
13. The cost of single 3-phase transformer in comparison to a bank of three 1-phase transformers of equal rating is
(*a*) 15% less (*b*) 15% more (*c*) 50% less (*d*) 50% more
14. Usually *HV* winding is placed
(*a*) next to the core (*b*) over the low voltage winding
(*c*) both *a* and *b* (*d*) none of these

SHORT ANSWER QUESTIONS

1. What is a magnet ?
2. What do you understand by magnetic field ?
3. Define magnetic flux density.
4. Explain the term MMF.
5. Define relative permeability.
6. Define reluctance in a magnetic circuit and give its formula.
7. Explain B-H characteristics of a magnetic materiel.
8. What is a transformer ?
9. What are step-up and step-down transformers ?
10. What are the applications of step-up and step-down transformers ?
11. Which are the two winding present in a transformer ?

12. Explain the working principle of a transformer.
13. What are the functions of a transformer ?
14. Define voltage transformation ratio.
15. From construction point of view, name different types of transformers ?
16. Why the core of a transformer is laminated ?
17. Why is a transformer also called the static device ?
18. Why do we use iron - core in a Transformer ?
19. Explain with reasons what happens when a power transformer is connected to dc supply of the same voltage rating ?
20. Why is the efficiency of a transformer maximum among electrical equipments ? Explain.
21. What is the difference between an ideal and practical transformers ?
22. How does leakages flux occur in a transformer ?
23. Write a short note on a star-Delta transformer.

TEST QUESTIONS

1. Why the B-H curve of iron is not a straight line?
2. Define the terms: permeability, reluctance and permeance.
3. How is B-H curve of ferromagnetic material different from that of non-magnetic material ? Name all the salient regions of B-H curve of magnetic material.
4. What is a transformer ? What is its necessity in the power system ?
5. Explain the working principle of a transformer.
6. State why the core of a transformer is laminated ?
7. State why silicon steel in selected for the core of a transformer ?
8. Give the constructional details of a core-type transformer.
9. In a transformer explain how power is transferred from one winding to the other.
10. Show that $(E_1/E_2) = (I_2/I_1) = (T_1/T_2)$ in a transformer.
11. What happens when d.c. voltages is applied to the primary of a transformer ?
12. Derive an expression for the e.m.f. induced in a transformer winding. Show that the e.m.f. induced per turn in primary is equal to the e.m.f. per turn in secondary.
13. Explain what is meant by regulation of a transformer.
14. What are the various losses in a transformer ? Where do they occur and how do they vary with load ?
15. Define efficiency of a transformer and find the condition for obtaining maximum efficiency.
16. In what ways does an auto transformer differ from a conventional two winding transformer? What are its application?

ANSWER TO MULTI-CHOICE QUESTIONS

Multi-choice questions:

1. (*c*)	2. (*c*)	3. (*b*)	4. (*c*)	5. (*c*)
6. (*d*)	7. (*b*)	8. (*b*)	9. (*c*)	10. (*c*)
11. (*b*)	12. (*b*)	13. (*a*)	14. (*b*)	

✪✪✪✪✪

ELECTRICAL MACHINES

LEARNING OBJECTIVES

- How a 3-phase induction motor operates and how its direction of rotation can be reversed?
- What are the various losses in an induction motor and how these are affected by slip of an induction motor?
- What is the significance of torque-slip characteristics of an induction motor?
- What are various starters used to start 3-phase induction motor?
- How to vary the speed of 3-phase induction motor?
- How to make 1-phase induction motors self starting?
- How a shaded pole motor works?
- Why we use universal motor in mixer-grinder?
- What are separately excited dc motors and how their speed is controlled?
- What are the various parts of an alternator?

INTRODUCTION

Induction machines or induction motor are the backbone of any industry. Induction motors may be single-phase or three phase. The single phase induction motors are usually built in small sizes (upto 3 H.P) and are employed with light loads to handle domestic appliances. Three phase induction motors are the most commonly used a.c. motors in the industry because they have simple and rugged construction, low cost, high efficiency, reasonably good power factor, self starting torque and low maintenance. Almost more than 90% of the mechanical power used in industry is provided by three phase induction motors.

In this chapter, we shall deal with all the important aspects of a three phase induction motors as well as the functioning of 1-phase induction motors, dc motors and synchronous generators.

4.1 CONSTRUCTIONAL FEATURES OF A 3-PHASE INDUCTION MOTOR

A 3-phase induction motor consists of two main parts namely stator and rotor.

1. **Stator:** It is the stationary part of the motor. It has three main parts, namely. (*i*) Outer frame, (*ii*) Stator core and (*iii*) Stator winding.

   (*i*) ***Outer frame:*** It is the outer body of the motor. Its function is to support the stator core and to protect the inner parts of the machine. For small machines the fame is casted but for large machines it is fabricated.

To place the motor on the foundation, feet are provided in the outer frame as shown in fig.4.1.

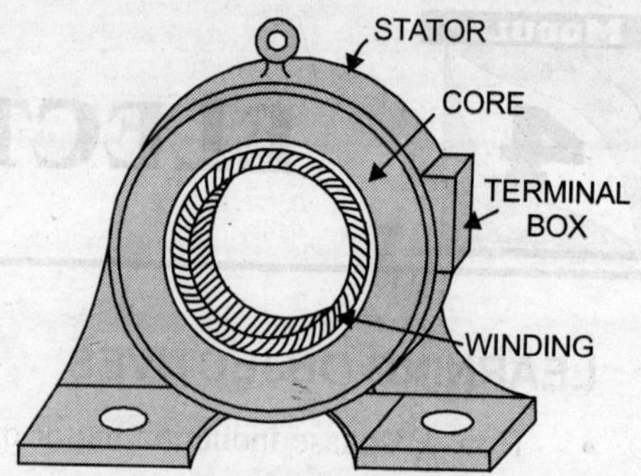

Fig.4.1: Stator

(ii) Stator core: The stator core is to carry the alternating magnetic field which produces hysteresis and eddy current losses, therefore, core is built up of high grade silicon steel stamping. The stampings are assembled under hydraulic pressure and are keyed to the frame. Each stamping is insulated from the other with a thin varnish layer. The thickness to the stamping usually varies from 0.3 to 0.5 mm. Slots are punched on the inner periphery of the stampings, as shown in fig.4.2, to accommodate stator winding.

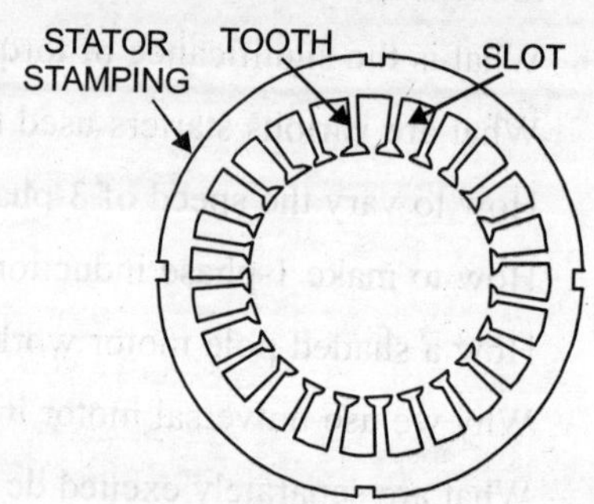

Fig.4.2: Stator stamping

***(iii) Stator winding*:** The stator core carries a three phase winding which is usually supplied from a three phase supply system. The six terminals of the winding (two of each phase) are connected in the terminal box of the machine. The stator of the motor is wound for definite number of poles, the exact number being determined by the requirement of speed. It will be seen that greater the number of poles, the lower is the speed and vice-versa, since $N_s \propto \frac{1}{P}\left(\because N_S = \frac{120f}{P}\right)$. The three- phase winding may be connected in star or delta externally through a starter.

2. **Rotor:** It is the rotating part of the motor. There are two types of rotors, which are employed in 3 phase induction motors.

(*i*) Squirrel cage rotor (*ii*) Phase wound rotor.

(*i*) Squirrel cage rotor: The motors employing this type of rotor are known as *Squirrel cage induction motors*. Most of the induction motors are of this type because of simple and rugged construction of rotor. A squirrel cage rotor consists of a laminated cylindrical core having semi-closed circular slots at the outer periphery. Copper or aluminium bar conductors are placed in these slots and short circuited at each end by copper or aluminium rings, called short circuiting rings, as shown in fig.4.3. Thus, the rotor winding is permanently short circuited and it is not possible to add any external resistance in the rotor circuit.

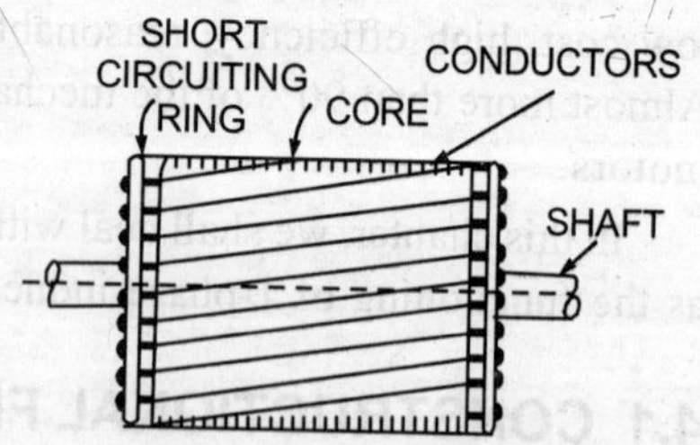

Fig.4.3: Squirrel cage rotor

The rotor slots are usually not parallel to the shaft but are skewed. Skewing of rotor has the following advantages:

(*a*) It reduces humming thus ensuring quiet running of a motor, (*b*) It results in a smoother torque curves for different positions of the rotor,

(*c*) It reduces the magnetic locking of the stator and rotor,

(*d*) It increases the rotor resistance due to the increased length of the rotor bar conductors.

(*ii*) ***Phase wound rotor:*** Phase wound rotor is also called slip ring rotor and the motors employing this type of rotor are known as *phase wound* or *slipring induction motors*. Slip ring rotor consists of a laminated cylindrical core having semi-closed slots at the outer periphery and carries a 3-phase insulated winding. The rotor is wound for the same number of poles as that of stator. The three finish terminals are connected together forming star point and the three start terminals are connected to three copper slipings fixed on the shaft (see fig.4.4).

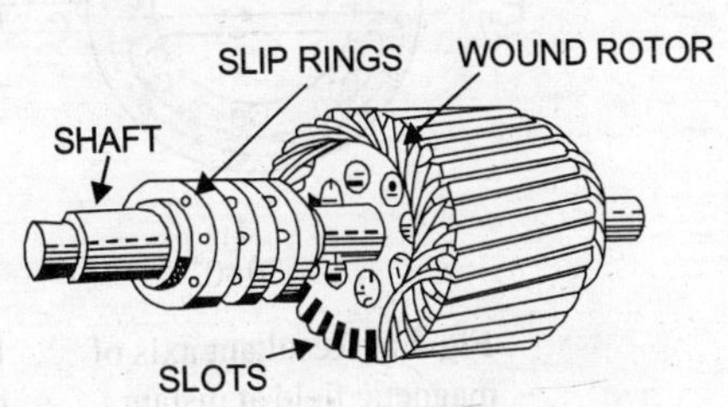

Fig.4.4: Phase wound rotor

In this case, depending upon the requirement any external resistance can be added in the rotor circuit. In this case also the rotor is skewed.

A mild steel shaft is passed through the centre of the rotor and is fixed to it with key. The purpose of shaft is to transfer mechanical power.

4.2 PRODUCTION OF REVOLVING FIELD

Consider a stator on which three different windings represented by three concentric coils a_1a_2, b_1b_2 and c_1c_2 respectively are placed 120° electrically apart.

Let a 3-phase supply, as shown in fig.4.5, is applied to the stator. Three phase currents will flow through the three coils and produce their own magnetic fields. The positive half cycle of the alternating current is considered as inward flow of current in the start terminals and negative half cycle is considered as outward flow of current in the start terminals. The direction of flow of current is opposite in the finish terminals of the same coil.

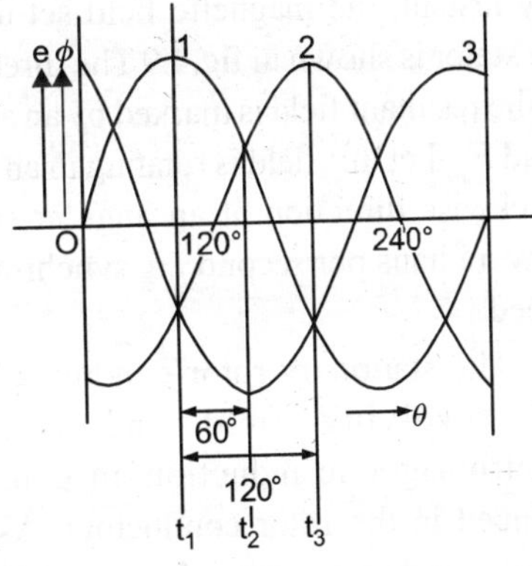

Fig.4.5: Wave diagram of magnetic flux produced in the stator core

Let at any instant t_1, current in coilside a_1 be inward and in b_1 and c_1 outward. Whereas, the current in the other sides of the same coils is opposite *i.e.* in coil side a_2 is outward and b_2 and c_2 is inward. The resultant field and its direction (F_m) is marked in fig.4.6.

At instant t_2 when θ is 60°, current in coil sides a_1 and b_1 is inward and in c_1 is outward. Whereas, the current in the opposite sides is opposite. The resultant field and its direction is shown in fig.4.7, which is rotated through an angle θ = 60° from its previous position.

At instant t_3 when is θ is 120° , current in coilside b_1 is inward and in c_1 and a_1 is outward. The resultant field and its direction is shown in fig.4.8. Which is rotated through an angle θ = 120° electrical from its first position.

Thus, in one cycle, the resultant field completes one revolution. Hence, we conclude that when 3-phase supply is given to a 3-phase wound stator, a resultant field is produced which revolves at a constant speed, called synchronous speed ($N_s = 120f/P$).

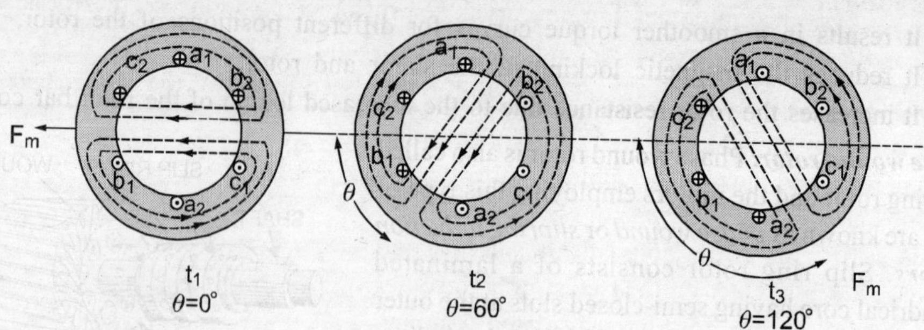

Fig.4.6: Resultant axis of magnetic field at instant t_1

Fig.4.7: Resultant axis of magnetic field at instant t_2

Fig.4.8: Resultant axis of magnetic field at instant t_3

In this case, we have seen that when supply from phase 1,2 and 3 is given to coil a_1a_2, b_1b_2 and c_1c_2, respectively, an anticlockwise rotating field is produced. If the supply to coil a_1a_2, b_1b_2 and c_1c_2 is given from phase 1, 3 and 2 respectively, the direction of rotating field is reversed. Thus, to reverse the direction of rotation of rotating field the connections of any two supply terminals are inter changed.

4.3. PRINCIPLE OF OPERATION

When 3-phase supply is given to the stator of a 3-phase wound induction motor, a revolving field is set up in the stator. At any instant, the magnetic field set-up by the stator is shown in fig.4.9 The direction of the resultant field is marked by an arrow head F_m. Let this field is rotating in an anti-clockwise direction at an angular speed of w_s radians per second *i.e.* synchronous speed.

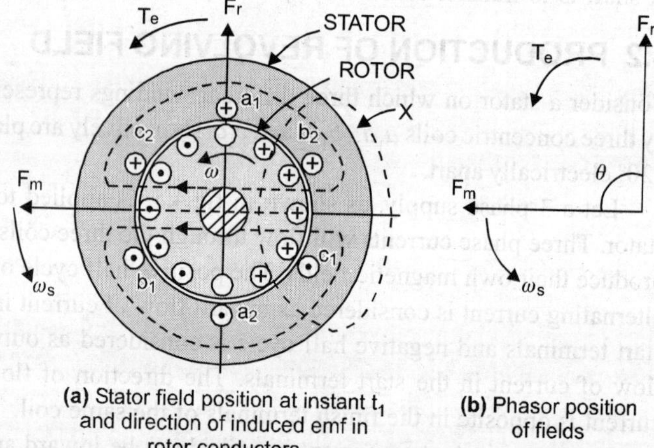

(a) Stator field position at instant t_1 and direction of induced emf in rotor conductors.

(b) Phasor position of fields

Fig.4.9: Working principle of 3-phase Induction motor

The stationary rotor conductors cut the revolving field and due to electromagnetic induction an e.m.f. is induced in the rotor conductors. As the rotor conductors are short circuited, current flows through them in the direction as marked in the figure. Rotor current carrying conductors set up a resultant field F_r. This field tries to come in line with the stator main field F_m. Due to this an electromagnetic torque T_e is developed in the anticlockwise direction. Thus, rotor starts rotating in same direction in which stator field is revolving.

Alternately: Reproducing section X of fig.4.9 (*a*) as shown in fig.4.10, when the revolving stator field see fig.4.10 (*a*) cuts the stationary rotor conductors, an e.m.f. is induced in the conductors by induction. As rotor conductors are short circuited, current flows through them, as marked in fig.4.10, (*b*) which sets up field around them. A resultant field is set up, as shown in fig.4.10 (*c*) which exerts force on the rotor conductors. Thus, the rotor starts rotating in the same direction in which stator field is revolving.

The rotor picks up speed and tries to attain the synchronous speed but fails to do so. It is because if the rotor attains the synchronous speed then the relative speed between revolving stator field and rotor will be zero, no e.m.f. will be induced in rotor conductors. No e.m.f. means no current, no rotor field F_r and hence no torque is produced. Thus, *an induction motors never runs at synchronous speed.* It always seems at a speed less than synchronous speed.

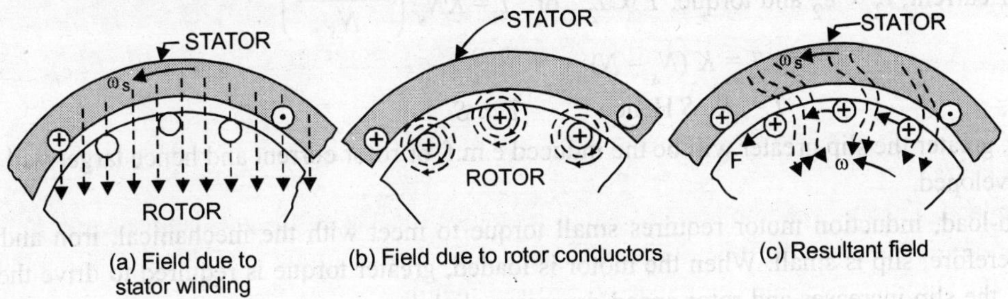

Fig.4.10: Field developed in the motor

Since, *the principle of operation of this motor depends upon electromagnetic induction, hence the name* ***induction motor***.

4.4 REVERSAL OF DIRECTION OF ROTATION OF 3-PHASE INDUCTION MOTORS

In Art. 2 it has been seen that a revolving field is set up in the stator of a 3-phase induction motor when 3-phase supply is given to its winding and the direction of rotation depends upon the supply sequence.

In Art. 3, it has been see that rotor of a three phase induction motor rotates in the same direction as that of the revolving field.

The direction of rotation of the revolving field or that of the rotor can be reversed if the sequence of supply is reversed. The supply sequence can be reversed by interchanging the connections of any two supply leads at the stator terminals.

Hence, *the direction of rotation of a 3-phase induction motor can be reversed by interchanging the connections of any two supply leads at the stator terminals.*

4.5 SLIP

The rotor of an induction motor always rotates at a speed less than synchronous speed. The difference between the flux speed (N_s) and the rotor speed (N) is called slip. It is usually expressed as a percentage of synchronous speed (N_s) and is represented by symbol S.

Mathematically,

% slip, $$\% S = \frac{N_S - N}{N_S} \times 100 \quad \text{or Fractional slip,} \quad S = \frac{N_S - N}{N_S}$$

Rotor speed, $$N = N_s (l - S)$$

The difference between synchronous speed and rotor speed is called *slip speed i.e.,*

$$\text{Slip speed} = N_s - N$$

The value of slip at full load varies from about 6% small motors to about 2% for large motors.

Importance of slip: Slip plays an important role in the operation of an induction motor. We have already seen that the difference between the rotor speed and synchronous speed of flux determine the rate at which the flux is cut by rotor conductors and hence the magnitude of induced e.m.f. *i.e.* $e_2 \propto N_s - N$.

Rotor current, $i_2 \propto e_2$ and torque, $T \propto i_2$ or $T = KN_s\left(\dfrac{N_S - N}{N_S}\right)$

$\therefore$ $T = K\,(N_s - N)$

or $T = K_1\,S$ Hence $T \propto S$

Thus, greater the slip greater will be the induced e.m.f. or rotor current and hence larger will be the torque developed.

At no-load, induction motor requires small torque to meet with the mechanical, iron and other losses, therefore, slip is small. When the motor is loaded, greater torque is required to drive the load, therefore, the slip increases and rotor speed decreases slightly.

Thus, it is seen that ship in an induction motor adjusts itself to such a value so as to meet the required driving torque under normal operation.

4.6 FREQUENCY OF ROTOR CURRENTS

The frequency of rotor currents depends upon the relative speed between rotor and stator field. When the rotor is stationary, the frequency of rotor currents is the same as that of the supply frequency. But once the rotor starts rotating, the frequency of rotor currents depends upon slip speed ($N_s - N$). Let at any speed N, the frequency of rotor currents be f_r. Then

$$f_r = \frac{(N_S - N)P}{120} = \frac{(N_S - N)}{N_S}\cdot\frac{N_S P}{120} = S \times f$$

Example 4.1.: *A 3-phase induction motor is wound for 6 poles and is supplied from a 50 Hz system. Calculate: (i) Synchronous speed. (ii) Actual speed of the motor when running at 3% slip (iii) Frequency of e.m.f. induced in rotor.*

Solution:

Synchronous speed, $N_s = \dfrac{120f}{P} = \dfrac{120 \times 50}{6} = \mathbf{1000\ r.p.m.}$ (*Ans.*)

Actual speed of motor, $N = N_s\,(1 - S)$ where, $S = 0{\cdot}03$

$N = 1000\,(1 - 0{\cdot}03) = \mathbf{970\ r.p.m.}$ (*Ans*)

Frequency of rotor e.m.f., $f_r = Sf = 0.03 \times 50 = \mathbf{1.5\ Hz}$ (*Ans*)

Example 4.2.: *A 5 HP, 230 V, 3-phase, 50 Hz, 4 pole squirrel cage induction motor operates at full load slip of 4% when rated voltage and rated frequency are applied. Determine (i) full loud speed. (ii) full load torque in Newton-metre. (iii) frequency of rotor current under this condition and (iv) speed of rotation of the stator m.m.f.*

Solution: Synchronous speed, $N_s = \dfrac{120f}{P} = \dfrac{120 \times 50}{4} = 1500$ r.p.m.

Full load speed, $N = N_s\,(1 - S) = 1500\,(1 - 0.04) = \mathbf{1440\ r.p.m.}$ (*Ans*)

Output $= \omega\,T = 5 \times 735.5$ W (where $\omega = 2\pi\,N/\,60$)

Or $T = \dfrac{5 \times 735{\cdot}5 \times 60}{2\pi \times 1440} = \mathbf{24.39\ Nm}$ (*Ans*)

Rotor current frequency, $f_r = Sf = 0.04 \times 50 =$ **2 Hz** *(Ans)*

Speed of rotation of stator m.m.f. = N_s = **1500 r.p.m.** *(Ans)*

Example 4.3.: *Power to an induction motor is supplied by an 8 pole, 3-phase, 750 r.p.m. alternator. The full load speed of the motor is 1440 r.p.m. Find the percentage slip and number of poles in the motor.*

Solution: Speed of the alternator, $N_{sa} = 750$ r.p.m.

No. of poles of the alternator, $P_a = 8$

Generated or supply frequency, $f = \frac{P_a N_{sa}}{120} = \frac{8 \times 750}{120} = 50$ Hz

Motor Speed, $N = 1440$ r.p.m.

No. of poles of motor, $P = \frac{120 f}{N} = \frac{120 \times 50}{1440} = 4 \cdot 16 = \mathbf{4}$ *(Ans)*

Synchronous speed, $N_s = \frac{120 f}{P} = \frac{120 \times 50}{4} = 1500$ r.p.m.

Percentage slip, $S = \frac{N_s - N}{N_s} \times 100 = \frac{1500 - 1440}{1500} \times 100 = \mathbf{4\%}$ *(Ans)*

Example 4.4.: *A 200 HP, 3-Phase, 400 V, 50 Hz induction motor has a speed of 1425 r.p.m at full load. The machine has 4 poles. Calculate the slip. How many complete alternations will the rotor e.m.f. make per minute.*

Solution:

Here, $f = 50$ Hz, $P = 6$, $N = 950$ r.p.m.

Synchronous speed, $N_s = \frac{120 f}{P} = \frac{120 \times 50}{4} = 1500$ r.p.m.

Slip, $S = \frac{N_s - N}{N_s}$

$= \frac{1500 - 1425}{1500} = \mathbf{0 \cdot 05}$ ***or* 5%** *(Ans)*

Frequency of rotor e.m.f. $f_r = Sf = 0.05 \times 50 = 2.5$ Hz or 2.5 c/s

Alternations of rotor e.m.f. per min $= 2.5 \times 60 =$ **150 c/ min** *(Ans)*

PRACTICE EXERCISE

Q.1 *A 3-phase induction motor is wound for 4 poles and is supplied from a 50 Hz system. Calculate: (i) Synchronous speed. (ii) Actual speed of the motor when running at 4% slip (iii) Frequency of e.m.f. induced in rotor.* ***(Ans.*** *1500 r.p.m.; 1440 r.p.m.; 2 Hz)*

Q.2 *A 10 HP, 230 V, 3-phase, 50 Hz, 6 pole squirrel cage induction motor operates at full load slip of 4% when rated voltage and rated frequency are applied. Determine (i) full loud speed. (ii) full load torque in Newton-metre . (iii) frequency of rotor current under this condition and (iv) speed of rotation of the stator m.m.f.* ***(Ans.*** *960 r.p.m.; 73·16 Nm; 2 Hz; 1000 r.p.m.)*

Q.3. *Power to an induction motor is supplied by a 12 pole, 3-phase, 500 r.p.m. alternator. The full load speed of the motor is 1440 r.p.m. Find the percentage slip and number of poles in the motor.*

(***Ans.*** *4; 4%*)

Q.4 *A 500 HP, 3-Phase, 440 V, 50 Hz induction motor has a speed of 950 r.p.m at full load. The machine has 6 poles. Calculate the slip. How many complete alternations will the rotor e.m.f. make per minute.*

(***Ans.*** *0·05 or 5%; 150 c/ min*)

4.7 SPEED OF ROTOR FIELD OR M.M.F.

When three-phase currents are supplied to the stator winding of a poly phase induction motor, a resultant field is set up which rotates at a constant speed called synchronous speed ($N_s = 120f/P$).

This rotating field induces poly phase e.m.fs. in the rotor winding and if rotor winding is closed, poly phase currents circulate in it. These currents set up a revolving field in the rotor which rotates at a speed $N_r = 120\, f_r/P$ with respect to rotor.

Now $N_r = 120 \times Sf/P = SN_s$

When rotor itself is rotating at a speed N r.p.m. in the space.

$\therefore$ Speed of rotor field in space $= N + N_r = (1 - S)\, N_s + SN_s = N_s - SN_s + SN_s = N_s$

Thus, rotor magnetic field also rotates, in space, at the same speed and in the same direction as that of stator field. Hence, the two fields are magnetically locked with each other and are stationary with respect to each other.

4.8 ROTOR E.M.F.

The revolving magnetic field set up in the stator by poly phase currents is common to both stator and rotor winding. This field induces e.m.fs. in both the windings. The stator induced e.m.f. per phase is given by the relation;

$$E_1 = 4.44\, k_{w1}\, T_1\, f\, \phi_m \quad \text{...}(i)$$

Where k_{w1} = winding factor *i.e.* product or coil span factor k_c and distribution factor k_d.

T_1 = No. of turns/phase of stator winding;

f = stator or supply frequency and

ϕ_m = maximum value of flux.

The rotor induced e.m.f/phase, $E_2 = 4.44\, kw_2\, T_2 f_r\, \phi_m$ ...(*ii*)

Where f_r is the rotor current frequency, and under stationary condition *i.e.*, at the start $f_r = f$.

Therefore, rotor induced e.m.f./phase at stand still or start, $E_{2s} = 4.44\, kw_2\, T_2 f\, \phi_m$

Dividing eq. (*ii*) by (*i*), we get,

$$\frac{E_{2s}}{E_1} = \frac{4{\cdot}44\, kw_2\, T_2 f\, \phi_m}{4{\cdot}44\, kw_1\, T_1\, f\, \phi_m} = \frac{T_2}{T_1} = K \quad (\textit{i.e.}\text{ transformation ratio})$$

From eq. (*ii*), induced e.m.f. in the rotor under running condition,

$$E_2 = 4{\cdot}44\, kw_2\, T_2\, (Sf)\, \phi_m = SE_{2s}$$

The induced e.m.f. in the rotor circuit is maximum at the start and varies according to the value of slip under running condition. Since, the value of normal slip under loaded condition is nearly 5% therefore, the rotor induced e.m.f. is nearly 5% of the maximum value.

4.9 ROTOR RESISTANCE

Since the rotor winding is made of some conducting material (copper or aluminium), it has a definite resistance ($R = \rho\, l/a$). Its value remains constant and is denoted by R_2.

4.10 ROTOR REACTANCE

Whole of the flux produced by the rotor currents does not link with the stator winding. The part of rotor flux which links the rotor conductors but not with the stator winding is called leakages flux and hence develops leakage inductance (L_2). The leakage flux and hence the inductance is very small if the rotor conductors are placed at the outermost periphery of the rotor as shown in fig.4.11. Depending upon the rotor current frequency, rotor reactance will be developed.

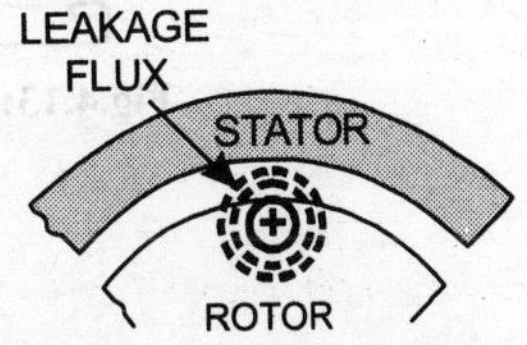

Fig.4.11: Rotor leakage flux

Rotor reactance, $X_2 = 2\pi f_r L_2 = 2\pi\, Sf L_2 = S\,(2\pi f L_2)$

When the rotor is standstill *i.e.*, at the start, when slip, $S = 1$

The value of rotor reactance $= X_{2s} = 2\pi f L_2$

Thus, under normal running, rotor reactance, $X_2 = SX_{2s}$

4.11 ROTOR IMPEDANCE

The total opposition offered to the flow of rotor current by the rotor circuit is called the rotor impedance.

Rotor impedance, $\overline{Z}_2 = R_2 + jX_2 = R_2 + j\,SX_{2s}$

Magnitude of rotor impedance, $Z_2 = \sqrt{(R_2)^2 + (SX_{2s})^2}$

4.12 ROTOR CURRENT AND POWER FACTOR

The rotor circuit diagram of an induction motor is shown in figure 4.12.

Under running condition;

Rotor induced e.m.f. $= E_2 = SE_{2s}$

Rotor impedance, $Z_2 = \sqrt{R_2^2 + X_2^2} = \sqrt{(R_2)^2 + (SX_{2s})^2}$

Rotor current, $I_2 = \dfrac{E_2}{Z_2}$

$$= \frac{E_2}{\sqrt{(R_2)^2 + (X_2)^2}} = \frac{SE_{2s}}{\sqrt{(R_2)^2 + (SX_{2s})^2}}$$

Rotor power factor, $\cos\phi_2 = \dfrac{R_2}{Z_2} = \dfrac{R_2}{(R_2)^2 + (SX_{2s})^2}$

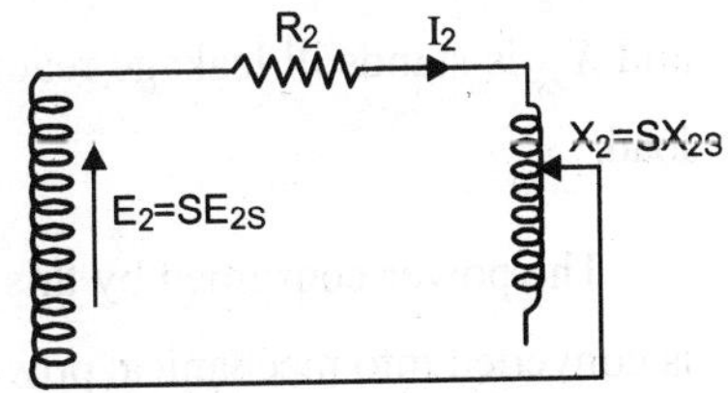

Fig.4.12: Rotor circuit $X_2 = SX_{2S}$

4.13 SIMPLIFIED EQUIVALENT CIRCUIT OF ROTOR

The various parameters and electrical quantities are represented on the circuit diagram, as shown in fig.4.13 The rotor current is given by the expression:

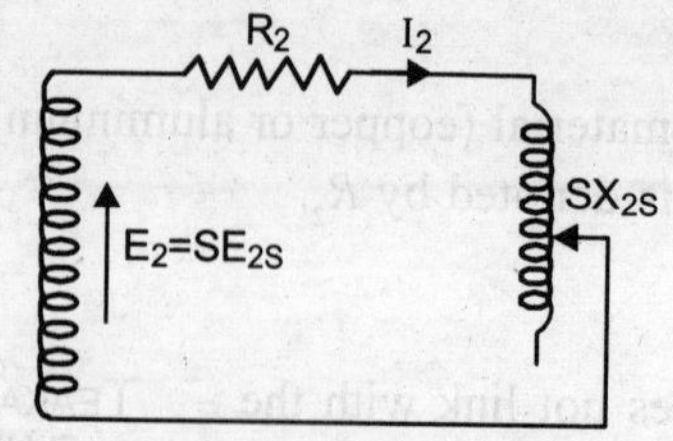

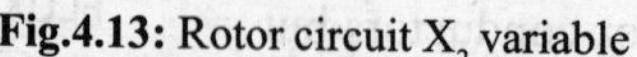
Fig.4.13: Rotor circuit X_2 variable

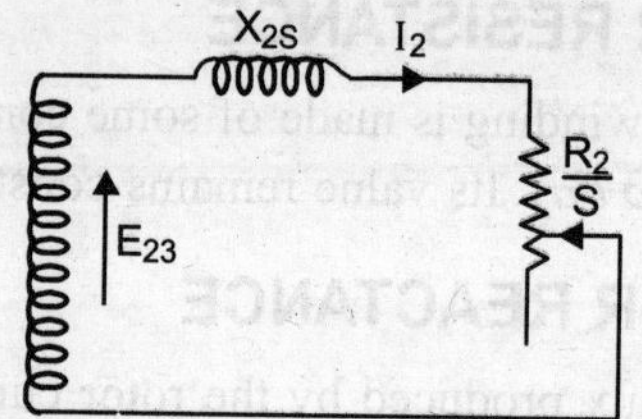

Fig.4.14: Rotor circuit X_2 is fixed to X_{2S}

$$I_2 = \frac{SE_{2s}}{\sqrt{(R_2)^2 + (SX_{2s})^2}}$$

The other expression for the rotor current is

$$I_2 = \frac{E_{2s}}{\sqrt{(R_2/S)^2 + (X_{2s})^2}} \text{ (dividing the numerator and denominator by } S\text{)}$$

This expression gives a convenient form of equivalent circuit as shown in fig.4.14.

The resistance is a function of slip and can be splitted into two parts; $\frac{R_2}{S} = R_2 + R_2\left(\frac{1-S}{S}\right)$

Where $R_2\left(\frac{1-S}{S}\right)$ represents electrical load on the rotor.

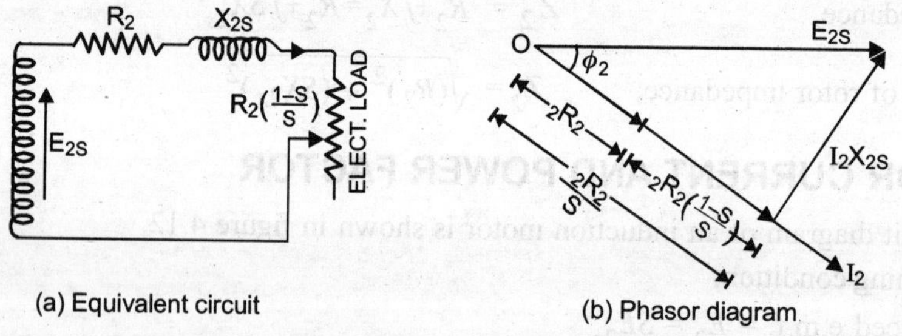

(a) Equivalent circuit (b) Phasor diagram

Fig.4.15: Rotor equivalent circuit and phasor diagram

Thus, the final simplified equivalent rotor circuit is shown in fig.4.15 (*a*). Where R_2 is rotor resistance and X_{2s} is standstill leakage reactance. The resistance $R_2\left(\frac{1-S}{S}\right)$ is fictitious resistance representing load.

The power consumed by this fictitious resistance *i.e.* $I_2^2 R_2\left(\frac{1-S}{S}\right)$ is the electrical power which is converted into mechanical power to pick the load. After subtracting the mechanical losses, we get the output power available at the shaft.

Thus, electrical power converted into mechanical power, $= I_2^2 R_2\left(\frac{1-S}{S}\right)$ watt

From the simplified equivalent circuit the phasor diagram of rotor circuit is drawn as shown in fig.4.15 (*b*).

Rotor current I_2 lags behind the rotor standstill induced e.m.f E_{2s} by an angle ϕ.

The voltage drop across R_2 *i.e.* I_2R_2 and across $R_2\left(\frac{1-S}{S}\right)$ *i.e.* $I_2R_2\left(\frac{1-S}{S}\right)$ are in phase with current I_2, whereas the voltage drop in X_{2s} *i.e.* I_2X_{2s} leads the current I_2 by 90°.

The vector sum of all the three drops is equal to E_{2s} *i.e.*

$$E_{2s} = I_2\sqrt{(R_2/S)^2 + (X_{2s})^2}$$

Power factor of rotor circuit,

$$\cos\phi = \frac{R_2/S}{\sqrt{(R_2/S)^2 + (X_{2s})^2}}$$

4.14 LOSSES IN AN INDUCTION MOTOR

The major losses in an induction motor are:

1. **Stator losses:** The losses which occur in the stator of an induction motor are called stator losses.
   (*i*) Stator copper losses – $I_1^2 R_1$ (per phase)
   (*ii*) Stator iron losses- These are the hysteresis and eddy current losses.
2. **Rotor losses:** which occur in the rotor of an induction motor are called rotor losses.
   (*i*) Rotor copper losses – $I_2^2 R_2$ (per phase)
   (*ii*) Rotor iron losses- Since under normal running condition rotor frequency is very small, therefore. These losses are so small that they are neglected.
3. **Mechanical losses:** The sum of windage and friction losses are called mechanical losses.

4.15 POWER FLOW DIAGRAM

Electrical power input is given to the stator. There are stator copper and iron losses and the remaining power *i.e* stator output is transferred to the rotor through magnetic flux called rotor input. In the rotor there are rotor copper losses and the remaining power is converted into mechanical power called mechanical power developed in the rotor. Then there are mechanical losses and the remaining power is available at the shaft called mechanical power output.

The power flow diagram is shown in fig.4.16.

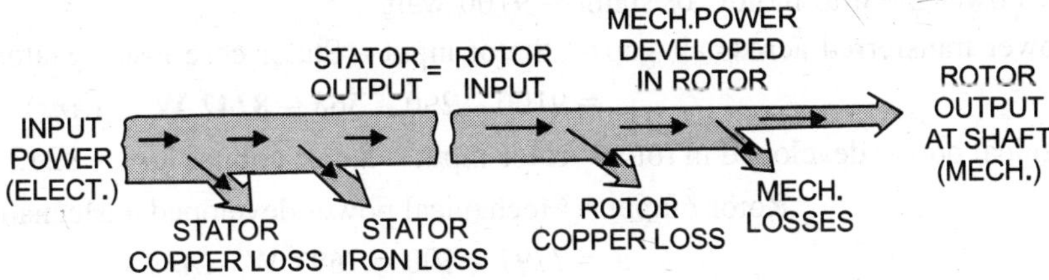

Fig.4.16: Power flow diagram

4.16 RELATION BETWEEN ROTOR COPPER LOSS, SLIP AND ROTOR INPUT

We have seen that the electrical power developed in the rotor is converted into mechanical power which is given by the relation:

$$\text{Mech. power developed in the rotor} = I_2^2 R_2 \left(\frac{I-S}{S}\right) \quad ...(i)$$

$$\text{The rotor copper losses} = I_2^2 R_2 \quad ...(ii)$$

From power flow diagram:

$$\text{Rotor input} = \text{Mech. power developed + rotor copper losses}$$

$$= I_2^2 R_2 \left(\frac{I-S}{S}\right) + I_2^2 R_2 = \left(\frac{I_2^2 R_2}{S}\right) \quad ...(iii)$$

From eq. (*i*) and (*ii*), we get, $\dfrac{\text{Rotor copper loss}}{\text{Mech. power developed}} = \dfrac{I_2^2 R_2}{I_2^2 R_2\left(\dfrac{1-S}{S}\right)}$

$\therefore$ **Rotor copper loss =** $\left(\dfrac{S}{1-S}\right)$ **Mech. power developed**

From. eq. (*ii*) and (*iii*), we get, $\dfrac{\text{Rotor copper loss}}{\text{Rotor input}} = \dfrac{I_2^2 R_2}{I_2^2 R_2 / S}$

$\therefore$ **Rotor copper loss = S × Rotor input**

Note: *All the values are the phase values.*

4.17 ROTOR EFFICIENCY

The ratio of rotor output (*i.e.,* mechanical power developed in rotor neglecting mechanical losses) to the rotor input is called the rotor efficiency.

$$\text{Rotor efficiency} = \frac{\text{Mech. Power developed}}{\text{Rotor input}} = \frac{I_2^2 R_2\left(\dfrac{1-S}{S}\right)}{I_2^2 R / S} = (1-S)$$

Example 4.5: *A 10 H.P., 4 pole, 25 Hz, 3-phase, wound rotor induction motor is taking 9100 watt from the line. Core loss is 290 watt, stator copper loss is 568 watt, rotor copper loss in 445 watt, friction and windage losses are 100 watt. Determine; (a) power transferred across air gap; (b) mechanical power in watt developed by rotor; (c) mechanical power output in watt; (d) efficiency; (e) slip.*

Solution: Power input to motor or stator = 9100 watt

Power transferred across air gap = Stator input – Stator core loss – Stator copper loss

= 9100 – 290 – 568 = **8242 W** (*Ans*)

Mechanical power developed in rotor = rotor input – Rotor copper loss = 8242 – 445 = **7797**

Rotor output = Mechanical power developed – Mechanical loss

= 7797 – 100 = **7697 W** (*Ans.*)

$$\text{Motor efficiency} = \frac{\text{Output}}{\text{input}} \times 100 = \frac{7697}{9100} \times 100 = \mathbf{84{\cdot}58\ \%} \quad (Ans)$$

$$\text{Slip,} \quad S = \frac{\text{Rotor copper loss}}{\text{Rotor input}} = \frac{445}{8242} = \mathbf{0{\cdot}05399} \quad (Ans)$$

Example 4.6: *A 4 pole, 3-phase, 50 Hz, 400 V induction motor has a delta connected stator and a star connected rotor. Each phase of rotor winding has one-fourth the number of turns on each phase of stator winding. The full load speed is 1455 r.p.m. The rotor resistance is 0·3 ohm, and rotor standstill reactance is 1 ohm per phase. The stator and rotor windings are similar. Stator losses equal 100 W. Friction and windage losses are equal to 50 W. Calculate.*

(i) blocked rotor voltage per phase *(ii) rotor current per phase at full load*
(iii) total rotor power input at full load *(iv) rotor power loss at full load*
(v) efficiency.

Solution: Here, $P = 4; f = 50$ Hz; $V_2 = 400$ V; $\dfrac{N_2}{N_1} = \dfrac{1}{4}$;

$N = 1455$ r.p.m; $R_2 = 0{\cdot}3\ \Omega$; $X_{2s} = 1\ \Omega$; Stator loss = 100 W; Mech loss = 50 W

Stator induced e.m.f. per phase, $E_1 = V_L = 400$ V (*delta connected*)

Now, $$\frac{E_{2s}}{E_1} = \frac{N_2}{N_1} = \frac{1}{4}$$

∴ Blocked rotor voltage per phase, $E_{2s} = \dfrac{400}{4} = \mathbf{100\ V}$ (*Ans*)

$$\text{Synchronous speed, } N_s = \frac{120f}{P} = \frac{120 \times 50}{4} = 1500 \text{ r.p.m.}$$

$$\text{Slip, } S = \frac{N_s - N}{N_s} = \frac{1500 - 1455}{1500} = 0 \times 03$$

$$\text{Rotor current, } I_2 = \frac{SE_{2s}}{\sqrt{(R_2)^2 + (SX_{2s})^2}} = \frac{0\cdot 03 \times 100}{\sqrt{(0\cdot 3)^2 + (0\cdot 03 \times 1)^2}} = \mathbf{9\cdot 95\ A} \quad (Ans)$$

Rotor copper loss $= 3I_2^2$ R2 $= 3 \times (9{\cdot}95)^2 \times 0{\cdot}3 =$ **89 W** (*Ans*)

$$\text{Power input to rotor} = \frac{\text{Rotor copper loss}}{\text{Slip}} = \frac{89}{0\cdot 03} = \mathbf{2967\ W} \quad (Ans)$$

Input to the motor = rotor input + stator losses = 2967 + 100 = 3067 W

Output at the shaft = rotor input – rotor copper loss – mech. loss = 2967 – 89 – 50 = 2828 W

$$\text{Efficiency} = \frac{\text{output}}{\text{Input}} \times 100 = \frac{2828}{3067} \times 100 = \mathbf{92\cdot 2\ \%} \quad (Ans)$$

Example 4.7: *A 60 H.P., 6-pole, 3-phase induction motor delivers load output at 960 r.p.m. at 0·8 power factor when supplied with 400 V. 50 Hz supply. Losses due to windage and friction come to 3 H.P. and the stator losses are 2kW. Find out:* (*a*) *Total rotor copper loss*; (*b*) *efficiency, and* (c) *line current.*

Solution: Rotor output = 60 H.P = 60 × 735·5 = 44130 W

Mechanical losses = 3 H.P. = 3 × 735·5 = 2206 W

Mechanical power developed = Rotor output + Mechanical losses = 44130 + 2206 = 44336 W

$$\text{Synchronous speed, } N_s = \frac{120f}{P} = \frac{120 \times 50}{6} = 1000 \text{ r.p.m.}$$

$$\text{Slip, } S = \frac{N_s - N}{N_s} = \frac{1000 - 960}{1000} = 0\cdot04$$

$$\text{Rotor copper loss} = \frac{S}{1-S} \text{ mech. power developed} = \frac{0\cdot04}{1-0\cdot04} \times 46336 = 1931 \text{ W}$$

$$\text{Stator losses} = 2\text{kW} = 2000\text{W}$$

$$\text{Input to stator or motor} = \text{Mech. power developed} + \text{Rotor } Cu \text{ loss} + \text{Stator losses}$$

$$= 46336 + 1931 = 2000 = 50267\text{W}$$

$$\text{Efficiency, } \eta = \frac{\text{Output}}{\text{Input}} = \frac{44130}{50267} \times 100 = \mathbf{87\cdot79\%} \; (Ans.)$$

$$\text{Line current, } I_2 = \frac{\text{Input power}}{\sqrt{3} \times \text{Line voltage} \times \text{p.f.}} = \frac{50267}{\sqrt{3} \times 400 \times 0\cdot8} = \mathbf{90\cdot67\ A} \; (Ans.)$$

Example 4.8: *A 4-pole, 50 Hz, 3-phase induction motor has an efficiency of 85% for useful output power at the shaft of 17 kW. For this load the total stator losses are 900 W and the windage and friction losses are 1100 W. Calculate the slip, torque developed by the rotor and torque available at the rotor shaft.*

Solution: No. of poles, $P = 4$; Supply frequency, $f = 50$ Hz

Motor efficiency, $\eta = 85\% = 0\cdot85$; Output power $= 17$ kW $= 17000$ W.

Stator losses $= 900$ W; Mechanical losses $= 1100$ W

$$\text{Input power} = \frac{\text{Output power}}{\eta} = \frac{17000}{0\cdot85} = 20000 \text{ W}$$

$$\text{Stator output or rotor input} = \text{Input power} - \text{Stator losses} = 20000 - 900 = 19100 \text{ W}$$

$$\text{Mech. power developed} = \text{Output power} + \text{Mech. losses} = 17000 + 1100 = 18100 \text{ W}$$

$$\text{Rotor Cu losses} = \text{Rotor input} - \text{Mech. power developed} = 19100 - 18100 = 1000 \text{ W}$$

$$\text{Slip, } S = \frac{\text{Rotor Cu loss}}{\text{Rotor input}} = \frac{1000}{19100} = 0\cdot05235$$

$$\text{Synchronous speed, } N_s = \frac{120f}{P} = \frac{120 \times 50}{4} = 1500 \text{ r.p.m.}$$

$$\text{Rotor speed, } N = N_s(1-S) = 1500\,(1 - 0\cdot05235) = 1421\cdot47 \text{ r.p.m.}$$

$$\text{Angular speed, } \omega = \frac{2\pi N}{60} = \frac{2\pi \times 1421\cdot47}{60} = 148\cdot85 \text{ rad/sec.}$$

Mech. power developed in the rotor $= \omega T$

$$\therefore \quad \text{Torque developed, } T = \frac{\textit{Mech. power developed}}{\omega} = \frac{18100}{148\cdot85} = \mathbf{121\cdot6\ Nm} \; (Ans.)$$

$$\text{Torque at the shaft, } T_m = \frac{\textit{Rotor output}}{\omega} = \frac{17100}{148\cdot85} = \mathbf{114\cdot2\ Nm} \; (Ans.)$$

Example 4.9: *A 3-phase induction motor has an efficiency of 90% and runs at a speed of 480 r.p.m. The motor is supplied from 400 V mains and it takes a current of 75 A at 0·77 p.f. Calculate the B.H.P. (metric) of the motor and pull on the belt when driving the line shaft through pulley of 0·75 m diameter.*

Solution: Supply voltage, V_L = 400 V; Rotor speed, N = 480 r.p.m.

Motor efficiency, η = 90% = 0·9; Current drawn from mains, I_L = 75 A

Motor p.f., $\cos\phi$ = 0·77 lag. Diameter of pulley, d = 0·75 m

$$\text{Radius of pulley, } r = \frac{0\cdot 75}{2} = 0\cdot 375 \text{ m}$$

$$\text{Input power} = \sqrt{3}\, V_L I_L \cos\phi = \sqrt{3} \times 400 \times 75 \times 0\cdot 77 = 40010 \text{ W}$$

$$\text{Output power} = \text{Input power} \times \eta = 40010 \times 0\cdot 9 = 36009 \text{ W}$$

$$\text{B.H.P. of the motor} = \frac{\text{Output power}}{735\cdot 5} = \frac{36009}{735\cdot 5} = \mathbf{48\cdot 958}\ (\textit{Ans.})$$

$$\text{Angular speed, } \omega = \frac{2\pi N}{60} = \frac{2\pi \times 480}{60} = 16\pi$$

$$\text{Torque at the shaft, } T_m = \frac{\text{Output power}}{\omega} = \frac{36009}{16\pi} = 716\cdot 376 \text{ Nm}$$

$$\text{Now, torque, } T_m = \text{Pull on the belt} \times \text{radius of pulley}$$

$$\therefore \quad \text{Pull on the belt} = \frac{T_m}{r} = \frac{716\cdot 376}{0\cdot 375} = 1910\cdot 34 \text{ N} = \frac{1910\cdot 34}{9\cdot 81} = \mathbf{194\cdot 73\ kg}\ (\textit{Ans.})$$

PRACTICE EXERCISE

Q.1 *The power input to a 3-phase induction motor is 80 kW. The stator losses total 1×5 kW. Find the total mechanical power developed if the motor is running with a slip of 4%.*

(**Ans.** *75.36 kW*)

Q.2 *A 50 H.P., 3-f, 6-pole induction motor delivers full load output at 960 r.p.m. at 0·8 p.f. when supplied with 400V, 50Hz supply. Losses due to windage and friction come out to be 2H.P. and stator losses are 2 kW. Find out. (a) total rotor Cu loss; (b) efficiency and (c) line current.*

(**Ans.** *1594 W; 87.9%; 75.49A*)

Q.3 . *A 400 V, 6 pole, 50 Hz, 3-phase induction motor develops 20 HP inclusive of mechanical losses when running at 965 r.p.m., the power factor being 0 × 87 lagging. Calculate (i) the slip (ii) rotor copper losses (iii) the total input if the stator losses are 1500 watt (iv) line current and (v) the number of cycles made per minute by the rotor e.m.f.*

(**Ans.** *3.5%; 833.5W; 16743.5W; 27.78A; 1.75 Hz*)

Q.4 . *A 4-pole, 3-phase, 50 Hz induction motor supplies a useful torque of 159 Newton-metre. Calculate at 4% slip: (i) The rotor input; (ii) Motor input; (iii) Motor efficiency, if the friction and windage losses total 500 watt and stator losses are 1000 watt.*

(**Ans.** *25497 W; 26497 W;90·49%*)

Q.5 *A 440 V, 50 Hz, 6-pole, 3-phase induction motor draws an input power of 81 kW from the mains. The rotor e.m.f. makes 120 complete cycle per minute. Its stator losses are 1 kW and rotor current per phase is 65 ampere. Calculate:*

(i) Rotor copper losses per phase; *(ii) Rotor resistance per phase;*

(iii) Torque developed. (**Ans.** *1067 W; 0.2525 ohm; 763.94 Nm*)

4.18 TORQUE DEVELOPED BY AN INDUCTION MOTOR

We have already seen that the electrical power of 3-phase induction motor converted into mechanical power is given by the relation;

$$P_0 = 3I_2^{\,2}R_2\left(\frac{1-S}{S}\right) \quad ...(i)$$

also,

$$P_0 = \omega\, T \quad ...(ii)$$

Where, ω = angular speed of the rotor in rad/sec. and

T = torque developed by an induction motor in Nm.

Equating eq. (*i*) and (*ii*), we get

$$\omega\, T = 3I_2^{\,2}R_2\left(\frac{1-S}{S}\right) = 3\frac{I_2^2R_2}{S}\cdot\frac{1-S}{\omega}$$

or

$$T = \frac{3}{\omega_2}\cdot\frac{I_2^2\,R_2}{S} \qquad [\text{since } \omega = \omega_s\,(1-S)]$$

where ω_s = angular synchronous speed in rad/sec.

As

$$I_2 = \frac{SE_{2s}}{\sqrt{(R_2)^2+(SX_{2s})^2}}$$

∴

$$T = \frac{3}{\omega_2}\left(\frac{SE_{2s}}{\sqrt{(R_2)^2+(SX_{2s})^2}}\right)^2\cdot\frac{R_2}{S}$$

or

$$T = \frac{3}{\omega_s}\,\frac{SE_{2s}^2R_2}{[(R_2)^2+(SX_{2s})^2]} = \frac{3}{\omega_s}\,\frac{E_{2s}^2R_2\,/\,S}{[(R_2\,/\,S)^2+(X_{2s})^2]}$$

This is the expression for *full load torque.*

4.19 CONDITION FOR MAXIMUM TORQUE AND EQUATION FOR MAXIMUM TORQUE

The full load torque developed in an induction motor is given by the relation:

$$T = \frac{3}{\omega_2}\,\frac{SE_{2s}^2R_2}{[(R_2)^2+(SX_{2s})^2]}$$

or

$$T \propto \frac{SR_2}{R_2^2+S^2X_{2x}^2} \qquad \left(\text{since } \frac{3}{\omega_s}E_{2s}^2 \text{ is constant}\right)$$

The torque developed will be maximum at a particular value of slip. As, slip (S) is a variable quantity, therefore, to obtain the condition for maximum torque, the above expression for torque is differentiated with respect to S and equated to zero.

$$\frac{dT}{dS} = \frac{(R_2^2 + S^2 X_{2s}^2) R_2 - SR_2(0 + 2SX_{2s}^2)}{(R_2^2 + S^2 X_{2s}^2)^2} = 0$$

or $$(R_2{}^2 + S^2 X_{2s}^2) R_2 = 2R_2 S^2 X_{2s}{}^2$$

or $$R_2{}^2 = (SX_{2s})^2 \text{ or } R_2 = SX_{2s}$$

or $$S = R_2/X_{2s} \text{ is the slip at which torque is maximum.}$$

To obtain the expression for maximum torque substitute the value of $R_2 = SX_{2s}$ in the expression for full load torque, we get,

$$\text{Maximum torque, } T_m = \frac{3}{\omega_s} \frac{SE_{2s}^2 (SX_{2s})}{[(SX_{2s})^2 + (SX_{2s})^2]} = \frac{3E_{2s}^2}{2\omega X_{2s}}$$

Thus, the maximum torque is independent of rotor resistance but it is inversely proportional to rotor reactance at standstill (*i.e.*, X_{2s}). Therefore, to achieve higher value of maximum torque, the leakage reactance of the rotor should be kept minimum. This is achieved (*i*) *by placing the rotor conductors very near to the outer periphery of the rotor and* (*ii*) *by reducing the air gap between stator and rotor to smallest possible value.*

4.20 STARTING TORQUE

At start rotor is stationary and the value of slip is one *i.e.*, $S = 1$.

Thus, to obtain the expression for starting torque, substitute the value of slip, $S = 1$ in the expression of full load torque;

$$\therefore \quad \text{Starting torque, } T_s = \frac{3}{\omega_s} \frac{E_{2s}^2 R_2}{[(R_2)^2 + (X_{2s})^2]}$$

Sometimes maximum torque is required at start. In that case, in the condition for maximum torque substitute the value of $S = 1$.

$$R_2 = SX_{2s} = X_{2s} \text{ (since } S = 1 \text{ at start)}$$

Thus, to obtain maximum torque at start, the value of rotor resistance must be equal to rotor leakage reactance at standstill. Therefore, at start some external resistance is added in the rotor circuit. This is only possible in case of slip ring induction motors. This is the reason, why slip ring induction motors are applied where heavy loads are required to be picked up at start such as in lifts, crains, elevators etc. Once the motor picks up the load the external resistance is gradually reduced to zero.

In case of squirrel cage induction motors, the rotor resistance is fixed and is kept quite low in comparison to rotor reactance, otherwise the rotor copper losses would be high and the efficiency of the motor would fall to low value. However to obtain higher starting torque in case of squirrel cage induction motors another cage is embedded in the rotor and the motor is called a *double cage induction motor.*

4.21 TORQUE-SLIP CURVE

The full load torque developed by an induction motor is given by the expression;

$$T = \frac{3}{\omega_s} \frac{SE_{2s}^2 R_2}{[R_2^2 + (SX_{2s})^2]}$$

To draw the torque-slip or torque-speed curve the following points are considered:

(*i*) At synchronous speed (N_s); slip, $S = 0$ and torque $T = 0$.

(*ii*) When rotor speed is very near to synchronous speed *i.e.*, when the slip is very low the value of the term $(SX_{2s})^2$ is very small in comparison to R_2^2 [*i.e.*, $(SX_{2s})^2 << R_2^2$)] and is neglected.

Therefore, torque is given by the expression;

$$T = \frac{3}{\omega_s} \frac{SE_{2s}^2 R_2}{R_2^2} = KS, \text{ or } T \propto S$$

Thus, at low values of slip, torque is approximately proportional to slip S and the torque-slip curve is a straight line, as shown in fig.4.17.

(*iii*) As the slip increases torque increases and attains its maximum value when $S = R_2/X_{2s}$. *This maximum value of torque is also known as* **break down** *or* **pull out torque.**

(*iv*) With further increase in slip due to increase in load beyond the point of maximum torque *i.e.* when slip is high, the value of term $(SX_{2s})^2$ is very large in comparison to R_2^2 [*i.e.*, $(SX_{2s})^2 >> R_2^2$]. Therefore, R_2^2 is neglected as compared to $(SX_{2s})^2$ and the torque is given by the expression.

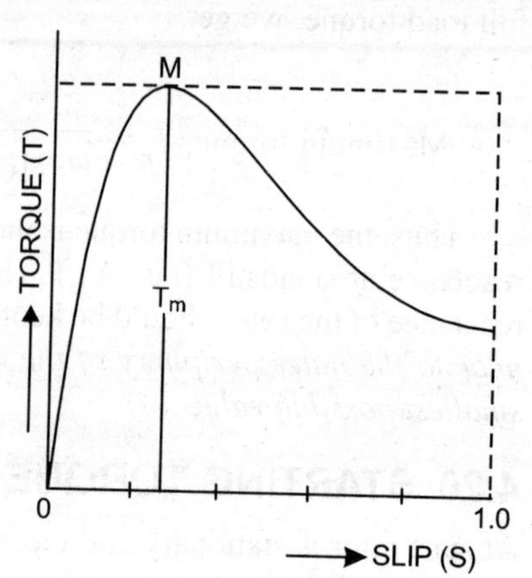

Fig.4.17: T-S curve

$$T = \frac{3}{\omega_s} \frac{SE_{2s}^2 R_2}{X^2 X_{2s}^2} = K' \frac{1}{S} \text{ or } T \mu \frac{1}{S}$$

Thus, at higher value of slip (*i.e.*, the slip beyond that corresponding to maximum torque), torque is approximately inversely proportional to slip S and the torque-slip curve is a rectangular hyperbola, as shown in fig.4.18.

Thus, with the increase of slip beyond the point of maximum torque, due to increase in load, torque decrease. The result is that the motor could not pick-up the load and slows down and eventually stops. This results in *blocked rotor or short circuited motor.*

4.22 TORQUE-SPEED CURVE AND OPERATING REGION

The torque-speed curve of an induction motor is shown in fig.4.18. It is the same curve which is already drawn, the only difference is that speed is taken on the abscissa instead of slip.

From the curve, it is clear that induction motor develops the same torque at point X and Y. However at point X the motor is unstable because with the increase in load speed decreases and the torque developed by the motor also decreases.

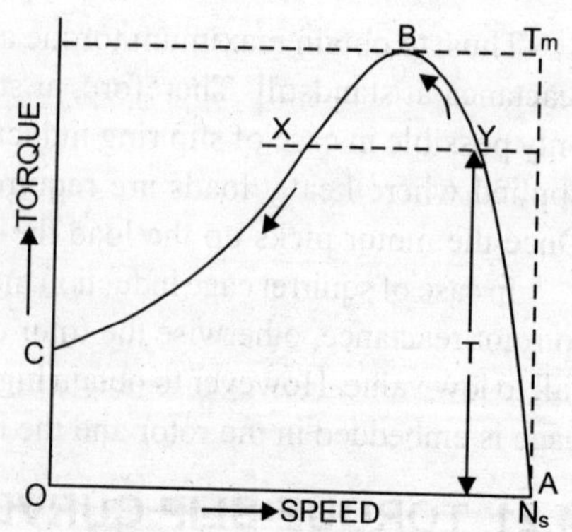

Fig.4.18: Operating region of T-S curve

Therefore, the motor could not pick up the load and the result is that the motor slows down and eventually stops. The miniature circuit breakers will be tripped open if the circuit has been so protected.

At point *Y*, the motor is stable because in this region with the increase in load speed decreases but the torque developed by the motor increases. Thus the motor will be in position to pick up the extra load effectively.

Thus, on the torque-speed curve region BC is the unstable region and **region AB is the stable or operating region** *of the induction motor as shown in fig.4.18.*

Example 4.10: *A 3-phase induction motor has a 4-pole star-connected stator winding. The motor runs at a line voltage of 400 V, 50 Hz supply. The rotor resistance and standstill reactance per phase are 0.15 and 1.0 ohm respectively. The ratio of rotor to stator turns is 0.7. Calculate the total torque at 4% slip.*

Solution: Here, $P = 4$; $f = 50$ Hz; $R_2 = 0{\cdot}15\ \Omega$; $X_{2s} = 1.0\ \Omega$; $S = 4\%$

Supply voltage (line value), $V_L = 400$ V

Stator induced voltage (phase value), $E_1 = \dfrac{V_L}{\sqrt{3}} = \dfrac{400}{\sqrt{3}} = 231$ V

Ratio of rotor to stator turns, $\dfrac{T_2}{T_1} = 0{\cdot}7 = \dfrac{E_{2s}}{E_1}$

$\therefore$ Rotor induced e.m.f. at standstill, $E_{2s} = 0{\cdot}7\ E_1 = 0{\cdot}7 \times 231 = 161{\cdot}7$ V

Synchronous speed, $N_s = \dfrac{120f}{P} = \dfrac{120 \times 50}{4} = 1500$ r.p.m.

Synchronous angular speed, $\omega_s = \dfrac{2\pi N_s}{60} = \dfrac{2\pi \times 1500}{60} = 50\ \pi$ rad/sec.

Torque developed, $T = \dfrac{3}{\omega_s} \dfrac{SE_{2s}^2\ R_2}{[R_2^2 + (SX_{2s})^2]}$

$$= \frac{3}{50\pi} \times \frac{0{\cdot}04 \times (161.7)^2 \times 0{\cdot}15}{[(0{\cdot}15)^2 + (0{\cdot}04 \times 1.0)^2]} = \mathbf{730.78\ Nm} \qquad (Ans)$$

Example 4.11: *A 6-pole, 50 Hz, 3-phase induction motor has a rotor resistance of 0·02 ohm per phase and standstill reactance of 0·8 ohm per phase. Calculate the speed at which maximum torque is developed.*

Solution: Here, $P = 6$; $f = 50$ Hz; $R_2 = 0{\cdot}02\ \Omega$; $X_{2s} = 0{\cdot}8\ \Omega$

Condition for maximum torque is; $R_2 = SX_{2s}$

$\therefore$ Torque will be maximum at a slip, $S = \dfrac{R_2}{X_{2s}} = \dfrac{0.02}{0.8} = 0{\cdot}025$

Synchronous speed, $N_s = \dfrac{120f}{P} = \dfrac{120 \times 50}{6} = 1000$ r.p.m.

Speed at which the torque will be maximum, $N = N_s (1 - S)$

$= 1000 (1 - 0{\cdot}025) =$ **975 r.p.m.** *(Ans)*

PRACTICE EXERCISE

Q.1 *A 3-phase induction motor has a 4-pole star-connected stator winding. The motor runs at a line voltage of 200 V, 50 Hz supply. The rotor resistance and standstill reactance per phase are 0.1 and 0.9 ohm respectively. The ratio of rotor to stator turns is 0.67. Calculate the total torque at 4% slip.*

(*Ans.* *40.48 Nm*)

Q.2 *A 4-pole, 50 Hz, 3-phase induction motor has a rotor resistance of 0·21 ohm per phase and standstill reactance of 0·7 ohm per phase. Calculate the speed at which maximum torque is developed*

(*Ans.* *1455 r.p.m.*)

4.23 NECESSITY OF A STARTER

The current drawn by a motor from the mains, depends upon the rotor current. The rotor current under running condition is given by the expression;

$$I_2 = \frac{SE_{2s}}{\sqrt{R_2^2 + (SX_{2s})^2}}$$

At start slip $S = 1$, therefore, rotor current. $I_{2s} = \dfrac{E_{2s}}{\sqrt{R_2^2 + X_{2s}^2}}$

This current is very large as compared to its full load current. Thus, when a squirrel cage induction motor is directly connected to the supply mains, it draws very large current (nearly 5 to 7 times of the full load current) from the mains. This heavy current may not be dangerous for the motor because it occurs for a short duration of time, but it causes the following affects:

(*i*) It produces large voltage drop in the distribution lines and thus affects the voltage regulation of the supply system.

(*ii*) It adversely affects the other motors and loads connected to the same lines.

Hence it is not advisable to start large capacity induction motors by direct switching. Rather, such motors should be started by means of some starting device known as *starter.*

The function of a starter is to limit the initial rush of current to a predetermined value. A starter also has some protective devices to protect the induction motors against over loading.

4.24 STARTING METHODS OF SQUIRREL CAGE INDUCTION MOTORS

The various starters which are employed to restrict the initial rush of current in squirrel cage induction motors are given below:

1. Direct On Line (D.O.L.) Starter; 2. Star/Delta Starter; 3. Auto-transformer Starter.

1. Direct On Line (D.O.L.) Starter: It is a starter by which the motor is switched *ON* direct to the supply mains by switching conductor. With normal industrial motors this operation results in a heavy rush of current of the order of five to seven times of the normal full load current. This high current rapidiy decreases as the motor picks up speed but it is at a very low power factor and thus tends to disturb the voltage of the supply in the distribution lines. For this reason, the supply authorities limit the size of motor upto 5 H.P. which can be started by this starter. An automatic *D.O.L.* Starter is shown in fig.4.19.

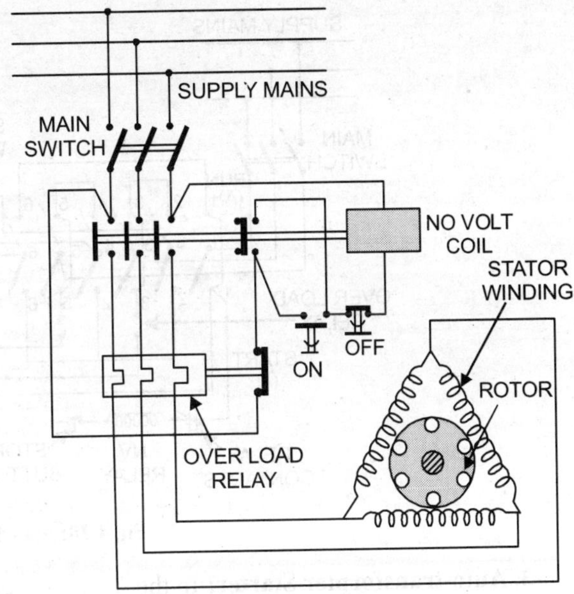

Fig.4.19: DOL starter

A direct on line starter essentially consists of a contactor having four normally open (*N.O.*) contacts and a contactor coil also known as no-volt coil or no volt release. There are two push buttons *ON* and *OFF* which are used to start and stop the motor. To protect motor against overload, thermal or magnetic over-load coils are connected in each phase.

To start the motor, the *ON* push button (green) is pressed which energies the no-volt coil by connecting it across two phases. The no-volt coil pulls its plunger in such a direction that all the normally open (*NO*) contacts are closed and motor is connected across supply through three contacts. The fourth contact serves as a hold on contact which keeps the no-volt coil circuit closed even after the *ON* push button is released. To stop the motor, *OFF* push button (red) is pressed momentarily which de-energises the no volt coil opening the main contacts.

When the motor is over loaded, the thermal overload relay contact, connected in the control circuit opens thus disconnecting the No-volt relay from the supply. Overload protection is achieved by thermal element overload relay.

2. Star-Delta Starter: This method is based upon the principle that is star connections, voltage across each winding is phase voltage *i.e.*, $1/\sqrt{3}$ times the line voltage, whereas the same winding when connected in delta will have full line voltage across it. So at start, connections of the motor are made in star fashion so that reduced voltage is applied across each winding. After the motor attains speed the same windings through a change-over switch, as shown in fig.4.20 are connected in delta across the same supply. The starter is provided with overload and under voltage protection devices also.

Since at start stator windings are connected in star connection, so voltage across each phase winding is reduced to $1/\sqrt{3}$ of line voltage, therefore, starting current/phase becomes equal to

$$I_{sc}/\sqrt{3} = \text{Starting line current}$$

Starting line current by direct switching with stator winding connected in delta = $\sqrt{3}\, I_{sc}$

$$\therefore \quad \frac{\text{Line current with star delta starter}}{\text{Line current with direct switching}} = \frac{I_{sc}/\sqrt{3}}{\sqrt{3} I_{sc}} = \frac{1}{3}$$

Thus, it concludes that when a 3-phase motor is started by a star/delta starter, the current drawn by it is limited to 1/3rd of the value that it would draw without starter.

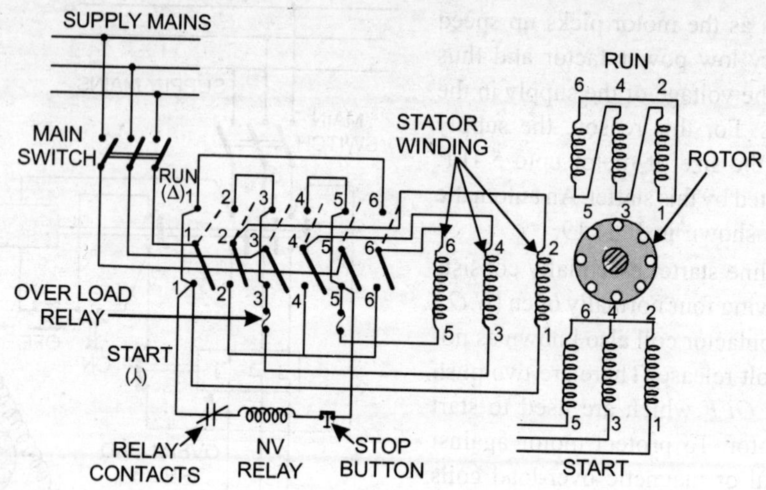

Fig.4.20: Star-delta starter

3. Auto-transformer Starter: In the previous method, the current can only be reduced to 1/3 times the short circuit current. Whereas, in this method, the voltage applied across the motor and hence current can be reduced to a very low value at the time of start. At the time of start, the motor is connected to supply through auto-transformer by a 6 pole double throw switch. When the motor is accelerated to about full speed, the operating handle is moved to run position. By this, motor is directly connected to the line as shown in fig.4.21

Fig.4.21: Auto-transformer starter

Overload protection and under voltage protection is provided as explained in the first method.

Although this type of starter is expensive but is most suitable for both the star-connected and delta-connected induction motors. It is most suitable for starting of large motor.

Large size motors draw huge amount of current from the mains if they are connected to mains without starter. However, if they are connected to the mains through star/delta starter, the current is limited to 1/3rd value which is still, so large that it would disturb the other loads connected to the same lines. Hence, to limit the initial rush of current to low values *auto-transformer starters* are preferred. With the help of auto-transformer starters, we can limit the starting current to any predetermined value as explained below:

Let the motor be started by an auto transformer having transformation ratio K.

If I_{sc} is the starting current when normal voltage is applied.

Applied voltage to stator at start = KV

Then motor input current $I_s = KI_{sc}$

Supply current = Primary current of Auto transformer
$= K \times$ Secondary current of Auto-transformer $= KKI_{sc} = K^2 I_{sc}$

If 20% (*i.e.* 1/5 th) voltage is applied to the motor through auto-transformer starter, the current drawn from the mains is reduced to $\left(\frac{1}{5}\right)^2$ *i.e.* 1/25th times.

4.25 STARTING METHOD OF SLIP RING INDUCTION MOTORS

To start a slip ring induction motor, a 3-phase rheostat is connected in series with the rotor circuit through brushes as shown in fig.4.22. This is called *rotor rheostat starter*. This is made of three separate variable resistors joined together by means of a 3-phase armed handle which forms a star point. By moving the handle equal resistance in each phase can be introduced.

At start, whole of the rheostat resistance is inserted in the rotor circuit and the rotor current is reduced to

$$I_{2s} = \frac{E_{2s}}{\sqrt{(R_2 + R)^2 + (X_{2s})^2}}$$

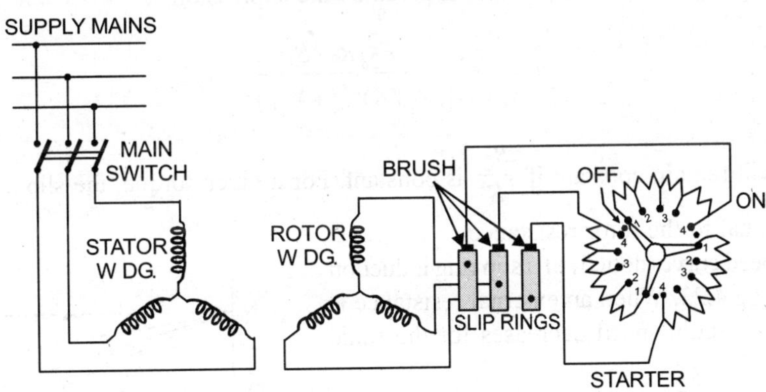

Fig.4.22: Rotor rheostat starter

Correspondingly it reduces the current drawn by the motor from the mains at start.

When the motor picks up speed the external resistance is reduced gradually and ultimately whole of the resistance is taken out of circuit and slip rings are short-circuited.

By inserting external resistance in the rotor circuit, not only the starting current is reduced but at the same time starting torque is increased due to improvement in power factor:

At starts:

Power factor without starter, $\cos\phi_s = \dfrac{R_2}{\sqrt{(R_2)^2 + (X_{2s})^2}}$

Power factor with starter, $\cos'\phi_s = \dfrac{(R_2 + R)}{\sqrt{(R_2 + R)^2 + (X_{2s})^2}}$

Hence, $\cos'\phi_s >> \cos\phi_s$

4.26 SPEED CONTROL OF INDUCTION MOTORS

The speed of an induction motor is given by the relation

$$N = N_s\,(1 - S) \ \text{ or } \ N = \frac{120 f}{P}(1 - S)$$

Hence the speed of the motor depends upon three factors *i.e.,* frequency, slip and number of poles for which the motor is wound. Thus, the speed of an induction motor can be controlled by changing or controlling any one of these three quantities.

4.27 SPEED CONTROL BY CHANGING THE SLIP

The speed of an induction motor can be changed by changing its slip, and the slip can be changed (1) by changing the rotor circuit resistance (2) by changing the supply voltage and (3) by injecting voltage in the rotor circuit.

1. Speed control by changing the Rotor Circuit Resistance

In the wound type motor the slip may be changed by introducing resistance in the rotor circuit and hence speed is changed.

Torque developed in an induction motor is given by the expression:

$$T = \frac{3}{\omega_s}\,\frac{E_{2s}^2 R_2 / S}{[(R_2/S)^2 + (X_{2s})^2]}$$

The torque will remain constant if $\frac{R_2}{S}$ is constant. For a given torque, the slip at which a motor works is proportional to the rotor resistance.

The torque-speed curve (dotted) of a slip ring induction rotor is shown in fig.4.23. When an external resistance is added in the rotor circuit, speed decreases for the same torque T, so that ratio $\frac{R_2}{S}$ remains constant.

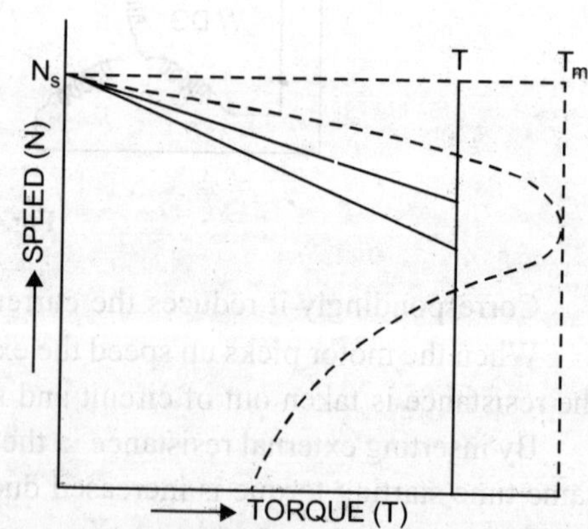

Fig.4.23: Graph between speed and torque

The disadvantages of this method of speed control are:

(*i*) *Poor efficiency:* By the introduction of external resistance in the rotor circuit, there is extra power loss (I_2^2R) in the rotor circuit which reduces the overall efficiency of the motor.

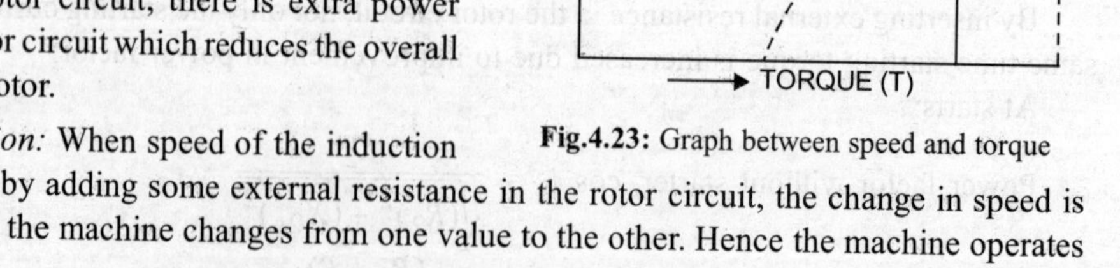

(*ii*) *Poor speed regulation:* When speed of the induction motor is controlled by adding some external resistance in the rotor circuit, the change in speed is larger when load on the machine changes from one value to the other. Hence the machine operates at a *poor regulation.*

For illustration, refer to fig 4.24. When the rotor resistance is R_1 and the load on the machine changes from half-load to full-load, the speed of the motor decreases from N_1 to N_2. However, when some resistance is added in the rotor circuit so that its value becomes R_2 (*i.e.,* $R_2 > R_1$), then the speed changes from N_3 to N_4 when load on the motor changes from half-load to full-load. It is very clear that $N_3 - N_4$ is larger than $N_1 - N_2$. Hence the machine operates at a *poor regulation.*

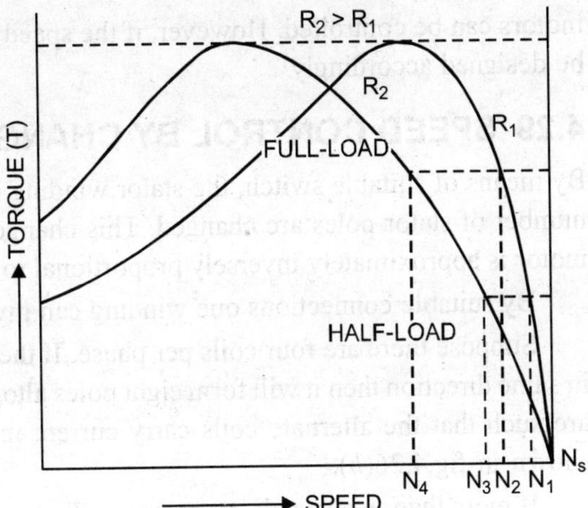

Fig.4.24: Speed-torque curve keeping R_2/S constant

2. Speed Control by Controlling the supply voltage: Slip or speed of a motor can also be changed by controlling the voltage fed to the motor. We have already seen that the torque developed by the motor is directly proportional to the square of the supply voltage. If the supply voltage is decreased, the torque developed by the motor decreases rapidly ($T \propto V^2$) and to pick-up the load slip increases or speed decreases.

For illustration, look at fig.4.25. At rated voltage and given load, the speed of the motor is N_1. If the supply voltage is reduced (say to 90%), the speed of the motor decreases to N_2 to pick-up the given load.

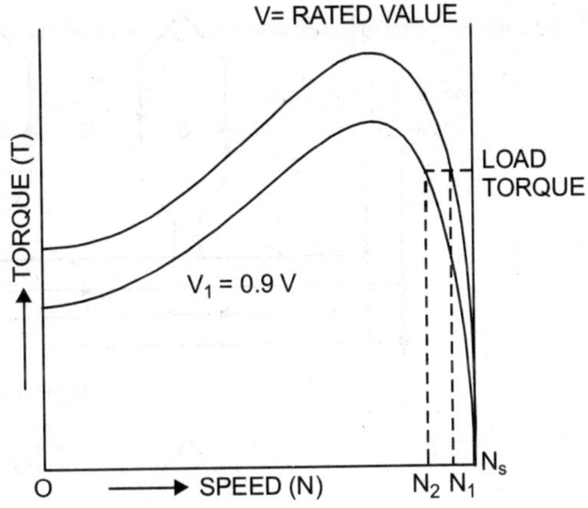

Fig.4.25: Speed-torque characteristies change in supply voltage

This method is never used for the speed control of three-phase large induction motors because the voltage control devices are very costly and bulky. However, this method is usually employed with single-phase induction motors *e.g.* ceiling fans etc.

3. Speed Control by Injecting voltage in the rotor circuit: The speed of an induction motor can also be controlled by injecting a voltage at slip frequency directly into the rotor circuit. This method was first of all introduced by K.H. Schrage of Sweeden and the motor in which this method is employed is called *Schrage motor*. If the injected e.m.f. has a component directly opposite to the rotor induced e.m.f., the motor speed decreases. On the other hand, if the injected e.m.f. has a component in phase with the rotor induced e.m.f. the motor speed increases and may rises beyond the synchronous speed.

Now a days, schrage motors are not preferred because of their heavy cost and bulky construction but these are still employed in large printing presses like newspaper printing.

4.28 SPEED CONTROL BY CHANGING THE SUPPLY FREQUENCY

The frequency of the power supply is constant, therefore, to control the speed of an induction motor by this method, the induction motor is connected to the alternator operating independently. To control the speed, the frequency of the alternator is changed. This is a costly affair.

Recent improvements in the capabilities of controlled rectifiers (*SCR*) and continued decreases in the cost of their manufacturing, it has made it possible to control the speed of induction motor by controlling the supply frequency fed to the motor. By this method 5 to 10% of rated speed of induction motors can be controlled. However, if the speed is to be controlled beyond this value, the motor has to be designed accordingly.

4.29 SPEED CONTROL BY CHANGING THE POLES

By means of suitable switch, the stator winding connections can be changed in such a manner that the number of stator poles are changed. This changes the actual speed of motor since actual speed of the motor is approximately inversely proportional to the number of poles.

By suitable connections one winding can give two different speeds.

Suppose there are four coils per phase. If these are connected in such a way that they carry current in same direction then it will form eight poles altogether as shown in fig.4.26 (*a*). Now if the connections are such that the alternate coils carry current in opposite directions, we get four poles altogether as shown in fig.4.26(*b*).

If more than two speeds are required. Two separate winding are housed in same slots and if each is arranged to give two speeds then two windings can give four different speeds.

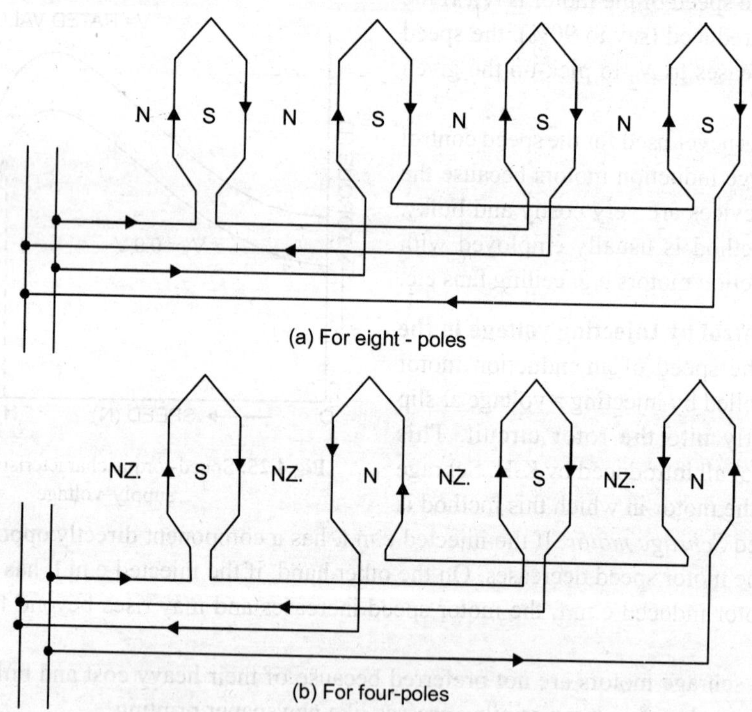

Fig.4.26: Speed control by changing number of poles

In squirrel cage motors, the rotor poles are adjusted automatically. However, in wound type motors, care has to be taken to change the rotor poles accordingly.

4.30 SPEED CONTROL BY CASCADE METHOD

A method of speed control of induction motor involving two or more motors is known as *cascade method of speed control.* The two machines are mechanically coupled with each other, one of them must be a slip ring induction motor. Mostly both slip ring motors are used having transformation ratio equal to unity. In this case supply is connected to the stator of one of the induction motor and the induced e.m.f. of the rotor from slip rings is fed to the stator or rotor of the second induction motor.

If P_1 and P_2 are the number of poles of the two machines and f is the supply frequency, then the set can given the following different speeds:

(*i*) When machine I works alone:

$$\text{The synchronous speed} = \frac{120f}{P_1}$$

(*ii*) When machine II works alone;

$$\text{The synchronous speed} = \frac{120f}{P_2}$$

(*iii*) When machine I and II are connected in cumulative cascade *i.e.*, the torque of the two motors are in same direction;

$$\text{The synchronous speed of the set} = \frac{120f}{P_1 + P_2}$$

(*iv*) When machine I and II are connected in differential cascade *i.e.*, the torque of the motors are in opposite direction;

$$\text{The synchronous speed of the set} = \frac{120f}{P_1 - P_2}$$

4.31 APPLICATIONS OF THREE-PHASE INDUCTION MOTORS

The applications of squirrel cage induction motors and slip-ring (phase wound) induction motors are given below:

1. Squirrel cage induction motors: These motors are mechanically robust and are operated almost at constant speed. These motors operate at high power factor and have high over load capacity. However, these motors have low starting torque. (*i.e.*, these motors can not pick-up heavy loads) and draw heavy current at start. On the bases of these characteristics, these motors are best suited for:

(*i*) Printing machinery
(*ii*) Flour mills
(*iii*) Saw mills
(*iv*) Shaft drives of small industries
(*v*) Pumps
(*vi*) Prime-movers with small generators etc.

2. Slip-ring (or phase-wound) induction motors: These motors have all the important characteristics (advantage) of squirrel cage induction motors and at the same time have the ability to pick-up heavy loads at start drawing smaller current from the mains. Accordingly these motors are best suited for;

(*i*) Rolling mills
(*ii*) Lifts and hoists
(*iii*) Big flour mills
(*iv*) Large pumps
(*iv*) Line shafts of heavy industries
(*vi*) Prime-moves with medium and large generators.

4.32 NATURE OF FIELD PRODUCED IN SINGLE PHASE INDUCTION MOTORS

The field produced in a single-phase induction motor can be explained by *double revolving field* theory which is given below:

This theory is based on the "Ferraris Principle" that pulsating field produced in single phase motor can be resolved into two components of half the magnitude and rotating in opposite direction at the same synchronous speed.

Thus the alternating flux which passes across the air gap of single phase induction motor at stand still consists of combination of two fields of same strength which are revolving with same speed, one in clockwise direction and the other in anticlockwise direction. The strength of each one of these fields will be equal to one half of the maximum field strength of the actual alternating field as shown in fig.4.27 (*a*).

Let ϕ_m be the pulsating field which has two components each of magnitude $\phi_m/2$. Both are rotating at the same angular speed ω_s rad/sec but in opposite direction as shown in fig.4.27 (*a*). The resultant of the two fields is ϕ_m cosθ. This shows that resultant field varies according to cosine of the angle θ. The wave shape of the resultant field is shown in fig.4.27 (*b*).

Thus an alternating field can be represented by the two fields each of half the magnitude rotating at same angular speed of ω_s *radians/sec but in opposite direction.*

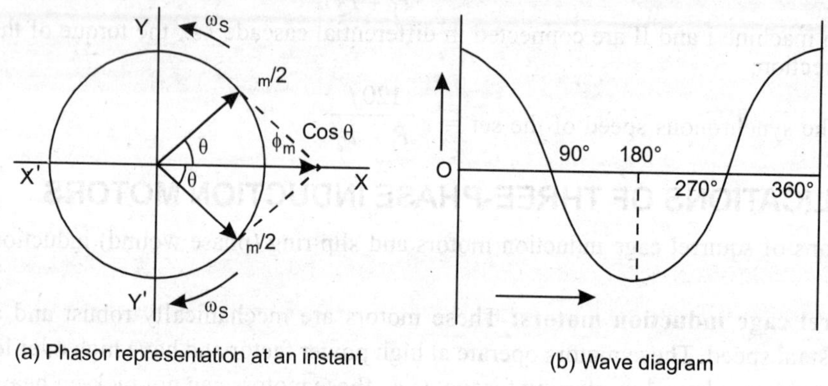

Fig.4.27: Two- field revolving theory

4.33 TORQUE PRODUCED BY 1-PHASE INDUCTION MOTOR

The two revolving fields will produce torques in opposite directions. Let the two revolving fields be field No. 1 and field No. 2 revolving in clockwise and anticlockwise direction produces torque in clockwise direction, whereas, the anticlockwise field produces torque in anticlockwise direction. The clockwise torque is plotted as positive and anticlockwise as negative. At stand still, slip for both fields is one. Synchronous speed in clockwise direction will give condition of zero slip for field 1 but it will give slip = 2 for field No. 2. Similarly synchronous speed in a counter clockwise direction will give condition of zero slip for field 2 but slip = 2 for field No. 1. The resultant torque developed in the rotor is shown by the curve passing through zero position as shown in fig.4.28. Now if we look at the resultant torque we see that the starting torque (torque at slip = 1) is zero. And except at starting there is always some magnitude of resultant torque, which shows if this type of motor once started in any direction it will develop torque in that direction and rotor will pick-up the required speed.

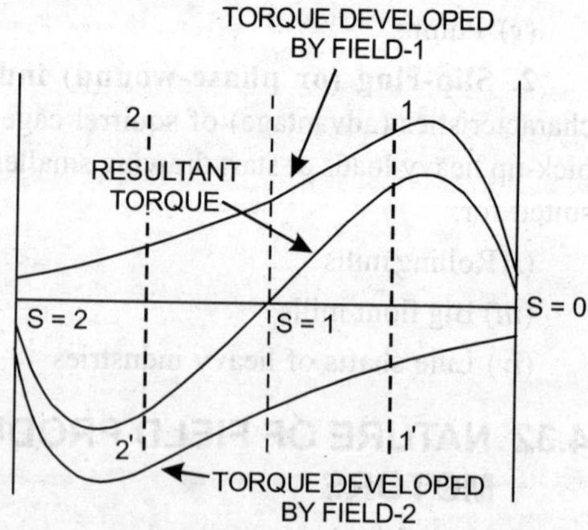

Fig.4.28: Torwue developed by two fields

The above analysis shows that single phase induction motor with single winding develops no starting torque but if the rotor is rotated in any direction by some auxiliary means in will develop torque in the same direction in which it has been rotated to start. So the problem is to find out the auxiliary means to give the starting torque to the motor.

4.34 TYPES OF 1-PHASE AC MOTORS

Various methods (means) are employed to obtain the starting torque in single-phase induction motors. Accordingly they are classified as:

1. Split phase motors
2. Shaded pole motors
3. A.C. Series motors or Commutator motors.

4.35 SPLIT PHASE MOTORS

Construction

The outer frame and stator core of a split-phase motor is similar to the outer frame and stator core of a 3-phase induction motor. The starting and running windings are placed in the stator slots. Both the windings are put in parallel as shown in fig.4.29(*a*). The purpose is to get two different currents sufficiently displaced from each other so that a revolving field is produced. The main winding which is highly inductive is connected across the line in the usual manner. The auxiliary or starting winding has a greater resistances and lesser reactance as compared to main winding.

The current in the starting winding I_s lags the supply voltage by lesser angle f_s whereas the current in the main winding I_m being highly inductive lags the supply voltage by greater angle f_m as shown in figure 4.29(*b*). The two currents have a phase difference of $q°$ electrical. Thus, a revolving field is set-up in the stator and a starting torque is developed in the rotor. Consequently rotor starts rotating and picks up the speed. A centrifugal switch which is normally closed is incorporated in series with the starting winding. When the motor attains a speed about 75% of synchronous speed, the centrifugal switch is opened automatically with the help of centrifugal force and puts the starting winding out of circuit. It is important that the centrifugal switch should open otherwise the auxiliary winding being made of thin wire will be over heated and may damage.

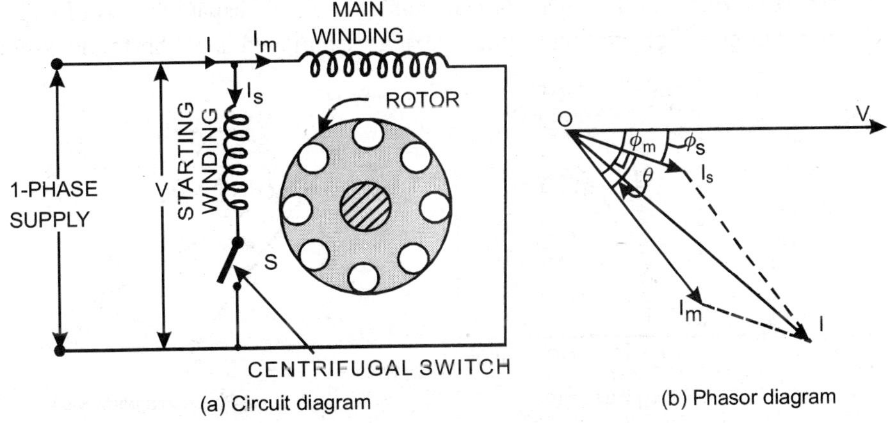

(a) Circuit diagram (b) Phasor diagram

Fig.4.29: Split-phase motor

Performance and characteristics: A typical torque speed characteristics are shown in fig.4.30, the starting torque is about twice the full load torque. The current at start is about 6 to 8 times. The speed falls with increase in load by only about 5% to 7% otherwise it is a constant speed motor. Speed is governed by the relation.

$$N_S = \frac{120\,f}{P} \text{ r.p.m.}$$

Actual speed is less than synchronous speed N_S. For the same weight its rating is about 60% to that of the poly phase induction motor. It has lower p.f. and lesser efficiency. P.f. is about 0.6 and efficiency is also about 60%.

Fig.4.30: Torque-speed curve

Applications: As starting torque is not so high so this machine is not used where large starting torque is required. It is used for smaller sizes about 0.25 h.p. It is used in washing machines, fans, blowers, wood working tools, grinders and various other low starting torque applications.

Reversal of direction of rotation: The direction of rotation of a 1-phase (split phase) induction motor can be reversed by reversing (interchanging) the connections of either starting winding or running winding.

4.36 CAPACITOR MOTORS

It is also a split phase motor. In this motor, a capacitor is connected in series with the starting winding. This is an improved form of the above said split phase motor. In these motors, the angular displacement between I_S and I_m can be made nearly 90° and high starting torques can be obtained since starting torque is directly proportional to sine of angle θ. The capacitor in the starting winding may be connected permanently or temporarily. Accordingly, capacitor motors may be

1. Capacitor start motors.
2. Capacitor run motors.
3. Capacitor start and capacitor run motors.

1. **Capacitor start motors:** In the capacitor start induction motor capacitor C is of large value such that the motor will give high starting torque. Capacitor employed is of short time duty rating.

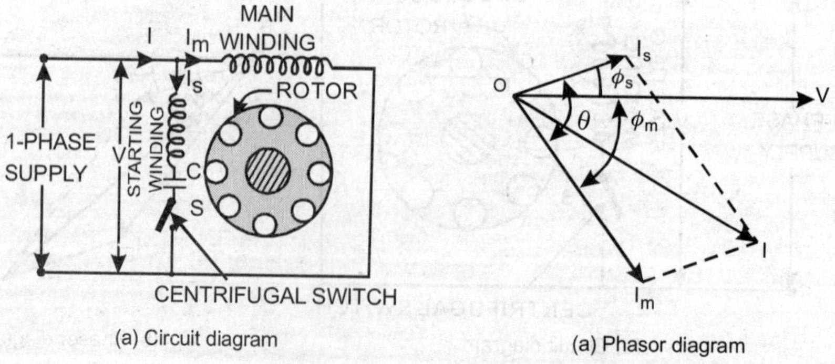

Fig.4.31: Capacitor start motor

Capacitor is of electrolytic type. Electrolytic capacitor C is connected in series with the starting winding along with centrifugal switch S as shown in fig.4.31 (*a*). When the motor attains the speed of about 75% of synchronous speed starting winding is cut-off. The construction of the motor and winding is similar to usual split phase motor. It is used where high starting torque is required such as refrigerators.

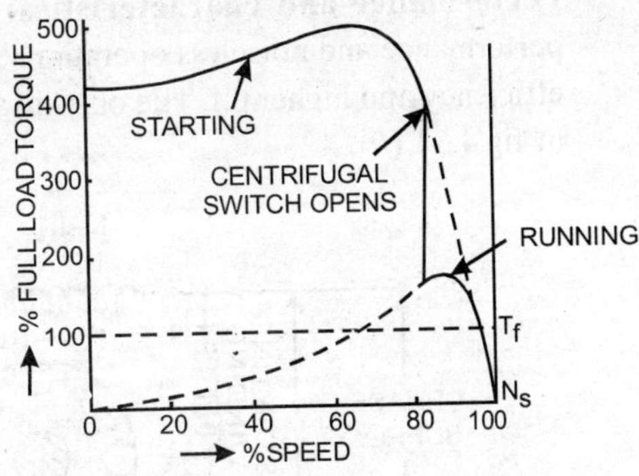

Fig.4.32: Torque-speed charactstics of

Performance and characteristics: Speed is almost constant with in 5% slip. This type of motor develops high starting torque about 4 to 5 times the full load torque. It draw low starting current. A typical torque speed curve is shown in fig.4.32. The direction of rotation can be changed by interchanging the connection of either starting or running winding.

2. **Capacitor run motors (fan motors):** In these motors, a paper capacitor is permanently connected in the starting winding, as shown in fig.4.33 (*a*). In this case, electrolytic capacitor can not be used since this type of capacitor is designed only for short time rating and hence can not be permanently connected in the winding. Both main as well as starting winding is of equal rating.

   Performance and characteristics. Starting torque is lower about 50 to 100% of full load torque. Power factor is improved may be about unity. Efficiency is improved to about 75%. A typical characteristics have been shown in fig.4.33 (*b*). It is usually used in fans, room coolers, portable tools and other domestic and commercial electrical appliances.

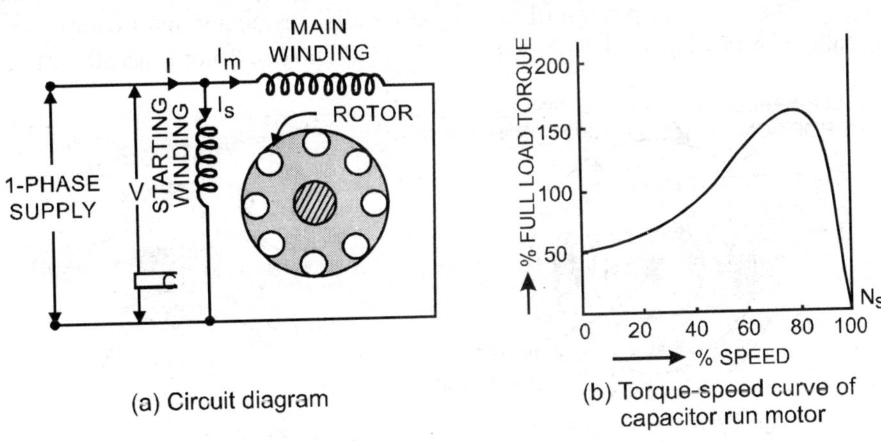

Fig.4.33: Capacitor run motor

3. **Capacitor start and capacitor run motors:** In this case, two capacitors are used one for starting purpose and other for running purpose as shown in fig.4.34 (*a*). The capacitor used for starting purpose C_S is of electrolytic type and is disconnected from the supply when the motor attains 75% of synchronous speed with the help of centrifugal switch S. Whereas, the other capacitor C_R which remains in the circuit of starting winding during running condition is a paper capacitor. This type of motor gives best running and starting operation. Starting capacitor C_S which is of higher value than the value of running capacitor C_R.

Performance and characteristics: Such motors operate as two phase motors giving best performance and noiseless operation. Starting torque is high, starting current is low and give better efficiency and higher p.f. The only disadvantage is high cost. A typical torque speed curve is shown in fig.4.34 (*b*).

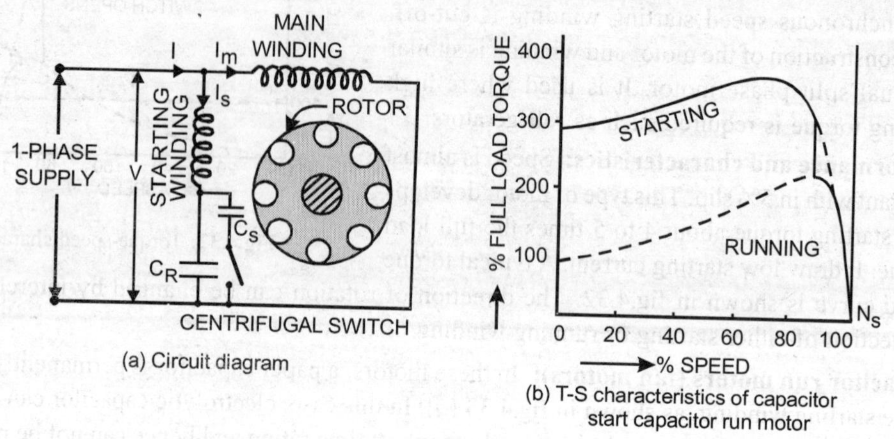

Fig.4.34: Capacitor start capacitor run motor

4.37 SHADED POLE MOTOR

Construction

Shaded pole motor is constructed with salient poles in stator. Each pole has its own exciting winding as shown in fig.4.35.(*a*). A 1/3rd portion of each pole core is surrounded by a copper strip forming a closed loop called the shading band as shown in fig.4.35 (*a*) and (*b*). Rotor is usually squirrel cage type.

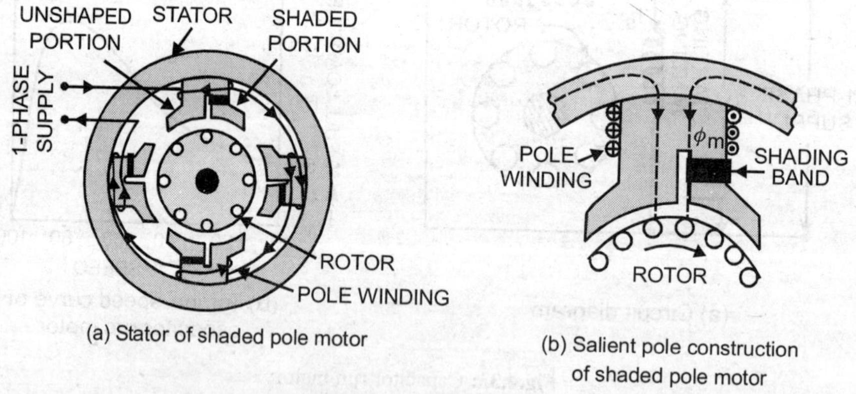

Fig.4.35: Shaded pole motor

When a single phase supply is given to the stator (exciting) winding, it produces alternating flux. When the flux is increasing in the pole, a portion of the flux attempts to pass through the shaded portion of the pole. This flux induces voltage and hence current in the copper ring, and by Lenz's law the direction of current is such that it opposes the causes *i.e. increase* of flux in shaded portion. Hence in the beginning, the greater portion of flux passes through unshaded side of each pole and resultant lies on unshaded side of the pole. When the flux reaches its maximum value, its rate of change is zero, thereby the e.m.f. and hence current in the shading coil becomes zero. Flux is uniformly distributed over the pole phase and the resultant field lies at the centre of the pole. After that the main flux tends to decrease, the current induced in he shading coil now tends to increase the flux on the shaded portion of the pole and resultant lies on the shaded portion of the pole as shown in fig.4.36.

Hence, a revolving field is set up which rotates from unshaded portion of the pole to the shaded portion of the pole as marked by the arrow head in fig.4.36. Thus, by electromagnetic induction, a starting torque develops in the rotor and the rotor starts rotating. After that its rotor picks up the speed.

Performance and characteristics: A typical speed torque characteristics is shown in fig.4.37. Starting torque is very small about 50% of full load torque. Efficiency is low because of continuous power loss in shading coil. These motors are used for small fans, electric clocks, gramophones etc.

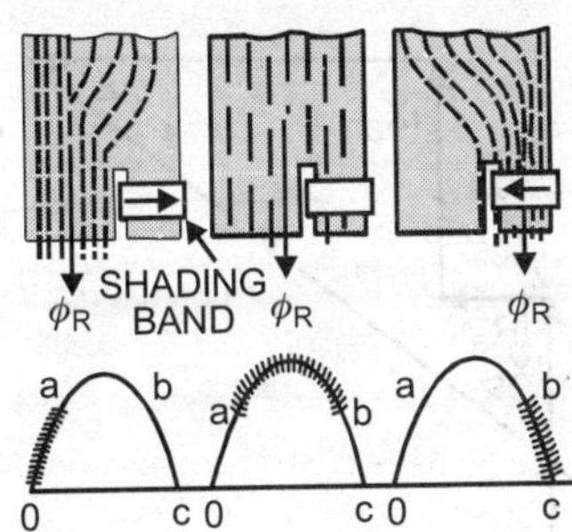

Fig.4.36: Flux distribution in the pole during half cycle

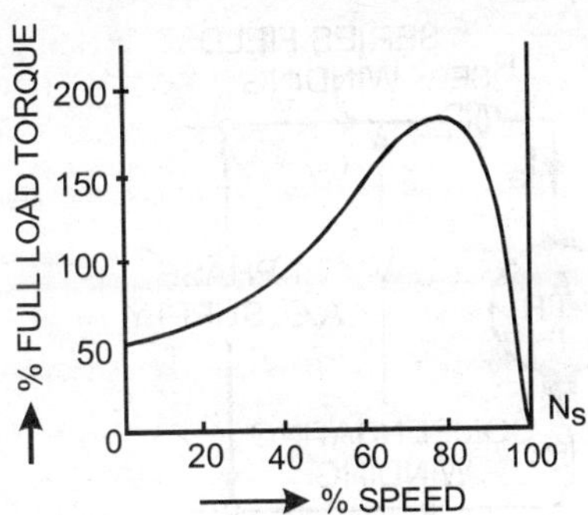

Fig.4.37: Torque-speed characteristics of shaded-pole motor

Its direction of rotation depends upon the position of the shading coil, *i.e.,* which portion of the pole is wrapped with shading coil. The direction of rotation is from unshaded portion of the pole to the shaded portion. Its direction of rotation cannot be reversed unless the position of the poles is reversed.

4.38 A.C SERIES MOTOR OR COMMUTATOR MOTOR

When a 1-phase a.c. supply is given to a d.c. series motor, a unidirectional torque is developed in it. In fact, during positive half cycle, same current flows in the series field winding and armature winding which develops a torque in one direction (say clockwise). During the negative half cycle, current flowing through series field winding is reversed and at the same time current flowing through the armature also reverses, therefore, torque is developed in the same direction (*i.e.* clockwise direction). Thus, a continuous rotation is obtained.

Mathematically we know torque in D.C. series motors, $T \propto \phi_{se} I_a$ where ϕ_{se} is the series field winding flux and I_a is the armature current.

Now when A.C. supply is given to series motor $T \propto \phi_{se} I_a$ for positive half cycle. For negative half cycle $T \propto (-\phi_{se})(-I_a) \propto \phi_{se} I_a$. Thus same torque is produced during positive and negative half cycle. However, some modifications are necessary in a d.c. series motor for its satisfactory operation on a.c. which are given below:

(1) The iron structure of field and yoke are laminated.

(2) The series field winding is so designed that it will produce relatively small m.m.f as compared to d.c. This is done by reducing the number of turns. The smaller field m.m.f would result in reduced

air gap flux. Therefore, in order to develop the necessary torque the number of armature conductors have to be increased proportionately.

(3) Increase in armature conductors would result in increased inductive reactance of the armature so the net inductive reactance may not be reduced. In order to over come this difficulty compensating winding is connected in series with armature as shown in fig.4.38. This completely neutralises the inductive effect of the armature winding.

(4) The reluctance of the magnetic circuit is reduced to have high flux with reduced m.m.f. So magnetic material used should be of high permeability and air gap should be small.

As shown in the vector diagram (fig.4.39) large voltage drop occurs in resistance and reactance of the armature and field winding. Voltage left for operation is only E.

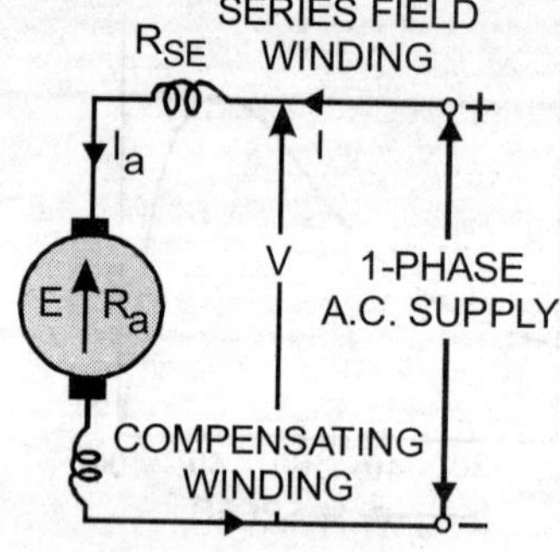

Fig.3.38: Circuit diagram of ac series motor

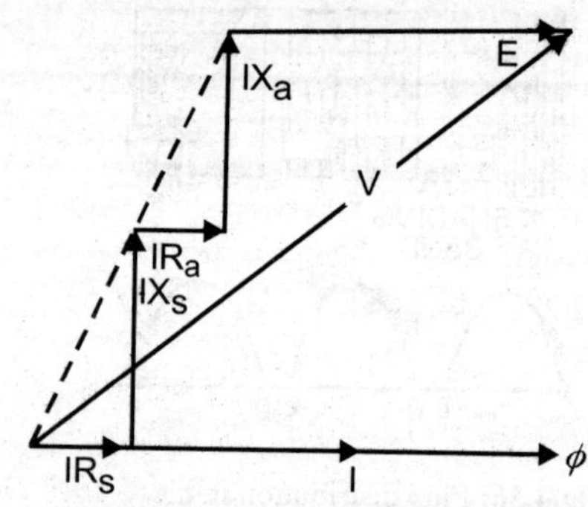

Fig.4.39: Voltage distribution

Where $\overline{E} = \overline{V} - (\overline{IR_S} + \overline{IX_S} + \overline{IR_a} + \overline{IX_a})$

Where $IR_S \longrightarrow$ voltage drop in series winding resistance

$IX_S \longrightarrow$ voltage drop in series winding reactance

$IR_a \longrightarrow$ voltage drop in armature resistance

$IX_a \longrightarrow$ voltage drop in armature reactance.

Performance and characteristics: The speed-torque characteristic for d.c. and a.c. series motors are shown in fig.4.40. The torque varies as square of the current and the speed varies inversely proportional to the current approximately. The efficiency will not be good as that of corresponding d.c. machine because of greater eddy current loss and effects of p.f.

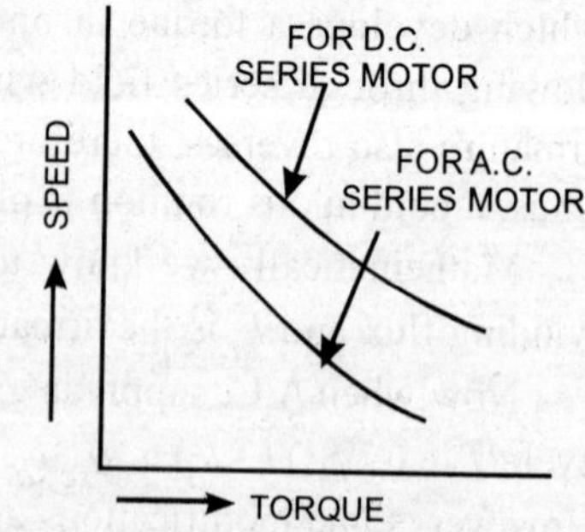

Fig.4.40: Torque-speed characteristics

These motors have their wide applications where high speed (20000 r.p.m.) is required *e.g.* mixer grinders, blowers, hair dryers etc.

4.39 UNIVERSAL MOTOR

A motor which can be operated on ac as well as on dc supply at the rated voltage is called **universal motor.**

Basically, universal motor is an ac series motor. It is just an improved from of a dc series motor. The core size of an universal motor is more than the core size of a dc series motor of the same rating.

Construction

The motor has two main parts namely.

1. Stator and 2. Rotor

1. **Stator.** It is the stationary part of the motor. It consists of *magnetic frame* (or yoke), *pole core* and *pole shoe* and *field* or *exciting winding* as shown in fig.4.41.

   The magnetic frame, pole core and pole shoe are made of silicon steel stampings. The stampings are insulated from each other by varnish layer. The hysteresis losses are very small in silicon steel and eddy current losses are reduced due to stampings. The field winding made of enameled copper is wound around the poles to produce the required flux.

2. **Rotor.** It is the rotating part of the motor. It consists of *shaft, armature, armature winding* and *commutator,* as shown in fig.4.42.

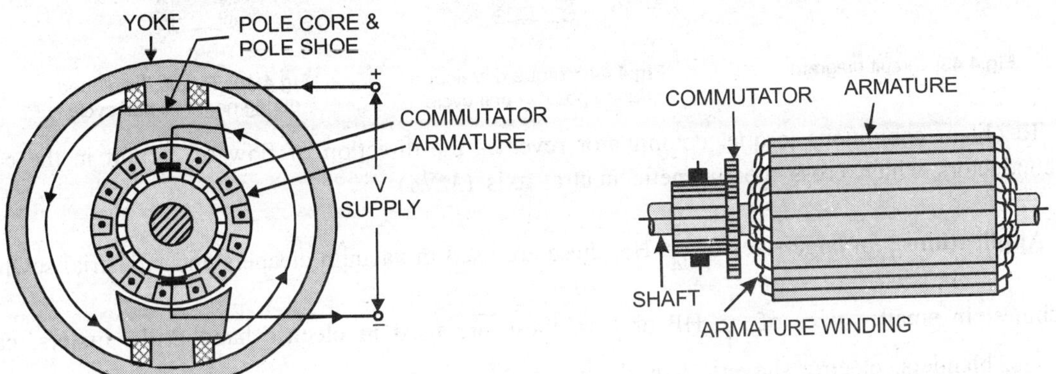

Fig.4.41: Universal motor

Fig.4.42: Rotor of a universal motor

Shaft is a part of rotor which transfers mechanical power or energy to the load. It is made-up of mild steel. Armature is made-up of stampings of silicon steel material since it is to carry the magnetic field. It is keyed to the shaft. Slots are cut at its outer periphery to accommodate armature winding. The ends of armature winding are braced to the commutator segments. Commutator is made-up of wedge shaped segments forming a ring. The wedges are insulated from each other by an insulating layer of micanite. The commutator is also keyed to the shaft.

Carbon brushes are pressed over the commutator surface to deliver current to the machine.

Principle

When a current carrying conductor is placed in the magnetic field, a force is exerted on it and torque develops. In other words, when the rotor field produced by the rotor current carrying conductors, tries to come in line with the main field, torque develops and rotor rotates.

Working

The armature winding and stator field winding both are connected in series, as shown in fig.4.43. When 1-phase, ac supply is given to the motor, current flows through the field winding and armature winding. The field winding sets up main stator field F_m, and the armature winding sets up rotor field F_r as shown in fig.4.44. Rotor field F_r tries to align itself with the main field F_m and an anticlockwise torque in produced.

During negative half cycle, the direction of flow of current in the field winding as well as in the armature winding is reversed, as shown in fig.4.45. The two windings set up their fields in the direction as shown in fig.4.45, again anticlockwise torque is produced in the rotor. Thus, unidirectional torque is produced in the motor.

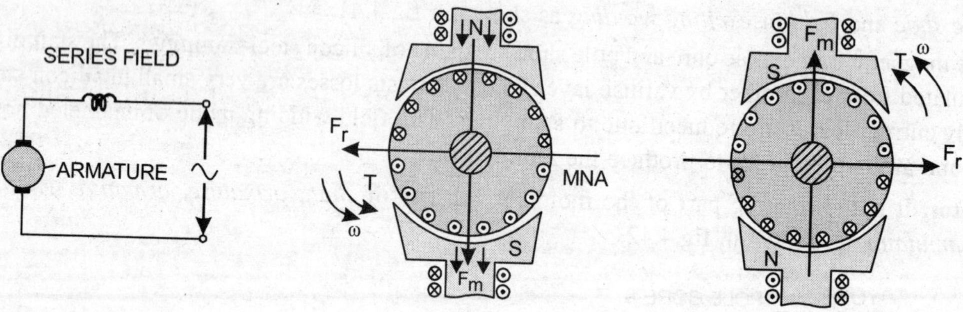

Fig.4.43: Circuit diagram

Fig.4.44: Torque developed during positive half cycle

Fig.4.45: Torque developed during negative half cycle

To obtain continuous torque, commutator reverses the direction of flow of current in the coil or conductors which cross the magnetic neutral axis (*MNA*)

Applications: In large sizes of $\frac{3}{4}$ HP, these are used in vacuum cleaners and industrial sewing machines. In smaller sizes of $\frac{1}{4}$ HP or less, these are used in electric hand drills, mixers, can openers, blenders, electric shavers, hair dryers etc.

4.40 D.C MOTOR

A machine which converts mechanical power into d.c. electrical power is called a d.c. generator. The same machine when used to convert d.c. electrical power into mechanical power is known as a d.c. motor. From construction point of view there is no difference between a d.c. generator and motor.

The d.c. motors are very useful where wide range of speeds and good speed regulation is required such as electric fraction..

4.41 MAIN CONSTRUCTIONAL FEATURES

The complete assembly of various parts in a scattered form of a d.c. machine is shown in fig.4.46. The essential parts of a d.c. machine are described below:

1. Magnetic frame or Yoke

The outer cylindrical frame to which main poles and inter poles are fixed and by means of which the machine is fixed to the foundation is called the *yoke*. It serves two purposes:

(*i*) It provides mechanical protection to the inner parts of the machine.

(*ii*) It provides a low reluctance path for the magnetic flux.

The yoke is made of cast iron for smaller machines and for larger machines, it is made of cast steel or fabricated rolled steel since these materials have better magnetic properties as compared to cast iron.

2. Pole core and pole shoes

The pole core and pole shoes are fixed to the magnetic frame or yoke by bolts. They serve the following purposes:

(*i*) They support the field or exciting coils.

(*ii*) They spread out the magnetic flux over the armature periphery more uniformly.

(*iii*) Since pole shoes have larger X-section, the reluctance of magnetic path is reduced.

Usually, the pole core and pole shoes are made of thin cast steel or wrought iron laminations which are riveted together under hydraulic pressure as shown in fig.4.47 (*a*).

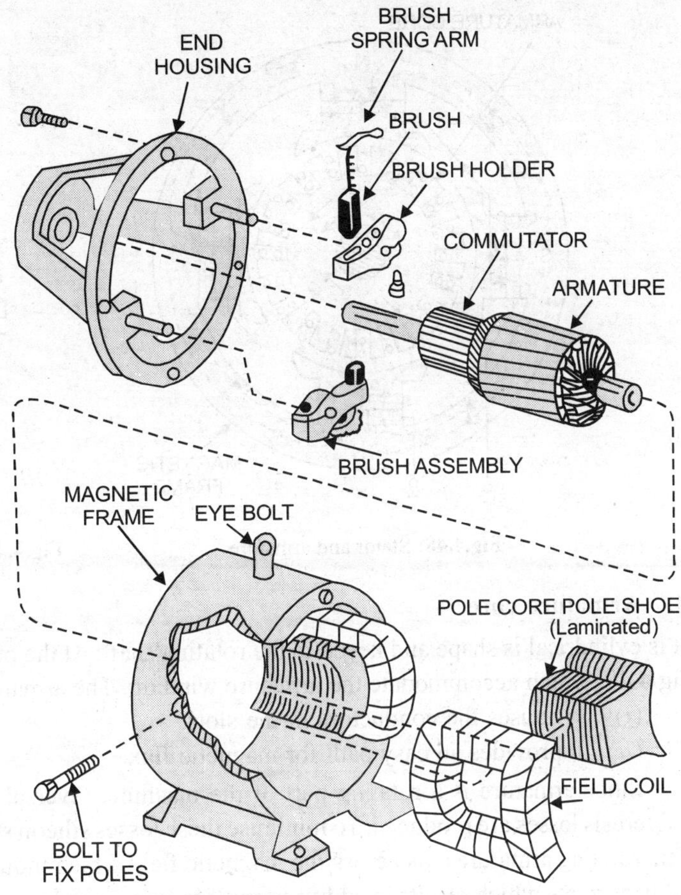

Fig.4.46: Dis-assembled view of a dc machine

3. Field or Exciting coils

Enamelled copper wire is used for the construction of field or exciting coils. The coils are wound on the former [see fig 4.47(*b*)] and then placed around the pole core as shown in fig 4.47(*a*). When direct current is passed through the field winding, it magnetises the poles which produce the required flux. The field coils of all the poles are connected in series in such a way that when current flows through them, the adjacent poles attain opposite polarity as shown if fig.4.48.

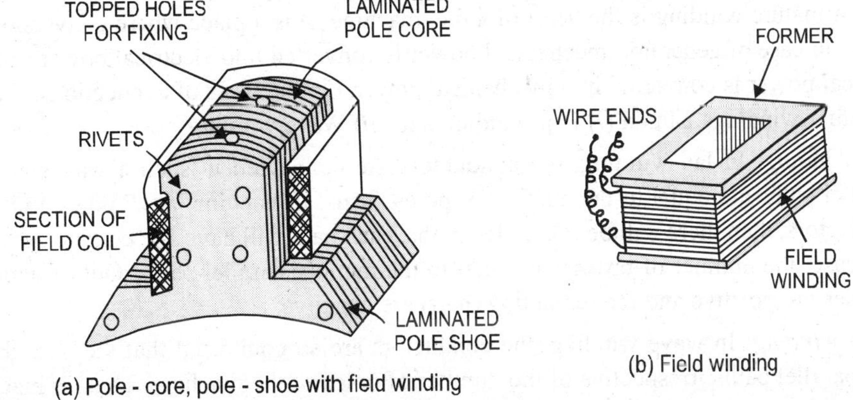

(a) Pole - core, pole - shoe with field winding

(b) Field winding

Fig.4.47: Pole with field winding

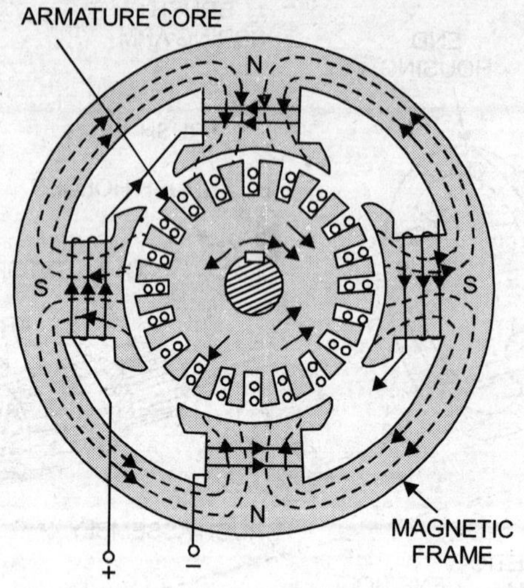

Fig.4.48: Stator and armature

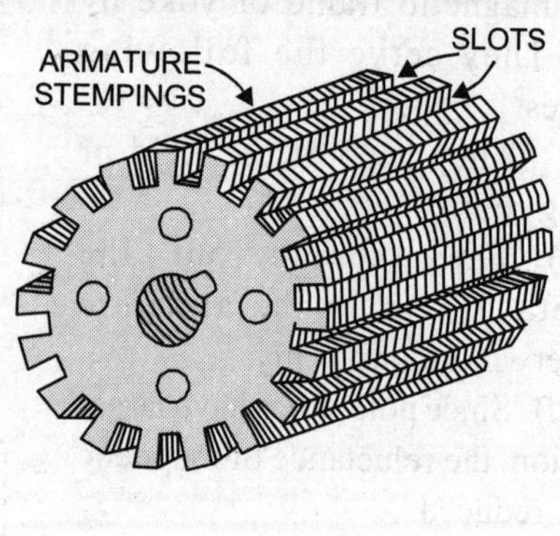

Fig.4.49: Armature of a dc machine

4. Armature core

It is cylindrical is shape and keyed to the rotating shaft. At the outer periphery slots are cut, as shown in fig.4.49, which accommodate the armature winding. The armature core serves the following purposes:

(*i*) It houses the conductors in the slots.

(*ii*) It provides an easy path for magnetic flux.

Since armature is a rotating part of the machine, reversal of flux takes place in the core, hence hysteresis losses are produced. To minimise these losses silicon steel material is used for its construction. The rotating armature cuts across the magnetic field which induces an e.m.f in it. This e.m.f. circulates eddy currents which results in eddy current loss in it. To reduce these losses, armature core is laminated, in other words we can say that about 0.3 to 0.5 mm thick stampings are used for its construction. Each lamination or stamping is insulated from the other by varnish layer (see fig.4.49).

5. Armature winding

The insulated conductors housed in the armature slots are suitably connected. This is known as armature winding. The armature winding is the heart of a d.c. machine. It is a place where conversion of power takes place *i.e.* in case of generator, mechanical power is converted into electrical power and in case of motor, electrical power is converted into mechanical power. On the basis of connections, there are two types of armature windings named (*i*) Lap winding and (*ii*) Wave winding.

(*i*) *Lap winding*: In lap winding, the conductors are connected in such a way that number of parallel paths are equal to the number of poles. Thus, if machine has P poles and Z armature conductors, then there will be P parallel, paths, each path will have Z/P conductors in series. In this case, the number of brushes is equal to the number parallel paths. Out of which half the brushes are positive and the remaining (half) are negative.

(*ii*) *Wave winding*: In wave winding, the conductors are so connected that they are divided into two parallel paths irrespective of the number of poles of the machine. Thus, if machine has Z armature conductors, there will be only two parallel paths each having $Z/2$ conductors in series. In this case, the number of brushes is equal to two *i.e.* number of parallel paths.

6. Commutator

It is the most important part of a d.c machine and serves the following purposes:

(*i*) It connects the rotating armature conductors to the stationary external circuit through brushes.

(*ii*) It converts the alternating current induced in the armature conductors into unidirectional current in the external load circuit in generator action, whereas, it converts the alternating torque into unidirectional (continuous) torque produced in the armature in motor action.

The commutator is of cylindrical shape and is made up of wedge-shaped hard drawn copper segments. The segments are insulated from each other by a thin sheet of mica. The segments are held together by means of two V-shaped rings that fit into the V-grooves cut into the segments. Each armature coil is connected to the commutator segment through riser. The sectional view of the commutator assembly is shown in fig.4.50

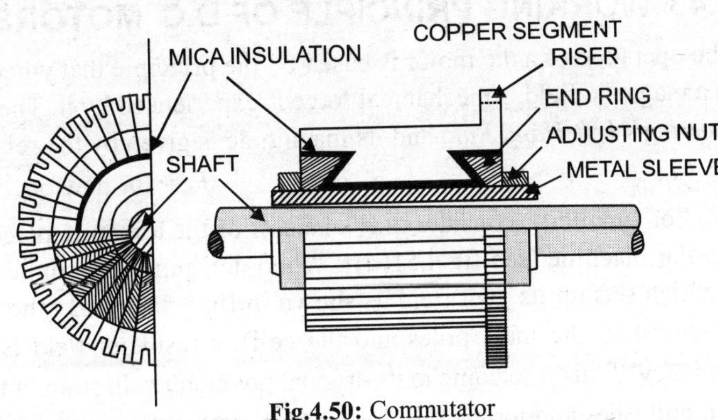

Fig.4.50: Commutator

7. Brushes

The brushes are pressed upon the commutator and form the connecting link between the armature winding and the external circuit. They are usually made of high grade carbon because carbon is conducting material and at the same time in powdered form provides imbricating effect on the commutator surface. The brushes are held in particular position around the commutator by brush holders and rocker.

8. Brush rocker

It holds the spindles of the brush holders. It is fitted on to the stationary frame of the machine with nut and bolts. By adjusting its position, the position of the brushes over the commulator can be adjusted to minimize the sparking at the brushes.

9. End housings

End housings are attached to the ends of the main frame and support bearings. The front housing supports the bearing and the brush assemblies whereas the rear housing usually supports the bearing only.

10. Bearings

The ball or roller bearings are fitted in the end housings. The function of the bearings is to reduce friction between the rotating and stationary parts of the machine. Mostly high carbon steel is used for the construction of bearings as it is very hard material.

11. Shaft

The shaft is made of mild steel with a maximum breaking strength. The shaft is used to transfer mechanical power from or to the machine. The rotating parts like armature core, commutator, cooling fan etc. are keyed to the shaft.

4.42 ARMATURE RESISTANCE

The resistance between the armature terminals is called *armature resistance.* It is generally represented by R_a The value of armature resistance is usually quite small (less than one ohm). Armature resistance depends upon the following factors:

(*i*) Length, area of cross-section and material of armature winding.

(*ii*) Type of armature winding *i.e* lap or wave winding. This will show the manner in which the conductors (*i.e.* their series-parallel combination) are connected.

4.43 WORKING PRINCIPLE OF D.C MOTORS

The operation of a d.c motor is based on the principle that when a current carrying conductor is placed in a magnetic field, a mechanical force is experienced by it. The direction of this force is determined by *Fleming's Left Hand Rule* and its magnitude is given by the relation:

$$F = Bil \text{ newton}$$

For simplicity, consider only one coil of the armature placed in the magnetic field produced by a bipolar machine [see fig 4.51(*a*)]. When d.c. supply is connected to the coil, current flows through it which sets up its own field as shown in fig 4.51 (*b*). By the interaction of the two fields (*i.e.* field produced by the main poles and the coil), a resultant field is set up as shown in fig 4.51 (*c*). The tendency of this is to come to its original position *i.e.* in straight line due to which force is exerted on the two coil sides and torque develops which rotates the coil.

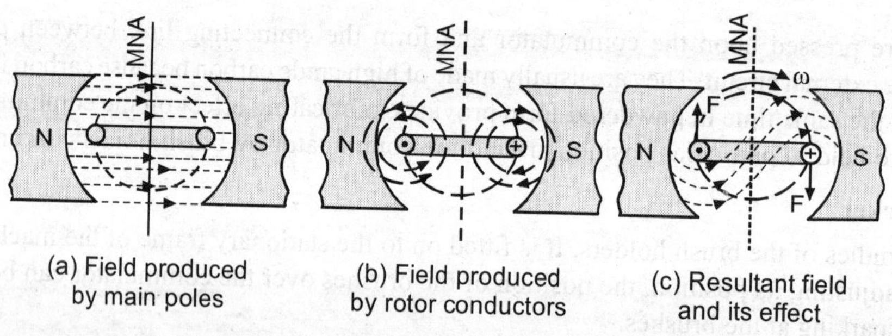

Fig.4.51: Magnetic field developed in a dc machine

Alternately, it can be said that the main poles produce a field F_m. Its direction is marked in fig 4.52. When current is supplied to the coil (armature conductors), it produces its own field marked as F_r. This field tries to come in line with the main field and an electromagnetic torque develops in clockwise direction as marked in fig.4.52.

In actual machine, a large number of conductors are placed on the armature. All the conductors, placed under the influence of one pole (say North pole) carry the current in one direction (outward). Whereas, the other conductors placed under the influence of other pole *i.e.* south pole, carry the current in opposite direction as shown in fig 4.53. A resultant rotor field is produced. Its direction is marked by the arrow-head F_r. This rotor field F_r tries to come in line with the main field F_m and torque (T_e) developes. Thus, rotor rotates.

It can be seen that to obtain a continuous torque, the direction of flow of current in each conductor or coil side must be reversed when it crosses the magnetic neutral axis (*MNA*). The is achieved with the help of a commutator.

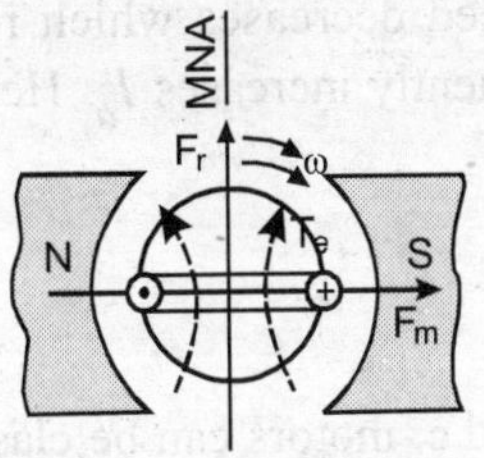

Fig.4.52: Position of axis of main field and rotor field

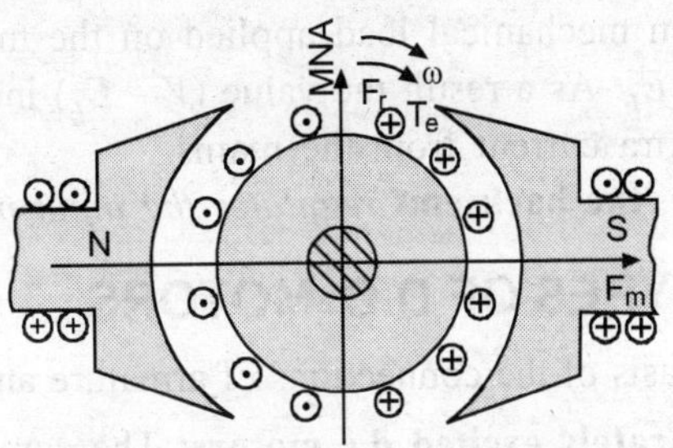

Fig.4.53: Torque development

Function of a commutator

The function of a commutator in d.c. motor is to reverse the direction of flow of current in each armature conductor when it crosses the M.N.A. to obtain continuous torque.

4.44 BACK E.M.F.

It has been seen that when current is supplied to the armature conductors placed in the main magnetic field, torque develops and armature rotates, the armature conductors cut across the magnetic field and an e.m.f. is induced in these conductors. The direction of this induced e.m.f. in the armature conductors is determined by Fleming's Right Hand Rule. It is marked in fig.4.54 (*b*).

It can be seen that the direction of this induced e.m.f. is opposite to the applied voltage. That is why this induced e.m.f. is called back e.m.f. (E_b). The magnitude of this induced e.m.f. is given by the relation; $E_b = PZ\phi N/60A$

A simple conventional circuit diagram of the machine working as motor is shown in fig.4.55. In this case, the supply voltage is always greater than the induced or back e.m.f. (*i.e* $V > E_b$). Therefore, current is always supplied to the motor from the mains and the relation among the various quantities will be; $E_b = V - I_a R_a$.

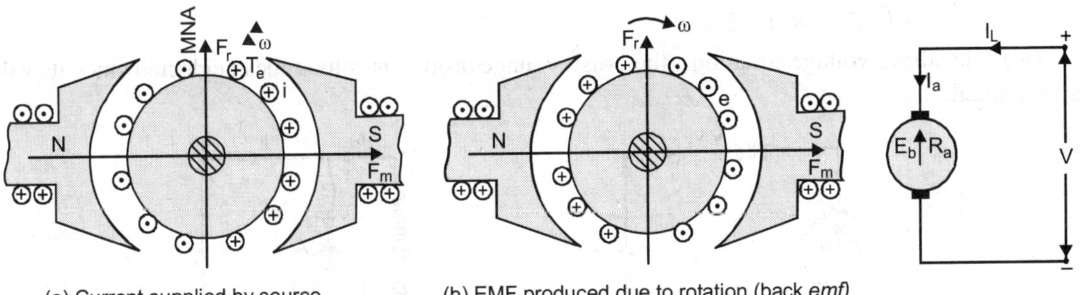

(a) Current supplied by source

(b) EMF produced due to rotation (back *emf*)

Fig.4.54:

Fig.4.55: Circuit diagram

Significance of back emf

The current flowing through the armature is given by the relation:

$$I_a = \frac{V - E_b}{R_a}$$

When mechanical load applied on the motor increases, its speed decreases which reduces the value of E_b. As a result the value $(V - E_b)$ increases which consequently increases I_a. Hence, motor draws extra current from the mains.

Thus, **the back emf** *regulates the input power as per the load.*

4.45 TYPES OF D.C MOTORS

On the basis of the connections of armature and their field winding, d.c. motors can be classified as;

1. **Separately excited d.c motors:** The conventional diagram of a separately excited d.c. motor is shown fig 4.56. Its voltage equation will be;
   $E_b = V - I_a R_a - 2n_b$
2. **Self excited d.c. motors:** These motors can be further classified as:
   (*i*) **Shunt motors:** Their conventional diagram is shown in fig 4.57.
   Important relations: $I_{sh} = V/R_{sh}$; $I_a = I_L - I_{sh}$
   $E_b = V - I_a R_a - 2u_b$

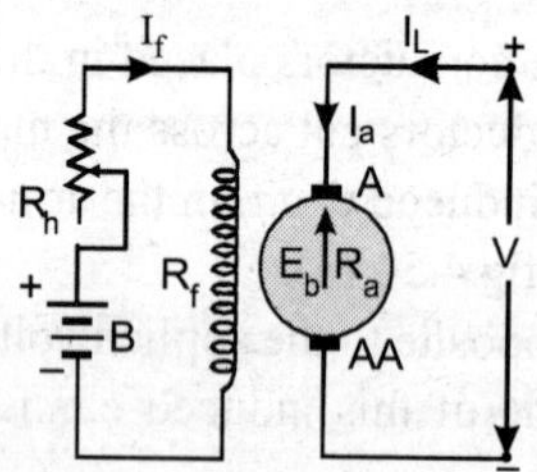

Fig.4.56: Circuit of separately excited motor

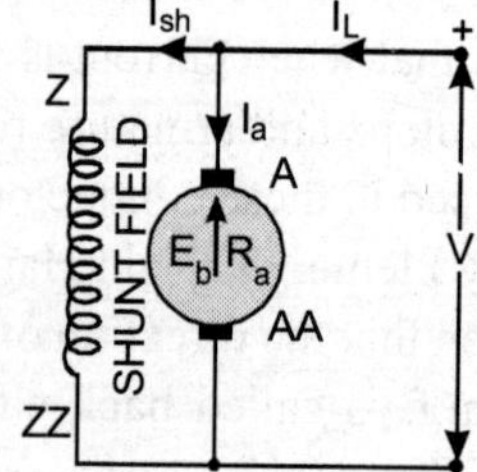

Fig.4.57: Circuit of shunt motor

(*ii*) **Series motor:** Its conventional diagram is shown in fig.4.58.
Important relations: $I_L = I_a = I_{se}$; $E_b = V - I_a (R_a + R_{se}) - 2\upsilon_b$

(*iii*) **Compound motor:** Its conventional diagram is shown in fig.4.59.
$E_b = V - I_a (R_a + R_{se}) - 2\upsilon_b$

In all the above voltage equations, the brush voltage drop v_b is some times neglected since its value is very small.

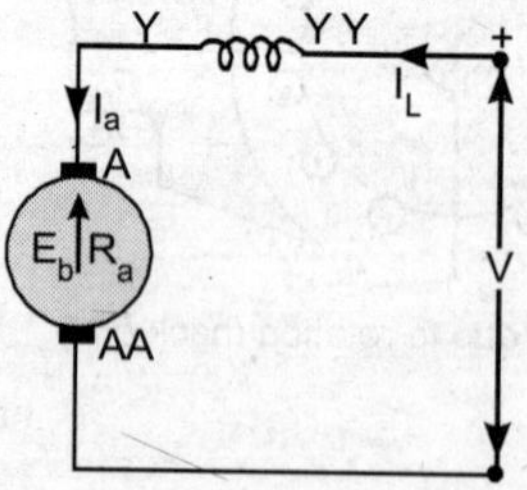

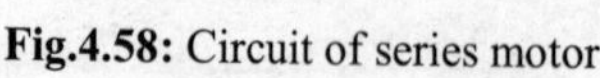

Fig.4.58: Circuit of series motor

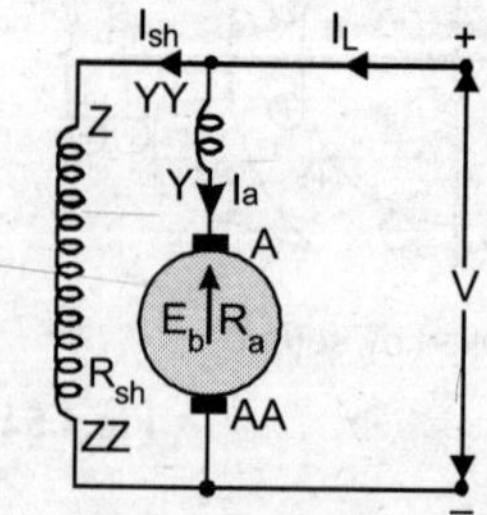

Fig.4.59: Circuit of compound motor

4.46 CHARACTERISTICS OF D.C MOTORS

The performance of a d.c. motor can be easily judge from its characteristic curves, known as motor *characteristics*. The characteristics of a motor are those curves which show relation between the two quantities. On the basis of these quantities, the following characteristics can be obtained:

1. **Speed and Armature current i.e. N – I_a Characteristics:** It is the curve drawn between speed N and armature current I_a. It is also known as *speed characteristics.*
2. **Torque and Armature current i.e. T – I_a Characteristics:** It is the curve drawn between torque developed in the armature T and armature current I_a. It is also known as *electrical characteristic.*
3. **Speed and Torque i.e. N – T characteristics:** It is the curve drawn between speed N and torque developed in the armature T. It is also known as *mechanical characteristics.*

The following important relations must be kept in mind while discussing the motor characteristics:

$$E_b \propto N\phi \quad \text{or} \quad N \propto \frac{E_b}{\phi} \quad \text{and} \quad T \propto \phi I_a$$

4.47 CHARACTERISTICS SEPARATELY EXCITED DC MOTORS

The conventional diagram of this motor is shown in fig 4.60. In these motors, the field current $I_f = V/R_f$ remains constant since the supply voltage V is constant. Hence, the flux in these motors is practically constant.

1. N – I_a characteristics

We know that, $N \propto \frac{E_b}{\phi}$ Since flux is constant; $N \propto E_b$ or $N \propto V - I_a R_a$

If the armature drop ($I_a R_a$) is negligible, the speed of the motor will remain constant for all values of load as shown by the dotted line AB in fig 4.61. But strictly speaking, as the armature current increases due to the increase of load, armature drop $I_a R_a$ increases and speed of the motor decreases slightly as shown by the straight line $A\,C$ in fig 4.61 (neglecting armature reaction). Moreover, the characteristic curve does not start from a point of zero armature current because a small current, no-load armature current I_{a0}, is necessary to maintain rotation of the motor at no load.

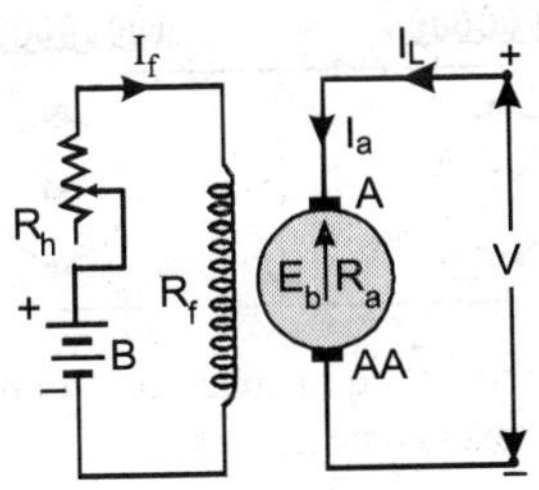

Fig.4.60: Separately excited dc motor

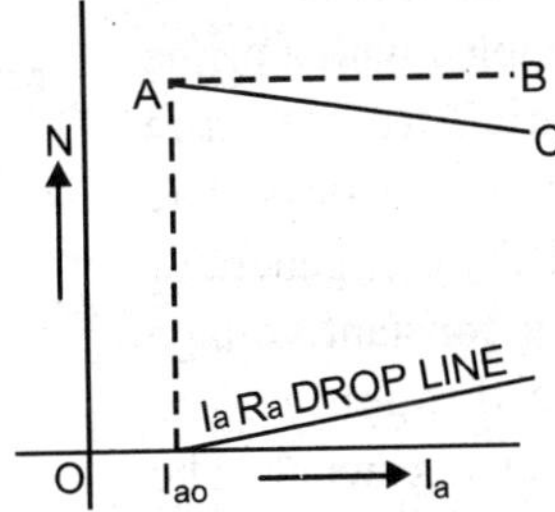

Fig.4.61: N-Ia characteristics

Since there is no appreciable change in the speed of a separately excited dc motor from no load to full load that is why it is considered to be a constant speed motor. This motor is best suited where almost constant speed is required and the load may be thrown off totally and suddenly.

2. T – I_a Characteristics

We know that, $T \propto \phi I_a$

Since flux is constant (independent to the voltage supplied to the armature), $T \propto I_a$

Hence, the electrical characteristic (*i.e.* $T - I_a$) is a straight line passing through the origin as shown in fig.4.62. It is clear from the characteristic curve that a large armature current is required at the start if machine is on heavy load. Hence, these motors are not started with heavy loads.

3. N – T Characteristics

The $N-T$ characteristic is derived from the first two characteristics. When load torque increases, armature current I_a increases but speed decreases slightly. Thus with the increase in load or torque, the speed decreases slightly as shown in fig 4.63.

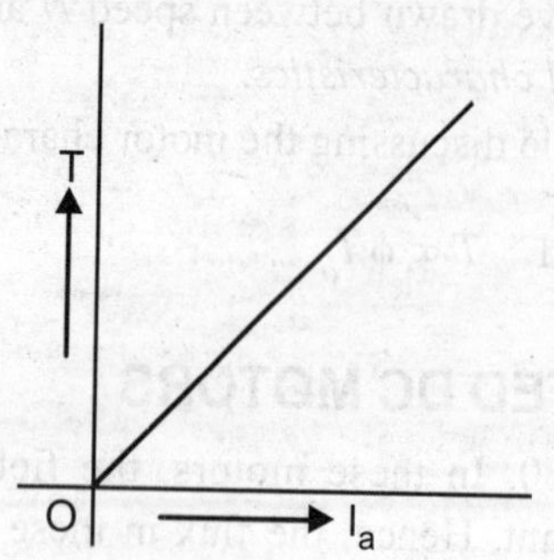

Fig.4.62: *T-I_0* charactoristics

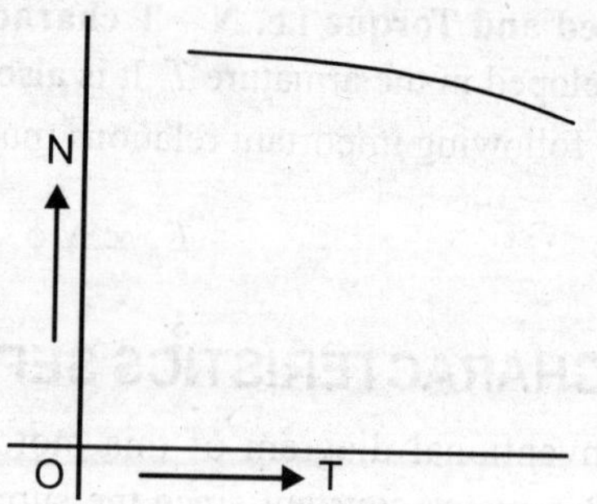

Fig.4.63: *N-T* characteristies

4.48 SPEED CONTROL OF SEPARATELY EXCITED MOTORS

The most common and accurate speed control method of separately excited d.c. motors is Ward-Leonard system as explained below:

Ward-Leonard System: This system is used to supply variable voltage to the motor. As shown in fig.4.64, a d.c. generator G is mechanically coupled with a prime mover PM which rotates the generator at constant speed. The field winding of the d.c. generator is connected to a constant voltage d.c. supply line through a field regulator and reversing switch. The d.c. motor M is fed from the generator G and its field winding is connected directly to a constant d.c. supply line.

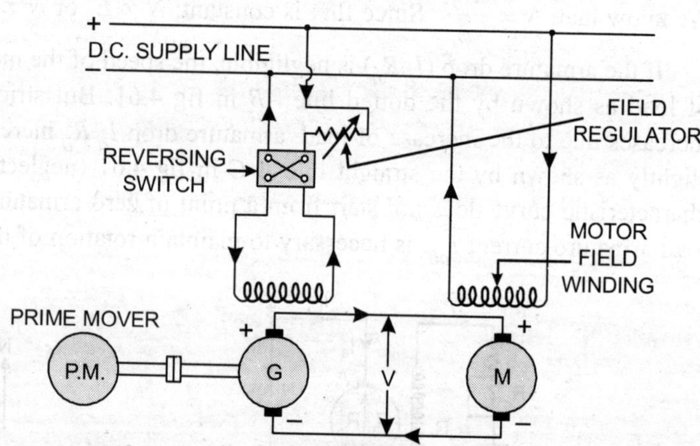

Fig.4.64: Ward-leonard method of speed control of resparately excited dc motors

The voltage of the generator fed to the motor, can be varied from zero to its maximum value by means of its field regulator.

By reversing the direction of the field current by means of the reversing switch, the polarity of the generated voltage can be reversed and hence the direction of rotation of motor M. Hence, by this method, the speed and direction of rotation both can be controlled very accurately.

The capital investment in this system is very high as two extra machines (a generator and a prime mover) are required. This system of speed control is best suited where almost unlimited speed control in either direction of rotation is required *e.g.* in steel rolling mills, paper industry, elevators, cranes, diesel-electric propulsion of ships etc.

4.49 SPEED REGULATION

The *speed regulation of a d.c. motor is defined as the change in speed from full load to no load and is expressed as a percentage of the full load speed.*

$$\therefore \quad \% \text{ Speed regulation} = \frac{\text{N.L.speed} - \text{F.L.speed}}{\text{F.L.speed}} \times 100 = \frac{N_0 - N}{N} \times 100$$

4.50 SYNCHRONOUS MACHINE

A synchronous machine is an a.c. machine whose satisfactory operation depends upon the maintenance of the following relationship:

$$N_S = \frac{120f}{P} \quad or \quad f = \frac{PN_S}{120}$$

where N_S is the synchronous sped in *r.p.m.*; *f* is the supply frequency and *P* is the number of poles of the machine.

When connected to an electric power system, a synchronous machine always maintains this relationship. If a synchronous machine working as a motor fails to maintain this average speed (N_S), the machine will not develop sufficient torque to maintain its rotation and will stop. Then the motor is said to be *pulled out of step*.

In case, the synchronous machine is operating as a generator, it has to run at a fixed speed called synchronous speed to generate power at a particular frequency since all the electrical equipment and machines are designed to operate at this frequency. In India the value of power frequency is 50 Hz.

4.51 BASIC PRINCIPLES

A *synchronous machine* is just an **electro-mechanical transducer** which converts mechanical energy into electrical energy or vice-versa. The fundamental phenomenon which make these conversions possible are:

(*i*) the law of electro-magnetic induction and (*ii*) law of interaction.

(*i*) *Law of electromagnetic induction:* This relates to the production of *e.m.f., i.e. e.m.f.* is induced in a conductor whenever it cuts across the magnetic field (see fig.4.65). This is called *Faraday's first law of electromagnetic induction.*

(*ii*) *Law of interaction:* This law relates to the phenomenon of production of *force.* or *torque i.e.* whenever a current carrying conductor is placed in the magnetic field, by the interaction of the magnetic fields produced by the current carrying conductor and the main field, force is exerted on the conductor and torque is developed (see fig.4.66).

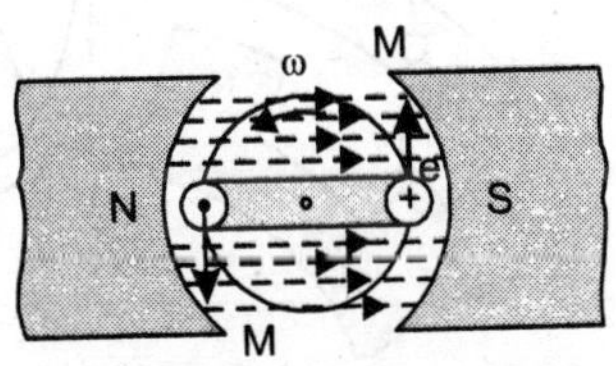

Fig.4.65: Generator action

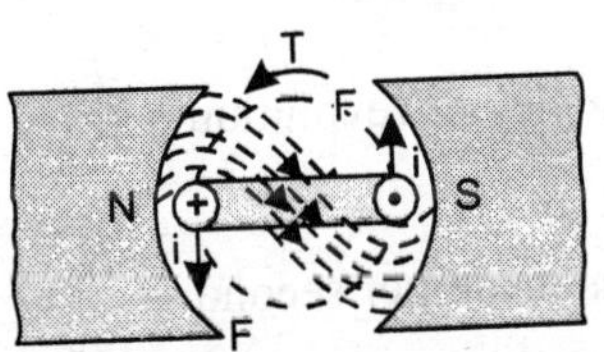

Fig.4.66: Motor action

4.52 PRODUCTION OF SINUSOIDAL ALTERNATING E.M.F.

When a conductor or coil cuts across the magnetic field an *e.m.f.* is induced in it by the phenomenon called electromagnetic induction. This can be achieved either by rotating a coil in the stationary magnetic field or by keeping the coil stationary and rotating the magnetic field. (The magnetic field can be rotated by placing the field winding on the rotating part of the machine).

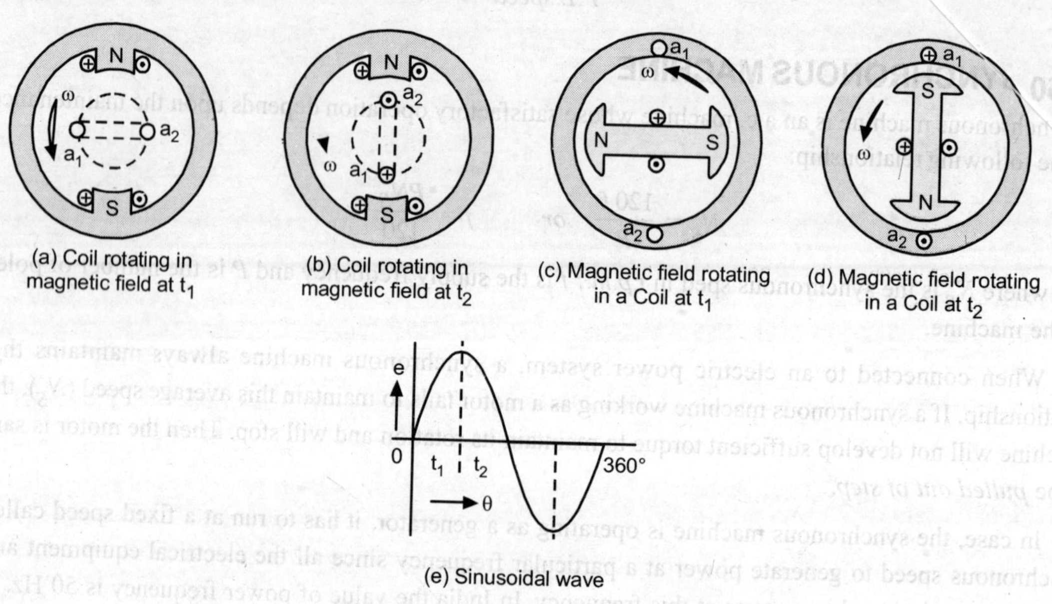

Fig.4.67: Generation of alternating *emf*

For illustration see fig.4.67 (*a*) and (*b*), two positions of a coil rotating in a stationary magnetic field is shown. Whereas, in fig.4.67 (*c*) and (*d*), two positions of a rotating electro-magnet in a coil placed on stationary armature is shown. At first instant, the *e.m.f.* induced in the coil is zero since flux cut by the coil is zero. However, at second instant, the *e.m.f.* induced in the coil is maximum (say positive). The two instants t_1 and t_2 are marked on the wave diagram shown in fig.4.67 (*e*). In one revolution the induced *e.m.f.* completes one cycle and its wave shape is shown in fig.4.67 (*e*).

4.53 RELATION BETWEEN FREQUENCY SPEED AND NUMBER OF POLES

In fig.4.68, a machine is shown having P number of poles on the rotor revolving at a speed at N_s *r.p.m.* When a conductor passes through a pair of poles one cycle of *e.m.f.* is induced in it.

$$\therefore \text{ No. of cycle made per revolution} = \frac{P}{2}$$

$$\text{No. of revolutions made per second} = \frac{N_s}{60}$$

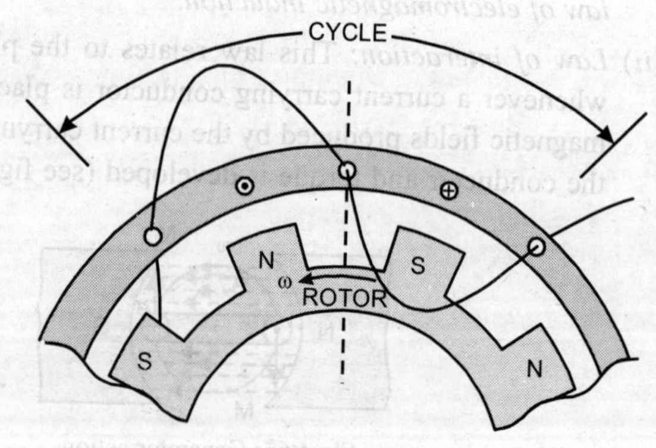

Fig.4.68: Rotor having number poles

∴ No. of cycles made per second

= No. of cycles/revolution × No. of revolutions/s

$$f = \frac{P}{2} \times \frac{N_s}{60} = \frac{PN_s}{120} \text{ cycles/s or Hz}$$

4.54 CONSTRUCTIONAL FEATURES OF SYNCHRONOUS MACHINES

Only in small synchronous machines the field system is placed on stator and armature winding on rotor, but in larger machines, the field winding is placed on the rotor and armature winding is placed on the stator, as shown in fig.4.69. The rotating field and stationary armature system is preferred over stationary field and rotating armature system.

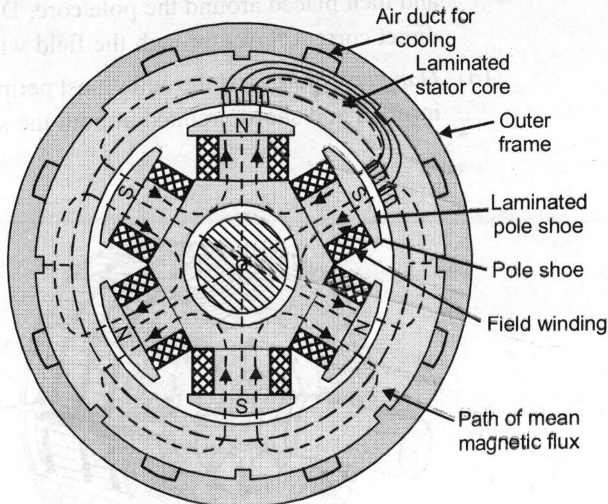

Fig.4.69: Constructional feactures of salient pole alternator

The important parts of a synchronous machine are given below:

1. Stator 2. Rotor 3. Miscellaneous

1. **Stator:** The outer stationary part of the machine is called stator, it has the following important parts:

(*i*) *Stator frame:* It is the outer body of the machine made of cast iron and it protects the inner parts of the machine. It can be also made of any other strong material since it is not to carry the magnetic field. Cast iron is used only because of its high mechanical strength.

(*ii*) *Stator Core:* The stator core is made of silicon steel material. It is made from number of stamping which are insulated from each other. Its function is to provided an easy path for the magnetic lines of force and accommodate the stator winding.

(*iii*) *Stator Winding:* Slots are cut on the inner periphery of the stator core in which 3-phase or 1-phase winding is placed. Enameled copper is used as winding material.

2. **Rotor:** The rotating part of the machine is called rotor. From construction point of view, there are two types of rotors named as

(*i*) Salient pole type rotor;

(*ii*) Non-salient pole type rotor.

(*i*) **Salient pole type rotor:** In this case, projected poles ar[illegible] type construction is suited for medium [illegible] electric and diesel p[illegible] (generators[illegible] po[illegible] with water turbines) the diameter of the machines is as large as 14 metre and length is only 1 metre. The salient pole type rotor has the following important parts

(*a*) *Spider:* Spider is made of cast iron to provide an easy path for the magnetic flux. It is keyed to the shaft and at the outer surface, pole core and pole-shoe are keyed to it (see fig.4.70).

(*b*) *Pole core and pole shoe:* It is made of laminated sheet material [see fig 4.70 (*b*)]. Pole core provides least reluctance path for the magnetic field and pole shoe distributes the field over the whole periphery uniformly to produce sinusoidal wave form of the generated *e.m.f.*

(*c*) *Field winding or Exciting winding:* Field winding is wound on the former [see fig.4.70 (*c*)] and then placed around the pole core. D.C. supply is given to it through slip rings. When direct current flows through the field winding, it produces the required magnetic field.

(*d*) *Damper winding:* At the outer most periphery, holes are provided in which copper bars are inserted and short-circuited at both the sides by rings forming damper winding.

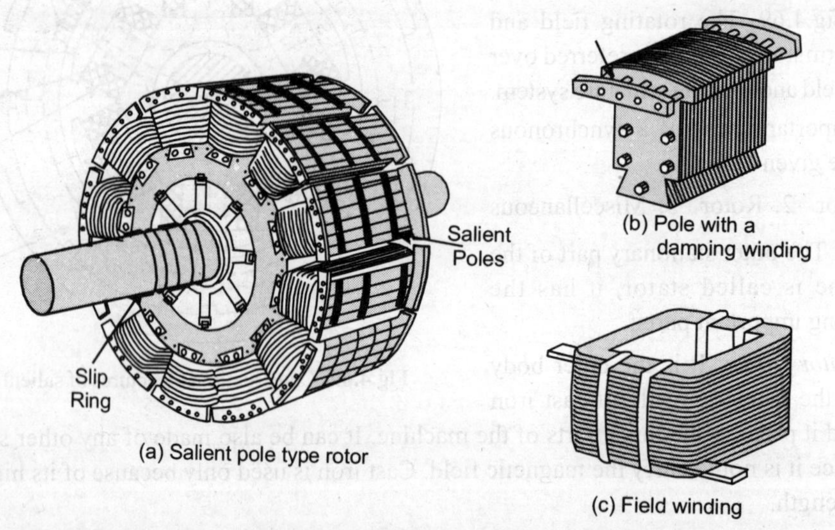

(a) Salient pole type rotor

(b) Pole with a damping winding

(c) Field winding

Fig.4.70: Parts of an alternator

(*ii*) **Non-salient pole type rotor:** In this case, there are no projected poles but the poles are formed by the current flowing through the rotor (exciting) winding. Non-salient pole type construction is suited for the high speeds. The steam turbines rotate at a high speed (3000 *r.p.m.*). When these turbines are used as prime-mover for this machine working as a generator, a small number of poles are required for given frequency. Hence, these machines have smaller ...eter and larger length. Non salient pole type rotors have the following parts:

... Rotor core is made of silicon steel stampings. It is keyed to the shaft. At the ... cut in which exciting coils are placed. It provides an easy path to

... laced in rotor slots and current is passed ... cording to the requirement

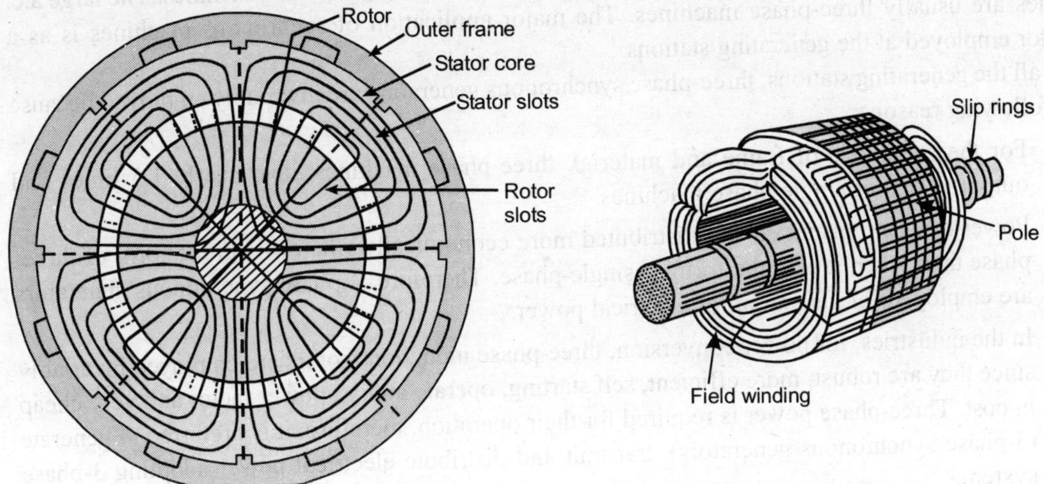

Fig.4.71: Non-salient pole type alternator

Fig.4.72: Non-salient pole type rotor

3. **Miscellaneous Part:** The following are few important miscellaneous parts;

   (*i*) *Brushes:* Brushes are made of carbon and these just slip over the slip rings. D.C. supply is given to the brushes. From brushes current flows to the slip rings and then to the exciting winding.

   (*ii*) *Bearings:* Bearings are provided between the shaft and outer stationary body to reduce the friction. The material used for their construction is high carbon steel.

   (*iii*) *Shaft:* Shaft is made of mild steel. Mechanical power is taken or given to the machine through shaft.

4.55 ADVANTAGES OF ROTATING FIELD SYSTEM OVER STATIONARY FIELD SYSTEM

Following are the important advantages of rotating field system over stationary field system:

(*i*) The armature winding is more complex than the field winding. Therefore, it is easy to place armature winding on stationary structure.

(*ii*) In the modern alternators (synchronous generators), high voltage is generated, therefore, heavy insulation is provided and it is easy to insulate the high voltage winding when it is placed on stationary structure.

(*iii*) The size of the armature conductors is much more to carry heavy current, therefore, high centrifugal stresses are developed. Thus, it is preferred to place them on stationary structure.

(*iv*) The size of slip rings depends upon the magnitude of flow of current, therefore, it is easy to deliver small current for excitation, through slip rings of smaller size when rotating field system is used.

(*v*) It is easier to build and properly balance high speed rotors when they carry the lighter field system.

(*vi*) The weight of rotor is small when field system is provided on rotor and as such friction losses are produced.

(*vii*) Better cooling system can be provided when the armature is kept stationary.

4.56 THREE-PHASE SYNCHRONOUS MACHINES

Only small *a.c. machines* employed in household applications are single-phase machines. The large a.c. machines are usually three-phase machines. The major application of sychronous machines is as a generator employed at the generating stations.

At all the generating stations, three-phase synchronous generators are invariably employed because of the following reasons:

(*i*) For the same size of frame and material, three-phase machines have nearly 1.5 times and output to that of single-phase machines.

(*ii*) Power can be transmitted and distributed more economically when it is in the form of three-phase than when it is in the form of single-phase. Therefore, 3-phase synchronous generators are employed for generation of electrical power.

(*iii*) In the industries, for power conversion, three-phase induction motors are employed invariably since they are robust, more efficient, self starting, operate at high power factor and very cheap in cost. Three-phase power is required for their operation, therefore, it is preferred to generate (3-phase synchronous generators), transmit and distribute electrical power adopting 3-phase system.

4.57 E.M.F EQUATION

Let P = No. of poles; ϕ = Flux per pole in webers;

N = Speed in *r.p.m.*; f = frequency in Hz;

Z_{ph} = No. of conductors connected in series per phase

T_{ph} = No. of turns connected in series per phase

$^*K_c$ = Coil span factor; $^{**}K_d$ = Distribution factor

Flux cut by each conductor during one revolution = $P\,\phi$ weber

Time take to complete one revolution = $\dfrac{60}{N}$ second

---

* Coil span factor (K_c)

The ratio of induced e.m.f. in a coil when the winding is short pitched to the induced e.m.f. in the same coil when it is full pitched is called a coil span factor or pitch factor or chorded factor. It is generally denoted by K_c and its value is always less than unity.

$\therefore$ Coil span factor, $K_c = \dfrac{2e\cos\beta/2}{2e} = \cos\beta/2$ where β is the angle through which the coil is short pitched.

** Distribution factor (K_d)

The ratio of induced e.m.f. in the coil group when the winding is distributed in number of slots to the induced e.m.f. in the coil group when the winding is concentrated in one slot is called a distribution factor or breadth factor. It is generally denoted by K_d and its value is always less than unity.

$\therefore$ Distrubution factor, $K_d = \dfrac{\sin\dfrac{m\,\alpha}{2}}{m\sin\dfrac{\alpha}{2}}$ where, m = No. of slots per pole per phase and

a = 180°/No. of slots per pole *i.e.* slot pitch.

Average *e.m.f.* induced per conductor $= \dfrac{P\phi}{60/N} = \dfrac{P\phi N}{60}$

Average *e.m.f.* induced per phase,

$$= \frac{P\phi N}{60} \times Z_{ph} = \frac{P\phi N}{60} \times 2T_{ph} \qquad \left(\because T_{ph} = \frac{Z_{ph}}{2}\right)$$

$$= 4 \times \phi \times T_{ph} \times \frac{PN}{120} = 4\,\phi f T_{ph}$$

R.M.S. values of *e.m.f.* induced per phase,

$$E_{ph} = \text{Average value} \times \text{form factor}$$

$$E_{ph} = 4\,\phi f T_{ph} \times 1{\cdot}11 = 4{\cdot}44\,\phi f T_{ph} \text{ volt}$$

Taking into consideration the coil span factor (K_C) and distribution factor (K_d) of the winding.

Actual *e.m.f.* induced per phase

$$\mathbf{E_{ph} = 4{\cdot}44\, K_c\, K_d\, \phi f\, T_{ph} \text{ volt}}$$

Example 4.12: *A three-phase 50 Hz, synchronous generator runs at 250 r.p.m. Find the number of poles of the machine ? What type of prime-mover would you expect for this machine ?*

Solution: Frequency, $f = 50$ Hz

Speed, $N_s = 250$ r.p.m.; Then $P = \dfrac{120f}{N_s} = \dfrac{120 \times 50}{250} = \mathbf{24}$ (*Ans.*)

Since the speed of the synchronous generator is very low the prime-mover would be a *water turbine (hydraulic-turbine)*. For such a large number of poles the machine would be a *salient-pole type*.

Example 4.13. *Calculate the no-load terminal voltage of a 3-phase, 6-pole, star connected alternator running at 1000 r.p.m. having following data: sinusoidally distributed flux per pole = 60 m Wb; Total No. of armature slots = 60; No. of conductors per slot = 10; Distribution factor; $K_d = 0{\cdot}96$. Assume full pitch windings.*

Solution: For full pitch winding; Coil span factor, $K_c = 1$

Distribution factor is given, therefore, it is not to be calculated.

No. of turns/phase, $T_{ph} = \dfrac{60 \times 10}{2 \times 3} = 100$

Supply frequency, $f = \dfrac{PN_S}{120} = \dfrac{6 \times 1000}{120} = 50\,\text{Hz}$

E.M.F. induced per phase, $E_{ph} = 4{\cdot}44\, K_c\, K_d f \phi\, T_{ph}$

$= 4{\cdot}44 \times 1 \times 0{\cdot}96 \times 50 \times 60 \times 10^{-3} \times 100 = 1278.7\text{V}$

Since the alternator is star connected;

No load terminal voltage, $E_L = \sqrt{3}\, E_{Ph} = \sqrt{3} \times 1278.7$

$= \mathbf{2215\ V}$ (*Ans.*)

Example 4.14: *Calculate the no-load terminal voltage of a 3-phase, 12-pole, star connected alternator running at 500 r.p.m. having following data:*

Sinusoidally distributed flux per pole = 55 m Wb

Total No. of armature slots = 72

Number of conductors/slot = 8

Distribution factor = 0·96

Assume full pitch windings.

Solution: No. of poles, $P = 12$;

Speed, $N_s = 500$ rpm

Flux, $\phi = 55 \times 10^{-3}$ Wb;

No. of slots = 72

No. of conductors/slot = 8;

Distribution factor, $K_d = 0{\cdot}96$

For full pitch winding,

Coil span factor, $K_c = 1$

Distribution factor is given, therefore, it is not to be calculated

$$\text{No. of turns/phase, } T_{ph} = \frac{72 \times 8}{2 \times 3} = 96$$

$$\text{Supply frequency, } f = \frac{PN_S}{120} = \frac{12 \times 500}{120} = 50 \text{ Hz}$$

$$\text{E.M.F. induced per phase, } E_{ph} = 4{\cdot}44\, K_c\, K_d\, f\, \phi\, T_{ph}$$

$$= 4{\cdot}44 \times 1 \times 0{\cdot}96 \times 50 \times 55 \times 10^{-3} \times 96 = 1125.3 \text{ V}$$

Since the alternator is star connected;

No load terminal voltage, $E_L = \sqrt{3}\, E_{ph} = \sqrt{3} \times 1125.3 =$ **1949 V** *(Ans.)*

PRACTICE EXERCISE

Q.1 *A three-phase 50 Hz, synchronous generator runs at 187·5 r.p.m. Find the number of poles of the machine ? What type of prime-mover would you expect for this machine ?* ***(Ans. 32)***

Q.2 *Calculate the no-load terminal voltage of a 3-phase, 4-pole, star connected alternator running at 1500 r.p.m. having following data: sinusoidally distributed flux per pole = 66 m Wb; Total No. of armature slots = 72; No. of conductors per slot = 10; Distribution factor; K_d = 0·96. Assume full pitch windings.* ***(Ans. 2924V)***

Q.3 *Calculate the no-load terminal voltage of a 3-phase, 8-pole, star connected alternator running at 750 r.p.m. having following data:*

Sinusoidally distributed flux per pole = 55 m Wb

Total No. of armature slots = 72

Number of conductors/slot = 10

Distribution factor = 0·96

Assume full pitch windings. ***(Ans. 2436·3 V)***

SUMMARY

1. *Three-phase induction motor:* A machine that converts 3-phase a.c. electric, power into mec. power by using an electromagnetic induction phenomenon is called a *3-phase induction motor.*
2. *Types of induction motors:* According to the construction of rotor there are two types of inductio motors namely *squirrel cage I.M.* and *phase-wound I.M.*
3. *Production of revolving field:* When a 3-phase supply is given to a 3-phase wound stator of an induction motor, a resultant field of magnitude $1{\cdot}5\ \phi_m$ is step-up which revolves in space at a constant speed called synchronous speed N_s ($N_s = 120f/P$).
4. *Working principle:* At start, stationary rotor conductors cut across the revolving magnetic field and an e.m.f. is induced in them by the electromagnetic induction phenomenon. Current flows through the rotor conductors as they are short circuited and produce rotor field. By the interaction of rotor and stator magnetic field, torque develops and rotor starts rotating in the same direction as that of the revolving field.
5. *Reversal of direction of rotation:* The direction of rotation of 3-phase induction motor can be reversed by interchanging the connections of any two supply leads at the stator terminals.
6. *Slip:* The rotor of a 3-phase motor never obtains synchronous speed because at that speed there would be no relative speed between rotor conductors and stator revolving field and induction phenomenon is not possible. Its speed is always less than synchronous speed.

   The difference between synchronous speed and rotor speed is called slip. It is generally denoted as a fraction of synchronous speed. $S = (N_s - N)/N_s$.
7. *Frequency of rotor currents:* $f_r = Sf$
8. *Rotor e.m.f.:* At stand still, $E_{2s} = 4 \cdot 44\ kw_2\ T_2 f \phi$ or $E_{2s} = KE_1$

   Under running conditions $E_2 = SE_{2s}$
9. *Rotor resistance:* R_2
10. *Rotor reactance:* At stand-still – X_{2s}; At running conditions, $X_2 = SX_{2s}$
11. *Rotor current:* $I_2 = E_2/Z_2$ where $E_2 = SE_{2s}$ and $Z_2 = \sqrt{R_2^2 + (SX_{2s})^2}$
12. *Rotor p.f.:* $\cos\phi_2 = R_2/Z_2 = R_2/\sqrt{R_2^2 + (SX_{2s})^2}$
13. *Power flow:* Input elect. power to stator–stator cu. loss–stator iron loss = stator output or rotor input.

    Rotor input–rotor cu. loss = mech. power developed in rotor.

    Mech. power developed in rotor – mech. loss = output mech. power at the shaft.
14. *Relation between rotor Cu loss and slip:* Rotor cu. loss = $S \times$ Rotor input.
15. Torque developed: $$T = \frac{3}{\omega_s}\frac{SE_{2s}^2 R_2}{[(R_2^2) + (SX_{2s})^2]} = \frac{3}{\omega_s}\frac{E_{2s}^2 R_2 / S}{[(R_s/S)^2 + (X_{2s})^2]}$$
16. *Condition for max. torque:* $R_2 = SX_{2s}$ or $S = R_2/X_{2s}$
17. *Max. Torque:* $T_m = \dfrac{3E_{2s}^2}{2\omega_s X_{2s}}$ and $T_m \propto \dfrac{1}{X_{2s}}$

18. *Starter:* A device used to limit the inrush flow of current at start is know as starter. It contains no volt coil and over-load coil for motor protection.

19. *Direct On Line (DOL) Starter:* This starter does not limit the inrush flow of starting current but it contains *No-volt* and *Over-load relays* to protect the motors. It is used with the motors of smaller size (less than 3HP).

20. *Star-delta Starter*: It connects the stator winding first in star and then in delta thus reduced the starting current to $\frac{1}{3}$rd value. It is employed with most of the 3-phase squirrel cage induction motors.

21. *Auto-transformer Starter:* It provides lower voltages to the stator winding at the start and has the ability to reduce the starting current to any predetermined value. It is used to start 3-phase squirrel cage induction motors of very large sizes.

22. *Starting of slip-ring induction motors:* In slip-ring induction motors, the starting current is limited by adding resistance in the rotor circuit at the start.

23. *Speed control of Induction Motors:* The speed of a 3-phase induction motor can be changed (*i*) by changing the slip, (*ii*) by changing the supply frequency and (*iii*) by changing the number of poles of an induction motor.

24. *Single-phase I.M.:* A machine that converts 1-phase a.c. elect. power into mech. power by using an electromagnetic induction phenomenon is called a 1-phase induction motor.

25. *Torque development in a 1-phase I.M.:* When 1-phase supply is given to the stator of a 1-phase I.M., two fields of magnitude $\phi_m/2$ are produced which rotate in opposite direction at synchronous speed. An equal and opposite torque is developed by the two fields and the resultant torque is zero at start. Thus, 1-phase I.M. is not a self starting motor. However, if the rotor is rotated in either direction by some external means, torque develops and rotor picks-up the speed in that direction.

26. *Split phase motor:* To obtain starting torque in 1-phase induction motors, the single winding is splitted into two parts having different resistance and inductance. They carry currents at different angles which produce a resultant field revolving in space at synchronous speed, this develops starting torque in the motor.

27. *Capacitor motors:* These may be (*i*) Capacitor start motors, (*ii*) Capacitor run motor and (*iii*) Capacitor start and capacitor run motors.

28. *Shaded pole motors:* These motors have projected poles with 1/3 pole part rapped with a copper strip. This portion of the pole is called shaded part of the pole. Because of this, a revolving field is set-up in the stator and torque is developed.

29. *AC series motors*: These motors have the ability to develop high starting torque and can rotate at high speeds. The torque varies as square of the current and the speed varies inversely proportional to the current approximately.

30. *Universal motors*: A motor which can be operated on ac as well as on dc supply at the rated voltage is called *universal motor.*

31. *D.C. motor*: A machine that converts d.c. electrical power into mechanical power is known as d.c. motor. Its working depends upon the basic principle that when a current carrying conductor is placed in the magnetic field, a force is exerted on it and torque develops.

32. *Constructional features of a dc machine:* The essential parts of a d.c. machine are: Magnetic frame or Yoke, Pole core and pole shoes, Field or Exciting coils, Armature core, Armature winding, Commutator, Brushes, Brush rocker, End housings etc.

33. *Functions of commutator:* Commutator reverses the direction of flow of current in the armature conductors when they cross the M.N.A. to obtain continuous torque.

34. *Back e.m.f.*: When d.c. supply is given to motor armature rotates. The armature conductors cut across the main magnetic field and an e.m.f. is induced in them in opposite direction to that of supply voltage called back e.m.f. (E_b).

$$E_b = PZN\ \phi/60\ A;\ E_b \propto N\phi \text{ or } N \propto E_b/\phi;\ E_b < V \text{ and } E_b = V - I_a R_a$$

35. *Types of d.c. motors and important relations:*

Separately excited: $I_a = I_L;\ E_b = V - I_a R_a - 2v_b$

Shunt motors: $I_{sh} = V/R_{sh};\ I_a = I_L - I_{sh};\ E_b = V - I_a R_a - 2v_b$

Series motors: $I_{se} = I_a = I_L;\ E_b = V - I_a (R_a + R_{se}) - 2\ v_b$

Compound motors: Cumulative $\phi_r = \phi_{sh} + \phi_{se}$; differential $\phi_r = \phi_{sh} - \phi_{se}$

36. *Speed control of d.c. motors*: The speeds of dc motors can be controlled very acurately by employing field control and armature control methods.

37. *Speed regulation* $= \dfrac{NL \text{ speed} - FL \text{ speed}}{FL \text{ speed}}$

38. *Synchronous machines:* A machines that rotates only at synchronous speed N_S is called a *synchronous machine.* Its satisfactory operation depends upon the relation.

$$N_S = \frac{120 f}{P}$$

39. *Alternator:* An *a.c.* machine that converts mechanical power or energy into *a.c.* electrical power or energy at a desired frequency (50 Hz in India) is called an *alternator.* It is also called as *synchronous generator* or simply *a.c.* generator.

Its basic principle of operation is *electro-magnetic induction.*

40. *Construction of Synchronous Machines:* Usually machines of large size have stationary armature and rotating field system because of economy and simple designing.

As per rotor construction, there are two types of synchronous machines namely,

(*i*) Salient pole type (*ii*) Non-salient pole type.

41. *Applications*: (*i*) *Salient pole type machines* (alternators) are operated at low speeds and are coupled with water turbines at *hydro-electric power plants.* These machines have large number of poles, larger diameter and smaller length. (*ii*) *Non-salient pole type machines* (alternators) are operated at high speeds and are coupled with steam turbines at *thermal power plants*. These machines have small number of poles, smaller diameter and larger length.

42. *Three-phase synchronous machines*: Larger *a.c.* machines are always 3-phase wound machines because of their high efficiency and economy.

43. *E.M.F. education*

$$E_{ph} = 4{\cdot}44\ K_c\ K_d\ \phi f\ T_{ph}$$

SHORT ANSWER QUESTIONS

1. Can the outer frame of an Inductor motor be made of plastic, state why ?
2. Can the outer frame of a dc machine be made of plastic, state why ?
3. The rotor of an inductor motor is skewed, why ?
4. Can we reverse the direction of rotation of a 3-phase induction motor by just interchanging the connections of any two terminals at the input, if so state why ?
5. In a squirrel cage rotor, no insulation is provided between the rotor conductors and slots, why?
6. An induction motor is also called asynchronous motor, why ?
7. Can an induction motor run at synchronous speed, state why ?
8. When 3-phase supply is given to the stator of a 3-phase wound induction motor, a revolving field is set-up. When the same supply is given to a 3-phase transformer no revolving field is set up, state why ?
9. The rotor conductors are placed at the outermost periphery of the rotor, Why ?
10. The air gap between stator and rotor of an inductor motor is kept as small as possible, why ?
11. The direct on line starter does not limit the starting current of an induction motor still it is called a starter not a switch, why ?
12. For very large 3-phase induction motors auto transformer starters are preferred over star delta starters, why ?
13. Large induction motors are not started direct on line, why ? *(PSB/HSB Dec. 1996)*
14. In lifts which type of 3-phase induction motor is employed, state why ?
15. An induction motor always draws power at lagging power factor, why ?
16. When load on an induction motor increases, its speed decreases, why ?
17. In case of 1-phase capacitor run motor if capacitor is damaged the motor does not start. However if it is rotated in either direction it picks up speed in that direction, why?
18. The direction of rotation of a shaded pole motor cannot be reversed, why?
19. The direction of rotation of a split-phase single-phase induction motor can be reversed, how?
20. Why universal motor is employed in mixer-grinder?
21. What are the functions of armature core ?
22. What is the function of field system ?
23. What are the main functions of yoke ?
24. What are the main functions of pole shoes ?

25. What are the advantages of carbon brushes ?

26. Why is armature of *dc* machines is made up of silicon steel ?

27. What is a *dc* Motor ?

28. What is the working principle of a *dc* Motor ?

29. What is Back *EMF* ?

30. What are the different types of *dc* Motors ?

31. What factors determine the number of poles of a synchronous generator ?

32. Why are the alternators rated in *kVA* ?

NUMERICAL FOR PRACTICE

1. If the e.m.f. in the stator of an 8 pole induction motor has a frequency of 50 Hz and that in rotor 1·5 Hz find the speed at which motor is running and its slip. **(Ans.** *727.5 r.p.m., 0.03)*

2. An 8-pole alternator runs at 750 r.p.m. and supplies power to a 6-pole induction motor which has full load slip of 3%. Find the full load speed of the induction motor and the frequency of its rotor e.m.f. **(Ans.** *970 r.p.m., 1·5 Hz)*

3. It is desired to obtain a speed of approximately 700 r.p.m. with a 3-phase induction motor. Determine number of poles for (*i*) 60 Hz motor; (*ii*) 25 Hz motor. If the rated load slip of each motor is 5%, determine rated speed for each motor.

(Ans. *10; 684 r.p.m., 4; 712 · 5 r.p.m.)*

4. Determine the number of poles, the slip and the frequency of rotor currents at rated load for a 3-phase, 3 · 7 kW induction motor rated at:

(*i*) 220 V, 50 Hz, 1440 r.p.m.

(*ii*) 120 V, 400 Hz, 3800 r.p.m. **(Ans.** *4, 4%, 2 Hz 12, 5%, 20 Hz)*

5. A balanced 3-phase, 50 Hz voltage is applied to a 3-phase 6-pole induction motor. When the motor delivers rated output the slip is found to be 0·04. Determine:

(*i*) The speed of the revolving field relative to the stator structure.

(*ii*) The frequency of the rotor currents.

(*iii*) The speed of the rotor m.m.f. relative to the rotor structure.

(*iv*) The speed of the rotor m.m.f. relative to the stator structure.

(*v*) The speed of the rotor m.m.f. relative to the stator field distribution.

(*vi*) Are the conditions right for the development of the net unidirectional torque ?

(Ans. *1000 r.p.m.; 2 Hz; 40 r.p.m.; 1000 r.p.m.; zero; yes)*

6. A 6-pole, 3-phase, 50 Hz induction motor has a star connected rotor. The rotor has resistance and standstill reactance of 0 · 25 ohm and 2 · 5 ohm per phase respectively. The induced e.m.f. between sliprings at start is 100 V. If the full load speed is 960 r.p.m. Calculate (*i*) the slip, (*ii*) rotor induced e.m.f. per phase (*iii*) the rotor current and power factor at standstill and (*iv*) the rotor current and power factor at rated load.

(Ans. *4%; 57 · 735 V; 22 · 98 A, 0 · 0995 lagging; 8 · 577 A, 0 · 9285 lagging)*

7. A 3-phase induction motor having a star-connected rotor has an induced e.m.f. of 80 V between sliprings at standstill on open circuit. The rotor has a resistance and reactance of 0 · 5 ohm and 2 ohm per phase respectively. Calculate the current per phase and p.f. when (*i*) sliprings are short circuited (*ii*) sliprings are connected to a star connected rheostat of 4 ohm per phase.

(**Ans.** *95 · 22 A, 0 · 2425 lagging; 9 · 38 A, 0 · 9138 lagging*)

8. The power supplied to a 3-phase induction motor is 40 kW and the corresponding stator losses are 1·5 kW. Calculate:

(*a*) The total mechanical power developed and the rotor I^2R losses when the slip is 0 · 04.

(*b*) The output power of the motor if the friction and windage losses are 0 · 8 kW.

(*c*) The efficiency of the motor. Neglect the rotor iron losses.

(**Ans.** *36 · 96 kW; 1 · 54 kW; 36 · 16 kW; 90 · 4%*)

9. The shaft output of a 3-phase induction motor is 75 kW. The friction and windage losses are 1000 watts. The stator core losses are 4000 W and the stator copper losses are 2500 W. If the slip is 3 · 5% what is the efficiency of the motor ?

(**Ans.** *87 · 97%*)

10. A 400 V, 6-pole, 50 Hz, 3-phase induction motor develops 20 H.P. inclusive of mechanical losses when running at 995 r.p.m., the power factor being 0 · 87. Calculate: (*i*) Slip (*ii*) The rotor copper losses (*iii*) The line current. The stator copper loss is 1500 W.

(**Ans.** *0 · 005; 73 · 92 W; 27 A*)

11. A 50 H.P., 6-pole, 3-phase induction motor delivers full load output at 955 r.p.m. and with 0 · 86 p.f. when connected to 500 V, 50 Hz mains. Friction and windage losses total 2 HP. and stator losses are 1 · 5 K.W. Determine for this load: (*i*) total rotor Cu losses; (*ii*) the efficiency; and (*iii*) the line current.

(**Ans.** *1 · 802 kW; 88 · 51%; 55 · 78 A*)

12. A 6-pole, 3-phase induction motor runs at a speed of 960 r.p.m. and the shaft torque is 135 · 7 Nm. Calculate the rotor copper loss if the friction and windage losses amount of 150 watt. The frequency of supply is 50 Hz.

(**Ans.** *574 · 67 W*)

13. An induction motor has an efficiency of 85% when loaded at 50 H.P. At this load stator copper loss and rotor copper loss each equal iron loss. Mechanical losses are one third of the no load loss. Calculate rotor copper loss, rotor input and slip.

(**Hints:** Input $= \dfrac{\text{Output}}{\eta}$; losses = Input – Output; If rotor copper loss = K.

Then total losses $= K + K + K + \frac{1}{3}K$)

(**Ans.** *1946·9 W, 39370·88 W, 0·04945*)

14. A 3-phase, 50 Hz, synchronous generator runs at 166·67 r.p.m. What is the number of poles ? What of prime-over would you except for this machine ?

(**Ans.** *36, Hydraulic turbine*)

15. The stator core of a 4-pole, 3-phase alternator has 36 slots. It carries a short pitch 3-phase winding with coil span equal to 8 slots. Determine the distribution and coil pitch factor.

(**Ans.** *0·9598; 0·9848*)

16. A 3-phase, 12-pole alternator has a star connected winding with 108 slots and 10 conductors per slot. The coils are full pitched. The flux per pole is 0·05 Wb sinusoidally distributed and speed is 600 rpm. Calculate the line voltage on open circuit. Assume distribution factor equal to 0·96.

(**Ans.** *3986·7 V*)

MULTI-CHOICE QUESTIONS

1. Stator core of an induction motor is made of
   (a) laminated cast iron (b) mild steel
   (c) silicon steel stampings (d) soft wood.
2. The stator winding of an induction motor can be designed for
   (a) any number of pole (b) any even number of poles
   (c) any odd number of poles (d) only for four pole's.
3. The rotor of squirrel cage induction motor is skewed because
   (a) it reduces humming thus ensures quite running of the motor.
   (b) it results in a smoother torque curves for different positions of the rotor.
   (c) it avoids the magnetic looking of the stator and rotor.
   (d) all of these.
4. Slip rings of phase-wound induction motor are made of:
   (a) wood (b) cast iron (c) steel (d) cooper
5. There is no electrical connection between stator and rotor, still power is transferred from stator to rotor through
   (a) magnetic flux (b) air (c) water (d) magnet.
6. In a large induction motor usually the value of full load slip is
   (a) 0 · 4% (b) 20% (c) 3 to 5% (d) 6 to 15%
7. At start, the slip of the induction motor is
   (a) zero (b) 0 · 5 (c) one (d) infinite.
8. Under running conduction, the rotor reactance is directly proportional to
   (a) induced e.m.f. (b) rotor current (c) slip (d) supply voltage.
9. At start the rotor power factor is
   (a) very high (b) very low (c) unity (d) zero.
10. The rotor copper losses of an induction motor are directly proportional to
    (a) input to the motor (b) output of the motor (c) rotor resistance (d) slip.
11. The condition for maximum torque is
    (a) $R_2 = SZ_{2s}$ (b) $X_{2s} = SR_2$ (c) $R_2 = X_{2s}$ (d) $R_2 = SX_{2s}$
12. The function of a starter is
    (a) to start the motor (b) to start and stop the motor
    (c) to limit the starting current (d) to limit the applied voltage
13. Stator core of a single-phase, split winding induction motor is made of
    (a) laminated cast iron (b) mild steel
    (c) silicon steel stampings (d) soft wood.
14. In a single phase induction motor at start, the two revolving fields produce
    (a) unequal torques in the rotor conductors

(*b*) no torque in the rotor conductors

(*c*) equal and opposite torque in the rotor conductors

(*d*) equal torques in same direction in the rotor conductors

15. During running condition if the starting winding of a split phase induction motor is disconnected.

(*a*) the motor will stop (*b*) the motor winding will burn

(*c*) the main winding will be damaged (*d*) the motor will continue to rotate

16. The starting torque of a capacitor start motor is

(*a*) more than a capacitor run motor (*b*) less than a capacitor run motor

(*c*) less than a shaded pole motor (*d*) less than a split-phase induction motor.

17. The direction of rotation of a split-phase induction motor can be reversed by

(*a*) reversing the connections of the supply terminals

(*b*) reversing the connections of the main winding only

(*c*) reversing the connections of the starting winding only

(*d*) either (*b*) or (*c*).

18. Which motor is best suited for domestic refrigerator ?

(*a*) 3-phase induction motor (*b*) DC motor

(*c*) Capacitor start motor (*d*) Shaded pole motor.

19. Which motor is best suited for domestic mixer-grinder?

(*a*) 3-phase induction motor (*b*) Universal motor

(*c*) Capacitor start motor (*d*) Shaded pole motor.

20. The yoke of a d.c. machine is made of

(*a*)copper (*b*) carbon (*c*) cast iron (*d*) sillicon steel

21. The armature core of a d.c. machine is made of

(*a*)copper (*b*) carbon (*c*) cast iron (*d*) silicon steel

22. The segments of the commutator of a d.c. machine are made of

(*a*)brass (*b*) copper (*c*) carbon (*d*) silicon steel

23. The segments of the commutator of a d.c. machine are insulated from each other by

(*a*) rubber (*b*) porcelain (*c*) mica (*d*) varnish

24. The brushes of a d.c. machine are made of

(*a*) iron (*b*) brass (*c*) mica (*d*) carbon

25. The speed of the d.c. motor can be varied

(*a*) by varying field current only, (b) by varying armature resistance only,

(*d*) stray losses are equal to copper losses.

26. *The frequency of voltage generated in large alternators in India is*

(*a*) in Megacycles (*b*) in kilocycles (*c*) 60 Hz (*d*) 50 Hz

27. *The rotors preferred for alternators coupled to hydraulic turbines are*

(*a*) salient pole type (*b*) cylindrical rotor type (*c*) solid rotor type (*d*) any of above.

28. *Salient pole type alternators are generally employed with*

(*a*) High speed prime movers (b) Low and medium speed prime movers

(*c*) Hydrogen cooled prime movers (*d*) Low voltages alternators.

29. *RMS value of voltage generated per phase in an alternator is given by*

(*a*) $E_{ph} = 4{\cdot}44\ K_c\ K_d\ N\ \phi f$ (*b*) $E_{ph} = 4{\cdot}44\ K_c\ K_d N\ \phi$

(*c*) $E_{ph} = 4{\cdot}44\ K_c\ K_d\ N^2\ \phi f$ (*d*) $E_{ph} = 1{\cdot}11\ K_c\ K_d\ N\ \phi f$

30. *The rating of alternators is usually expressed in*

(*a*) full load current (*b*) Horse power (*c*) kVA (*d*) kW

TEST QUESTIONS

1. Explain the construction of 3-phase squirrel cage induction motor.

2. Discuss how a rotating field is produced in a 3-phase induction motor. How does the rotating field help in the production of torque.

3. Derive the relationship between the frequency of rotor currents and supply frequency in case of a 3-phase induction motor.

4. If one of the phase of a 3-phase induction motor is blown off while running without load, what will happen to its rotation.

5. Can induction motor (3-phase) run at synchronous speed ? Explain your answer.

6. Explain the terms slip, slip frequency, wound rotor, cage rotor.

7. How can the direction of rotation of 3-phase induction motor be reversed ?

8. Explain the principle of working of a three-phase induction motor and give the expression of percentage slip.

9. How much torque does an induction motor develop at synchronous speed ? Explain your answer.

10. List the different types of losses in an induction motor.

11. What are the various losses in an induction motor ? On what factors do they depend ?

12. Derive an expression for the rotor copper loss in terms of slip and input to the rotor.

13. Obtain an expression for torque under running condition, for a 3-phase induction motor and then deduce the condition for maximum torque.

14. Derive the condition for maximum torque and also the expression for the maximum torque of a 3-phase induction motor.

15. Derive the simplified equation for torque for an induction motor.

16. Draw and explain the slip-torque characteristics of a 3-phase slip ring induction motor. Mark on it starting and maximum torque.

17. Why is starter necessary for starting 3-phase induction motors. Name various method of starting 3-phase squirrel cage induction motors and explain any one method in detail.

18. Name the various starters employed for the starting of 3-phase squirrel cage induction motor. Explain star-delta starter in detail.

19. Give various methods of starting large three-phase induction motors. Explain auto transformer starter in detail.
20. How many terminals do you expect to find on the terminal box of 3-phase squirrel cage induction motor to be used for starting by star-delta starter.
21. Explain how 3-phase wound type induction motor is started.
22. Explain how the speed of slip ring induction motor can be changed by changing the rotor circuit resistance. What are the limitations and disadvantages of this method ?
23. Name the different methods of speed control of a poly-phase squirrel cage type induction motor. Explain in brief the principle of pole changing method of speed control.
24. Explain that basically a single-phase induction motor is not a self-starting motor.
25. How will you make a single-phase induction motor self-starting ?
26. Show that a single-phase sinusoidal field can be replaced by two fields rotating around the air gap in opposite directions. Sketch a torque-slip curve due to each of these two fields. How can the fact that the single-phase induction motor has no starting torque be explained by these curves ? How do they explain the fact of the motor accelerating in the direction in which it is started ?

    State the manner in which the starting torque of a single-phase motor may be obtained by splitting the phase.
27. Explain the construction and working of a single-phase capacitor start induction motor.
28. Why is a capacitor employed with a ceiling fan ?
29. A ceiling fan when switched on to a single-phase a.c. supply does not start rotating. What may be the reasons ?
30. Explain the construction (with sketch) and working of a capacitor-start capacitor-run single-phase induction motor. What are its advantages and practical applications ?
31. How can you change the direction of rotation of a single-phase induction motor ?
32. Describe the working and construction of a single-phase shaded pole motor.
33. Describe the working of an ac series motor.
34. Describe the working and construction of a universal motor.
35. Name the various parts of a d.c. machine and give the function of each part.
36. Explain the principle of operation of a d.c. motor.
37. Explain with suitable diagram the working of a d.c. motor.
38. Explain the function of commutator in a d.c. motor.
39. What is back e.m.f.? Explain.
40. Mention the various types of d.c. motors and their uses.
41. With the help of speed-armature current characteristics, show that a separately excited dc motor runs at almost constant speed irrespective of the load.

42. Sketch the speed-torque curve of a separately excited dc motor and discuss its nature.

43. Sketch the speed-load and torque-load characteristics of a separately excited dc motor.

44. Write a general expression for the speed of a d.c. motor in terms of supply voltage and flux per pole.

45. Describe Ward-Leonard method of speed control of a separately excited dc motor.

46. Deduce the relation between number of poles frequency and speed of an alternator.

47. Name the various part of a synchronous machine. Give the function and material used for each of them.

48. Give the constructional details of cylindrical rotor alternator.

49. Explain the difference between salient pole and cylindrical pole type of rotor used in alternators. Mention their applications.

50. Explain why the stator core of an alternator is laminated.

51. List the advantages of making field system rotating and armature stationary in case of an alternator.

52. Derive an expression for induced e.m.f. for an alternator.

ANSWER TO MULTI-CHOICE QUESTIONS

1. (*c*)	2. (*b*)	3. (*c*)	4. (*d*)	5. (*a*)
6. (*c*)	7. (*c*)	8. (*c*)	9. (*b*)	10. (*d*)
11. (*d*)	12. (*c*)	13. (*c*)	14. (*c*)	15. (*d*)
16. (*a*)	17. (*d*)	18. (*c*)	19. (*b*)	20. (*c*)
21. (*d*)	22. (*b*)	23. (*c*)	24. (*d*)	25. (*d*)
26. (*d*)	27. (*a*)	28. (*b*)	29. (*a*)	30. (*c*)

✪✪✪✪✪

POWER CONVERTERS

LEARNING OBJECTIVES

- What are power converters?
- What is a simple linear voltage regulator?
- What is a buck converter and how it works?
- What is a boost converter and how it works?
- What are applications of buck and Boost converters?
- What is a buck–boost converter?
- What is a voltage source inverter and how it works?
- What is the necessity of modulation?
- What are various types of modulation?

INTRODUCTION

In electrical systems (power engineering), power conversion is considered as conversion of electric energy from one form to another, such as conversion from AC to DC, DC to AC, DC to DC or AC to AC. Sometimes, it may be just changing the voltage or frequency or combination of these.

A power converter is an electrical or electro-mechanical device for converting electrical energy. It can be simply a transformer to change the voltage level of AC power, but on the other hand, it may include far more complex systems. The term, power converter, can also be used for a class of electrical machinery that is used to convert one frequency of alternating current into another frequency as mentioned earlier.

Power conversion systems may be classified in different ways depending upon applications.

One way of classifying power conversion systems is according to whether the input and output are alternating current (AC) or direct current (DC).

DC to DC power conversion: DC-to-DC converters come under this category. These are usually employed to change the dc voltage level. A voltage regulator and linear regulator convert DC to DC

DC to AC power conversion: A power inverter is used to convert DC to AC.

AC to DC power conversion: The devices which are used to convert AC to DC are rectifier, mains power supply unit (PSU) and switched-mode power supply.

AC to AC power conversion: The devices which are used to convert AC to AC are transformer/ autotransformer, voltage converter, voltage regulator, cycloconverter and variable-frequency transformer

There are also some devices and methods to convert power for single and three-phase operation.

As we know that standard power voltage and frequency varies from country to country and sometimes within a country even. In America it is usually 120 volt, 60 Hz, but in Europe, Asia, Africa and many other parts of the world, it is usually 230 volt, 50 Hz. Moreover in *aircrafts*, usually power at 400 Hz is used, so 50 Hz/60 Hz to 400 Hz frequency conversion is needed for use in the ground power unit used to power the airplane while it is on the ground. Conversely, internal 400 Hz internal power may be converted to 50 Hz or 60 Hz for convenience power outlets available to passengers during flight.

It is also true that for some electronic gadgets AC *adapters* (a type of power supply) are used to convert mains-voltage AC to low-voltage DC suitable for consumption by microchips. Similarly, *voltage converters* (also known as "travel converters") are used when travelling between countries that use ~120 V versus ~240 V AC mains power. There are also consumer "adapters" which merely form an electrical connection between two differently shaped AC power plugs and sockets, these are changing neither voltage nor frequency.

In this chapter, we shall go through some of the important converters.

5.1 DC-DC POWER CONVERTERS

DC-DC power converters are employed in a variety of applications such as power supplies for personal computers, laptop, office equipment, spacecraft power systems, telecommunications equipment, dc motor drives etc. The input to a dc-dc converter is an unregulated dc voltage and it produces a regulated output voltage of different magnitude.

Classification of Converters

The dc-dc converters may be classified as:

Buck Converter: These are ***step down*** converters, their output voltage is less than the dc input voltage.

Boost converter: These are ***step up*** converters, their output voltage is more than the dc input voltage.

Buck-Boost converter: Their output voltage can be either higher or lower than the dc input voltage.

5.2 SIMPLE LINEAR VOLTAGE REGULATOR

DC voltage can be reduced or stepped down on the load side by using a simple linear voltage regulator, as shown in Fig.5.1. The most basic form of step down transition is to use a resistor as a potential divider or voltage dropper. In some cases a zener diode may also be used to stabilise the voltage. The issue with this form of voltage dropper or step down converter is that it is very wasteful in terms of power. Any voltage dropped across the resistor will be dissipated as heat, and any current flowing through the zener diode will also dissipate heat. Both of these elements result on the loss of valuable energy.

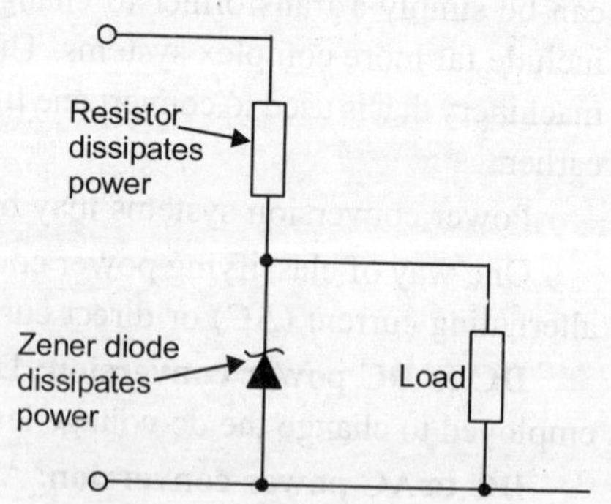

Fig.5.1: Linear voltage regulator

Moreover, in such cases, when voltage is decreased current is not increased. Whereas, in case of buck-converters, when voltage is stepped down, current is stepped up (*i.e.Pin = Pout*).

5.3 BUCK CONVERTERS

Buck converter is a DC-to-DC power converter which steps down voltage and steps up the current simultaneously such that input power supply remains the same as that of the output (load).

Conventional Circuit

The block diagram (conventional circuit) and pictorial view of a *buck converter* is shown in Fig.5.2 and 5.3 respectively. It essentially consists of;

- **At least two semiconductor devices:** a diode and a transistor, however, in modern buck converters usually diode is replaced with a second transistor and transistor is replaced with *MOSFET* or *IGBT*.
- **At least one energy storage element:** a capacitor, inductor or the two in combination.
- **Extra capacitors (or sometimes a combination of capacitors and inductors):** these are used to filter ripples. These components are normally added to output (load-side) and input (supply-side); not shown here for simplicity.

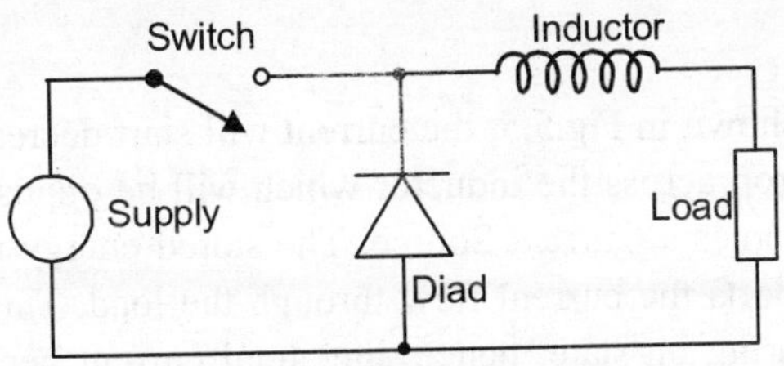

Fig.5.2: Buck converter circuit diagram.

Fig.5.3: Pictorial view of buck converter
The switch is typically a MOSFET, IGBT or BJT

Working

The two circuit configurations of a buck converter *i.e.* On-state, when the switch is closed, and Off-state, when the switch is open are shown in Fig.5.4 and 5.5 respectively. The arrows indicate the direction of flow of conventional current.

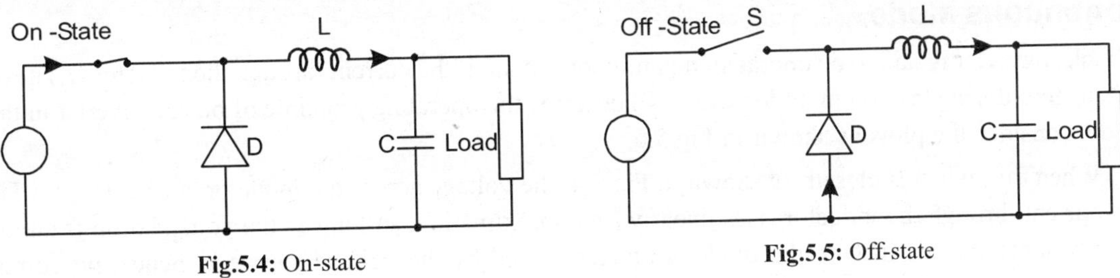

Fig.5.4: On-state

Fig.5.5: Off-state

For ideal operation, it is assumed that

- All the components are perfect specifically, the switch and the diode have zero voltage drop when on and zero current flow when off, and the inductor has zero series resistance
- The input and output voltages do not change over the course of a cycle. It implies that the output capacitance is being infinite.

The functioning of buck converter is best understood in terms of the relation between current and voltage of the inductor.

Off-state

When the switch is open (off-state), as shown in Fig.5.2, the current in the circuit is zero.

On-state

- When at first instant the switch is closed (on-state), as shown in Fig.5.4, the current will begin to increase, and the inductor will produce an opposing voltage across its terminals in response to the changing current. This voltage drop counteracts the voltage of the source and therefore reduces the net voltage across the load. As the rate of change of current decreases, the voltage across the inductor also decreases, this increases the voltage across the load. During this time, the inductor stores energy in the form of a magnetic field.
- If the switch is opened while the current is still changing, then there will always be a voltage drop across the inductor, so the net voltage at the load will always be less than the input voltage *i.e.* source voltage.

Off-state

- When the switch is opened again (off-state), as shown in Fig.5.5, the current will start decreasing. The decreasing current will produce a voltage drop across the inductor which will be opposite to the drop at on-state and hence, the inductor becomes a Current Source. The stored energy in the inductor's magnetic field is released which supports the current flow through the load. This off-state current is added to the current flowing during on-state, hence, the total current becomes greater than the average input current (being zero during off-state).
- Thus, during off-state, there is reduction in voltage but this reduction is compensated by the increase in average current. Ideally it preserves the power provided to the load. It all happens only because during off-state, the inductor is discharging its stored energy into the rest of the circuit.

If the switch is closed again before the inductor is fully discharged then the voltage at the load will always be greater than the input voltage.

The buck converters may work in continuous mode or discontinuous mode.

Continuous mode

A buck converter is said to be operated in continuous mode if the current through the inductor (I_L) never falls to zero during the commutation cycle. To describe the operating principle of buck converter in this mode, consider the plots as shown in Fig.5.6.

- When the switch is closed, as shown in Fig.5.4, the voltage across the inductor is $V_L = V_i - V_o$. The current through the inductor rises linearly (in approximation, so long as the voltage drop is almost constant). At this instant, the diode is reverse-biased by the voltage source V, hence, no current flows through it;
- When the switch is opened, as shown in Fig.5.5, the diode is forward biased. The voltage across the inductor is $V_L = -V_o$ (drop in diode is neglected). Current I_L decreases.

  The energy stored in magnetic field of inductor L is $E = \frac{1}{2} L I_L^2$

  Since I_L increases in L during on-time, the energy stored in it also increases and then decreases during the off-state. It shows that L is used to transfer energy from the input to the output of the converter.
- The rate of change of I_L can be determined from: $V_L = L\frac{dI_L}{dt}$

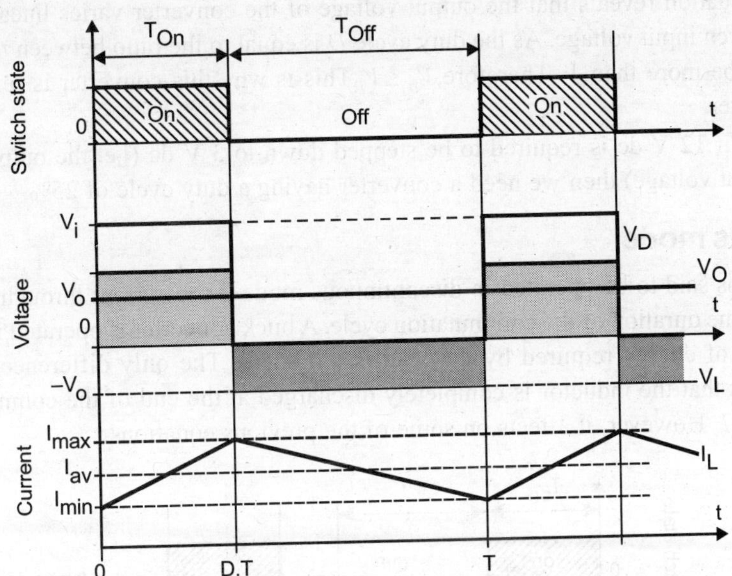

Fig.5.6: Evolution of the voltages and currents with time in an ideal buck converter operating in continuous mode.

- During the on-state, the value of V_L will be equal to $V_i - V_O$ and during the off-state it will be $-V_O$. Therefore, the increase in current during the on-state is given by the relation:

$$\Delta I_{Lon} = \int_0^{t_{on}} \frac{V_L}{L} dt = \frac{(V_i - V_o)}{L} t_{on},\ t_{on} = DT$$

Where D is called the *Duty Cycle* (a scalar quantity) which lies between 0 and 1.

- On the other hand, during off-state the current decreases. It is given by the relation:

$$\Delta I_{Loff} = \int_{t_{on}}^{T = t_{on} + t_{off}} \frac{V_L}{L} dt = \frac{V_o}{L} t_{off},\ t_{off} = (1 - D)T$$

- Assuming that the converter operates in the steady state condition, the energy stored in each component at the end of a commutation cycle T is equal to that at the beginning of the cycle. It shows that the magnitude of current I_L remains the same at $t = 0$ and at $t = T$ (see figure 5.6).
  From the above equations we get:

$$\Delta I_{L_{on}} + \Delta I_{L_{off}} = 0 \qquad \text{or} \qquad \frac{V_i - V_o}{L} t_{on} - \frac{V_o}{L} t_{off} = 0$$

- As per the above equations, the quantities can be represented graphically, as shown in Fig.5.6. Here $D L_{on}$ and $D L_{off}$ are represented by different rectangles (gray and dark gray). Under steady-state conditions, the area of rectangle for the quantity $(V_i - V_O)$ *ton* must be equal to the area of rectangle for the quantity $-V_O t_{off}$
  From the Fig.5.6, it is clear that $t_{on} = DT$ and $t_{off} = (1 - D)\,T$
  This yields; $(V_i - V_O)\,DT - V_O(1 - D)\,T = 0$ or $V_O - D\,V_i = 0$

$$\Rightarrow D = \frac{V_o}{V_i}$$

- The above equation reveals that the output voltage of the converter varies linearly with the duty cycle for a given input voltage. As the duty cycle D is equal to the ratio between *ton* and the period T, it cannot be more than 1. Therefore, $V_o \leq V_i$ This is why this converter is referred to as step-down converter.

For example, if 12 V dc is required to be stepped down to 3 V dc (i.e. the output voltage is one quarter of the input voltage) then we need a converter having a duty cycle of 25%.

Discontinuous mode

A buck converter is said to be operated in discontinuous mode if the current through the inductor (I_L) falls to zero for some duration of the commutation cycle. A buck converter is operated in this mode only when the amount of energy required by the load is too small. The only difference in the principle described above is that the inductor is completely discharged at the end of the commutation cycle as shown in figure 5.7. However, it affects on some of the previous equations.

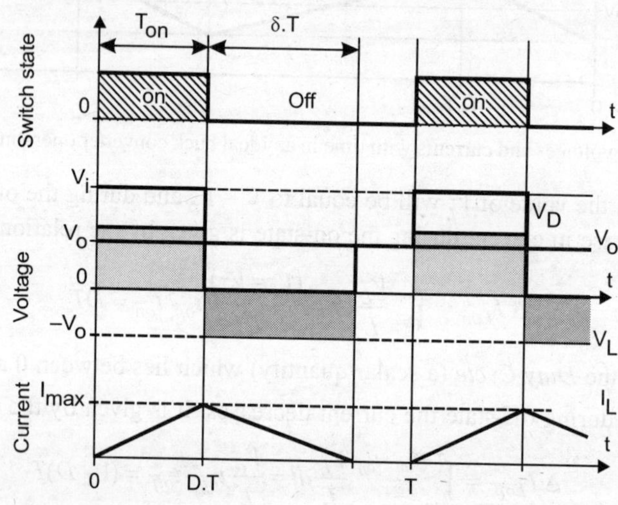

Fig.5.7: Evolution of the voltages and currents with time in an ideal buck converter operating in discontinuous mode.

- In these converters, in every commutation cycle, the current falls to zero. This results in the discharging of the output capacitor which causes more switching losses. To minimize these losses a different control technique known as *pulse-frequency modulation* is adopted.
- Still it is considered that the converter operates in steady state. Therefore, the energy in the inductor is considered to be the same at the beginning as well as at the end of the cycle (here in discontinuous mode, it is zero). This means that the average value of the inductor voltage (V_L) is zero; *i.e.* the area of the light and dark rectangles in Fig.5.7 are the same.

  This yields; $$(V_i - V_o)\,DT - V_o \delta T = 0$$

  So the value of δ is, $$\delta = \frac{V_i - V_o}{V_o} D$$

- The output current delivered to the load (I_o) is constant since we have considered that the output capacitor is large enough to maintain a constant voltage across its terminals during a commutation cycle. It implies that the current flowing through the capacitor has a zero average value.

Therefore, we have : $I_L' = I_o$ where I_L' is the average value of the inductor current

- It can be seen in Fig 5.7 that the inductor current waveform has a triangular shape. Therefore, the average value of *IL* can be sorted out geometrically as follow:

$$I_L' = \left(\frac{1}{2} I_{L\max} DT + \frac{1}{2} I_{L\max} \delta T\right)\frac{1}{T} = \frac{I_{L\max}(D+\delta)}{2} = I_o$$

- In the beginning, the inductor current is zero and rises during t_{on} up to I_{Lmax}. Therefore, I_{Lmax} will be equal to:

$$I_{L\max} = \frac{V_i - V_o}{L} DT$$

- Substituting the value of I_{Lmax} in the previous equation, we get,

$$I_o = \frac{(V_i - V_o)\, DT\,(D+\delta)}{2L}$$

- After substituting the value of δ in the above expression, it yields as:

$$I_o = \frac{(V_i - V_o)\, DT\left(D + \frac{V_i - V_o}{V_o} D\right)}{2L} \quad \text{or} \quad V_o = V_i \frac{1}{\frac{2LI_o}{D^2 V_i T} + 1}$$

Conclusion: From the above expression, it is clear that the output voltage of a buck converter operating in discontinuous mode is

- much more complicated than its counterpart of the continuous mode
- the output voltage is now a function not only of the input voltage (V_i) and the duty cycle *D*, but it also depends upon the inductor value (*L*), the commutation period (*T*) and the output current (I_o). Thus, continuous mode is always preferred over discontinuous mode.

Applications

Buck converters are remarkably efficient (often higher than 90%). These are very useful for converting a computer's main (bulk) supply voltage (often 12V) down to lower voltages needed by *USB, DRAM* and the *CPU* (1.8V or less).

5.4 BOOST CONVERTERS

Boost converter is a *DC-to-DC* power converter which steps up voltage and steps down the current simultaneously such that input power supply remains the same as that of the output (load).

Conventional circuit

The block diagram (conventional circuit) and pictorial view of a boost converter is shown in Fig.5.8 and 5.9 respectively. It essentially consists of:

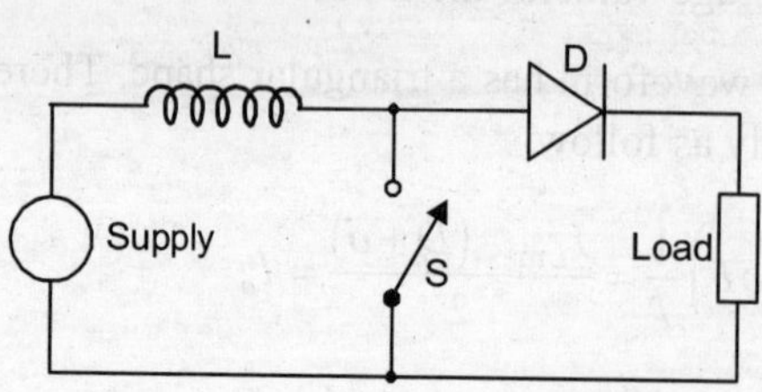

Fig.5.8: Boost converter circuit diagram.

Fig.5.9: Pictorial view of boost converter

The switch is typically a *MOSFET, IGBT* or *BJT*

- *At least two semiconductors* – a diode and a transistor, however, in modern buck converters usually diode is replaced with a second transistor and transistor is replaced with *MOSFET* or *IGBT.*
- *At least one energy storage element* – a capacitor, inductor or the two in combination.
- *Extra capacitors* (*or sometimes a combination of capacitors and inductors*) – these are used to filter ripples. These components are normally added to output (load-side) and input (supply-side); not shown here for simplicity.

Principle

The basic principle of a boost converter is the tendency of an inductor to oppose the changes in current by creating and destroying a magnetic field. In this converter, the output voltage is always higher than the input voltage. A schematic of a boost power stage is shown in Fig.5.10.

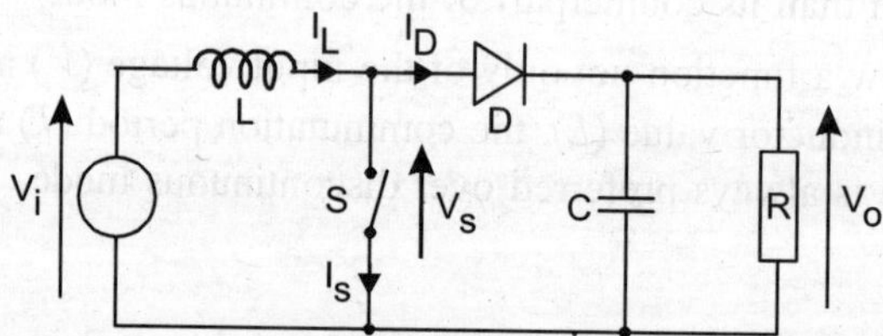

Fig.5.10: Boost converter schematic

Working

The two circuit configurations of a boost converter *i.e.* On-state, when the switch is closed, and Off-state, when the switch is open are shown in Fig.5.11 and 5.12 respectively. The arrows indicate the direction of flow of conventional current.

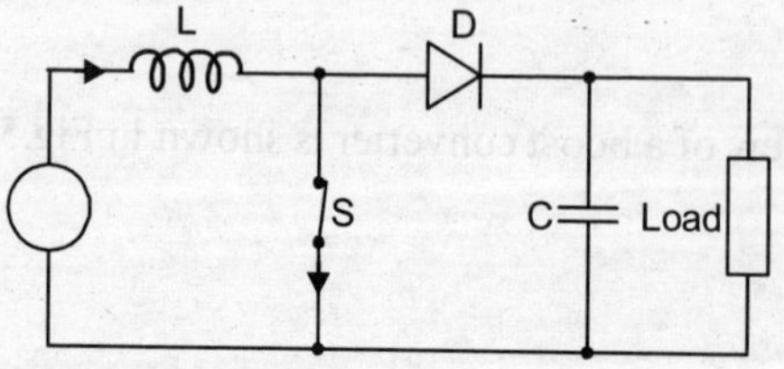

Fig.5.11: On-state (Boost converter)

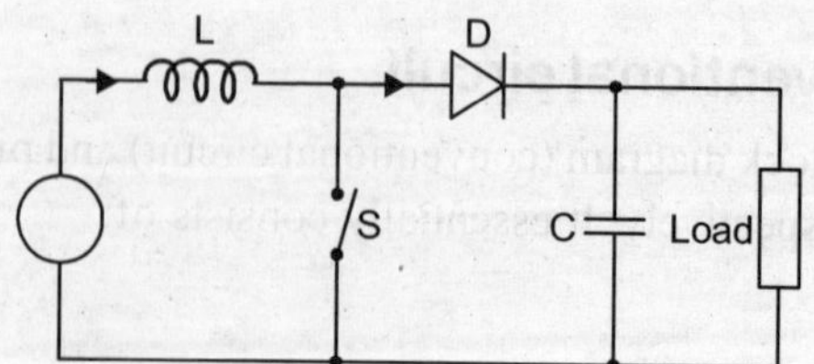

Fig.5.12: Off-state (Boost converter)

For ideal operation, it is assumed that

- All the components are perfect specifically, the switch and the diode have zero voltage drop when on and zero current flow when off, and the inductor has zero series resistance
- The input and output voltages do not change over the course of a cycle. It implies that the output capacitance is being infinite.

To understand its functioning proceed as follows:

On-state

- When the switch is closed (On-state), current flows through the inductor in clockwise direction, a magnetic field is developed and some energy is stored in it and the inductor attain a polarity.

Off-state

- When the switch is opened (Off-state), current will reduce since the impedance is higher. The magnetic field previously created will collapse to maintain the current towards the load. Thus, the *emf* of the two sources will be in series causing a higher voltage to charge the capacitor through the diode *D*.

Continuous mode

- If the switch is operating fast enough, the inductor will not discharge fully in between charging stages, and the load will always see a voltage greater than that of the input source alone when the switch is opened. Moreover, when the switch is in open position, the capacitor which is in parallel with the load is charged to this combined voltage.
- Again when the switch is closed, the right hand side is shorted out from the left hand side, in this stage the capacitor is able to provide the necessary voltage and energy to the load. During this time, the blocking diode prevents the capacitor from discharging through the switch. Obviously, the switch must be opened again fast enough to prevent the capacitor from discharging too much.

  It concludes that a Boost converter works under 2 distinct states:
- in the On-state (when the switch is closed), this results in an increase in the inductor current;
- in the Off-state (when the switch is open), in this case the only path offered to inductor current is through the flyback diode *D*, the capacitor *C* and the load *R*. This results in transferring of energy accumulated during the On-state into the capacitor.

  The boost converters may work in continuous mode or discontinuous mode.

Continuous mode

- A boost converter is said to be operated in continuous mode if the current through the inductor (I_L) never falls to zero during the commutation cycle. To describe the operating principle of boost converter in this mode, consider the plots as shown in Fig.5.13.
- When the switch is closed (on-state), as shown in Fig.5.11, the voltage which appears across the inductor is V_i. This causes a change in current (I_L) flowing through the inductor during a time period (t), it is given by the formula:

$$\frac{\Delta I_L}{\Delta t} = \frac{V_i}{L}$$ where L is the inductance of the inductor

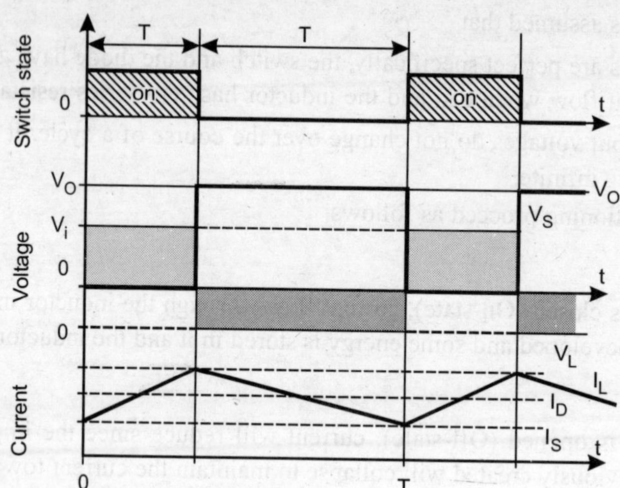

Fig.5.13: Waveforms of current and voltage in a boost converter operating in continuous mode.

At the end of the On-state, the current in the inductor increases to I_L, it changes as per the relation:

$$\Delta I_{L_{on}} = \frac{1}{L}\int_0^{DT} V_i dt = \frac{DT}{L} V_i \text{ where } D \text{ is the duty cycle.}$$

Duty cycle D represents the fraction of the commutation period T during which the switch is kept On. Therefore, D ranges between 0 (when S is off or never on) and 1 (when S is always on).

- When the switch is opened (off-state), as shown in Fig.5.12, the inductor current flows through the load. As it is Assumed that there is zero voltage drop in the diode, and a capacitor is large enough for its voltage to remain constant, the evolution of I_L is:

$$V_i - V_o = L\frac{dI_L}{dt}$$

Hence, the variation of I_L during the Off-period will be:

$$\Delta I_{L_{on}} = \int_{DT}^{T} \frac{(V_i - V_o)\,dt}{L} = \frac{(V_i - V_o)(1-D)\,T}{L}$$

Since the converter operates in steady-state conditions, the amount of energy stored in the inductor is given by:

$$E = \frac{1}{2} L I_L^{\,2}$$

For proper functioning, the inductor current has to be the same at the start and at the end of the commutation cycle. It means the overall change in the current (that is the sum of the changes) is zero:

Therefore, $\Delta I_{L_{on}} + \Delta I_{L_{off}} = 0$

Substituting the value of ΔL_{on} and ΔL_{off} the above expression yields to:

$$\Delta I_{L_{on}} + \Delta I_{L_{off}} = \frac{V_i DT}{L} + \frac{(V_i - V_o)(1-D)T}{L} = 0 \quad \text{or} \quad \frac{V_o}{V_i} = \frac{1}{1-D}$$

The above equation reveals that the output voltage is always higher than the input voltage (as the duty cycle varies from 0 to 1). Theoretically it may increases to infinity as D approaches 1.

That is why this converter is considered as a *step-up* converter.

Duty cycle can be determined by rearranging the above equation *i.e.*

$$D = 1 - \frac{V_i}{V_o}$$

Discontinuous mode

A boost converter is said to be operated in discontinuous mode if the inductor is completely discharged before the end of a whole commutation cycle. This commonly occurs under light loads. In this case, the current through the inductor falls to zero during part of the period (see waveforms in figure 5.14). Although the difference is slight but it has a strong effect on the output voltage equation. The voltage gain can be calculated as follows:

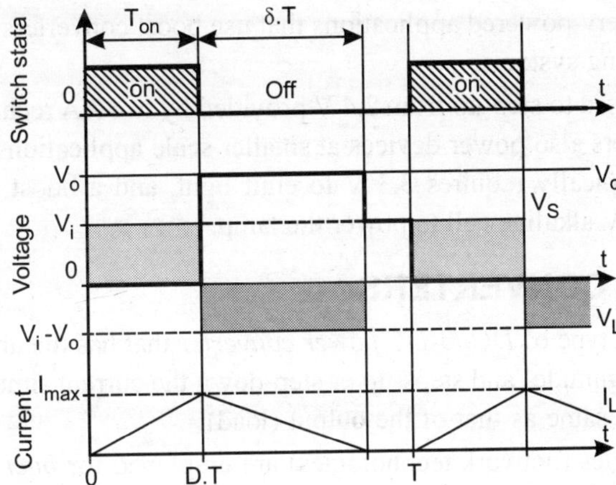

Fig.5.14: Waveforms of current and voltage in a boost converter operating in discontinuous mode.

As at the beginning of the cycle, the inductor current is zero, its maximum value I_{Lmax} (at $t = DT$) is

$$I_{L\max} = \frac{V_i DT}{L}$$

During the off-period, I_L falls to zero after δT; $I_{L\max} + \frac{(V_i - V_o)\,\delta T}{L} = 0$

Solving the above two equations, we can determine the value of δ, *i.e* $\delta = \frac{V_i D}{V_o - V_i}$

The load current I_o is equal to the diode current (I_D). It can be visualised in Fig.5.14. The diode current is equal to the inductor current during the off-state. Therefore, the output current can be written as:

$$I_o = I_D = \frac{I_{L\max}}{2}\delta$$

Substituting the value of I_{Lmax} and δ the expressions yields to:

$$I_o = \frac{V_i DT}{2L} \cdot \frac{V_i D}{V_o - V_i} = \frac{V_i^2 D^2 T}{2L\,(V_o - V_i)}$$

Thus, the output voltage gain: $\frac{V_o}{V_i} = 1 + \frac{V_i D^2 T}{2LI_o}$

This expression of the output voltage gain in comparison to the output voltage gain for continuous mode is much more complicated. Moreover, in discontinuous operation, the output voltage gain not only depends on the duty cycle (*D*), but it also depends upon the inductor value (*L*), the input voltage (*Vi*), the commutation period (*T*) and the output current (*Io*).

Applications

Boost converters are remarkably efficient (often higher than 90%). Battery power systems often stack cells in series to achieve higher voltage. However, sufficient stacking of cells is not possible in many high voltage applications due to lack of space. Boost converters can increase the voltage and reduce the number of cells. Two battery-powered applications that use boost converters are used in hybrid electric vehicles (*HEV*) and lighting systems.

Boost converter are used to step-up from 2.4 V provided by two AA rechargeable cells to 9 V for TI calculator. Boost converters also power devices at smaller scale applications, such as portable lighting systems. A *white LED* typically requires 3.3 V to emit light, and a boost converter can step up the voltage from a single 1.5 V alkaline cell to power the lamp.

5.5 BUCK-BOOST CONVERTERS

Buck–boost converter is a type of *DC-to-DC power converter* that has the ability to step-down or step-up the output voltage (magnitude) and steps-up or step-down the current simultaneously such that input power supply remains the same as that of the output (load).

Two different topologies (network technologies) are employed for *buck–boost converter*. Both of them can produce a range of output voltages, ranging from much larger (in absolute magnitude) than the input voltage, down to almost zero.

Conventional circuit

The block diagram (conventional circuit) and pictorial view of a buck-boost converter is shown in Fig.5.15 and 5.16 respectively.

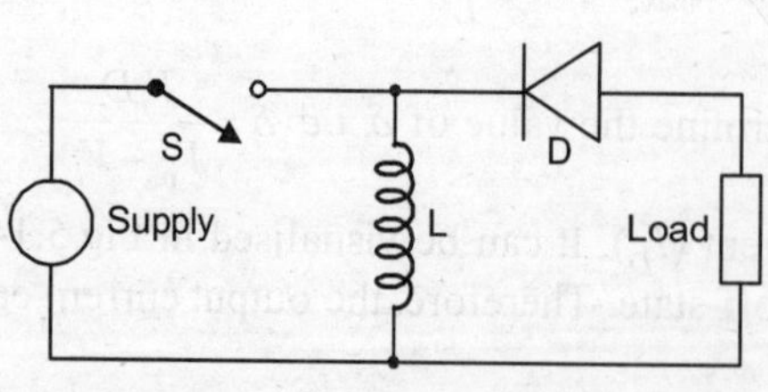

Fig.5.15: The basic schematic of an inverting buck–boost converter.

Fig.5.16: Pictorial view of a buck-boost converter

A detailed schematic diagram of a *buck-boost converter* is shown in Fig.5.17. It essentially consists of:

- *At least two semiconductors:* a diode and a transistor, however, in modern buck converters usually diode is replaced with a second transistor and transistor is replaced with *MOSFET* or *IGBT*.
- *At least one energy storage element:* a capacitor, inductor or the two in combination.
- *Extra capacitors (or sometimes a combination of capacitors and inductors)*: these are used to filter ripples. These components are normally added to output (load-side) and input (supply-side); not shown here for simplicity.

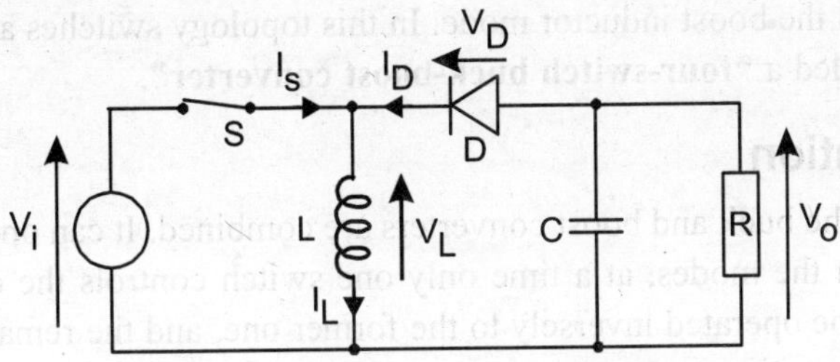

Fig.5.17: Schematic of a buck–boost converter.

The inverting topology or network technology

In this topology, the output voltage is of the *opposite polarity* than the input. This is a switched-mode power supply. It is a similar network technology which is used in boost converter and the buck converter. The output voltage is adjustable based on the duty cycle of the switching transistor.

The drawback in this network technology is that the switch of the converter is not grounded; this complicates the driving circuitry. However, this drawback is of no consequence if the power supply is isolated from the load circuit.

Principle of operation

The basic principle of the inverting buck–boost converter is fairly simple (see figure 2):

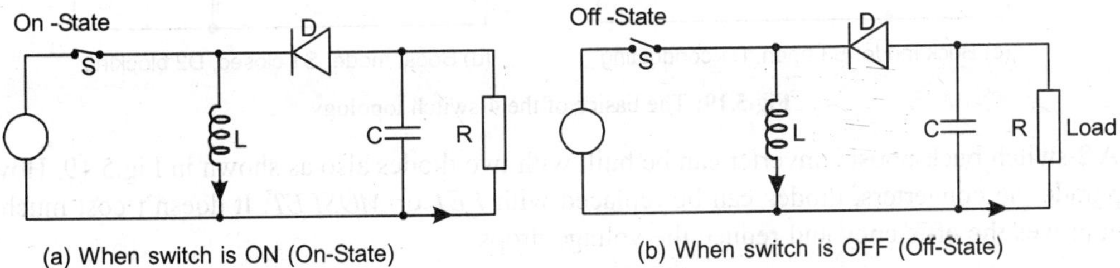

Fig.5.18: The two operating states of a buck–boost converter

- When the switch is in the On-state, the input voltage source is directly connected to the inductor (*L*). This results in accumulation of energy in *L* in the form of magnetic field. In this stage, the energy is supplies to the load by the capacitor.
- When the switch is in the Off-state, the inductor is connected to the load and capacitor. Therefore, energy is supplied by the inductor *L* to load *R* and capacitor *C*.

  In comparison to the buck and boost converters, the characteristics of the inverting buck–boost converter are mainly:

- polarity of the output voltage is opposite to that of the input, the output voltage can vary continuously from 0 to ∞ (for an ideal converter). The output voltage ranges for a buck and a boost converter are respectively Vi to 0 and Vi to ∞ .

A buck converter combined with a boost converter (four-switch topology)

The output voltage is typically of the same polarity of the input, and can be lower or higher than the input. Such a non-inverting buck-boost converter may use a single inductor which is used for both the buck inductor mode and the boost inductor mode. In this topology switches are used instead of diodes, that is why it is also called a **"four-switch buck-boost converter"**.

Principle of operation

In a 4-switch topology, the buck and boost converters are combined. It can operate in either the buck or the boost mode. In both the modes, at a time only one switch controls the duty cycle, another is for commutation and must be operated inversely to the former one, and the remaining two switches are in a fixed position.

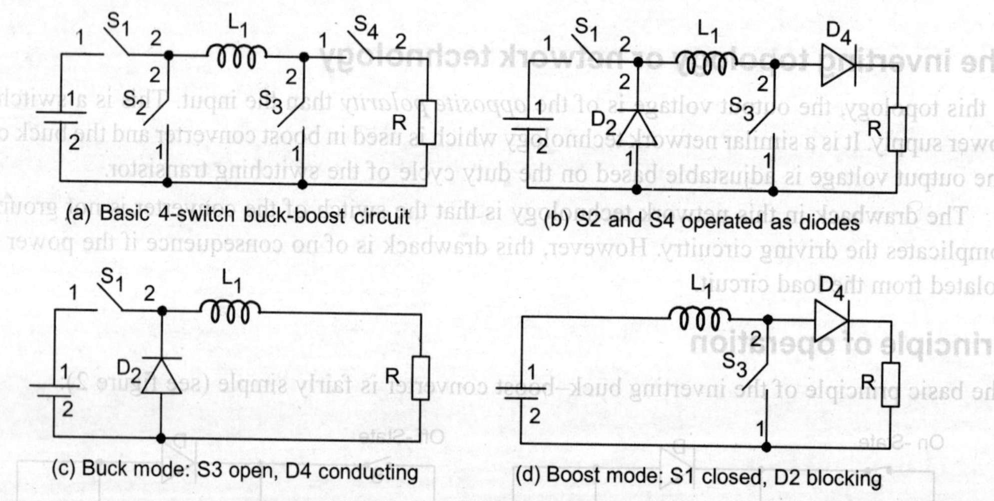

Fig.5.19: The basics of the 4-switch topology

A 2-switch buck-boost converter can be built with two diodes also as shown in Fig.5.19. However, to upgrade the converters, diodes can be replaced with *FET* or *MOSFET*. It doesn't cost much extra but improves the efficiency and reduce the voltage drops.

Continuous and discontinuous mode

The buck-boost converters can also be operated in continuous as well as discontinuous mode similar to buck converters and boost converters. When the current through the inductor (I_L) never falls to zero during the commutation cycle, it is called a continuous mode but if it falls to for some duration then it is called as discontinuous mode.

Applications of Buck boost converter

- It is used in the self regulating power supplies.
- It has consumer electronics.

- It is used in the Battery power systems.
- Adaptive control applications.
- Power amplifier applications.

Advantages of Buck Boost Converter

- It gives higher output voltage.
- Low operating duct cycle.
- Low voltage on *MOSFETs*

5.6 DUTY CYCLE

The period or duration in which a signal or system remains active is called duty cycle.

Duty cycle is commonly expressed as a percentage or a ratio. A period is the time duration that a signal takes to complete an on-and-off cycle.

As a formula, a duty cycle (%) may be expressed as:

$$D = \frac{PW}{T} \times 100\%$$

Similarly, a duty cycle (ratio) may be expressed as:

$$D = \frac{PW}{T}$$

where D is the duty cycle, *PW* is the pulse width (pulse active time), and *T* is the total period of the signal.

For a converter, $D = V_{out} / V_{in}$

Accordingly, a 60% duty cycle means the signal is on 60% of the time but off 40% of the time.

Note: The "on time" for a 60% duty cycle could be a fraction of a second, a day, or even a week, depending on the length of the period.

Duty cycle is used to describe the percent time of an active signal in an electrical device (power converter etc.).

5.7 VOLTAGE SOURCE INVERTERS

A circuit that operates from a stiff dc source and develops an ac output is usually known as ***voltage source inverter***.

If thc input dc is a voltage source, the inverter is called a voltage source inverter (*VSI*). One can similarly think of a current source inverter (*CSI*), where the input to the circuit is a current source. The *VSI* circuit has direct control over 'output (ac) voltage' whereas the *CSI* directly controls 'output (ac) current'. Shape of voltage waveforms output by an ideal *VSI* should be independent of load connected at the output.

The simplest dc voltage source for a *VSI* may be a battery bank, which may consist of several cells in series-parallel combination. Solar photovoltaic cells can be another dc voltage source. An ac voltage supply, after rectification into dc will also qualify as a dc voltage source. A voltage source is called stiff, if the source voltage magnitude does not depend on load connected to it.

Voltage source inverters (*VSI*) are used in uninterruptible power supply (*UPS*) units, adjustable speed drives (*ASD*) for ac motors, electronic frequency changer circuits etc.

Most of us are more familiar with commercially available inverter units used in our houses and offices as stand-by units to power some essential ac loads in case the power supply gets interrupted. In such inverter units, battery supply is used as the input dc voltage source and the inverter circuit converts the dc into ac voltage of desired power frequency (50Hz). The achievable magnitude of ac voltage is limited by the magnitude of input (dc bus) voltage. In ordinary household inverters the battery voltage may be just 12 volt and the inverter circuit may be capable of supplying ac voltage of around 10 volts (rms) only. In such cases the inverter output voltage is stepped up using a transformer to meet the load requirement of, say, 230 volt.

5.8 CONVERSION OF DC INPUT TO AC OUTPUT

Schematic circuit

For conversion of dc into ac, two schematic circuits, using transistor-switches, are shown in Figs. 5.20 (a) and 5.21 (a). In both the circuits, the transistors work in common emitter configuration and are interconnected in push-pull manner. In order to have a single control signal for the transistor switches, one transistor is of *npn* type and the other of pnp type and their emitters and bases are shorted as shown in the figures. A symmetrical bipolar dc supply is required for both the circuits. Collector of *npn* transistor is connected to positive dc supply (+*E*), whereas, the collector of *pnp* transistor is connected to negative dc supply of same magnitude (-*E*). A resistive load is connected between the emitter shorting point and the power supply ground.

Working

When the transistors work in active (amplifier) mode

The circuit for the transistors working in active (amplifier) mode is shown in Fig.5.20 (a). When applied base signal is positive, the *pnp* transistor is reverse biased and the *npn* transistor conducts the load current. Similarly for negative base voltage the *pnp* transistor conducts while *npn* transistor remains reverse biased. A suitable resistor connected in series with the base signal will limit the base current and keeps it sinusoidal provided the applied (sinusoidal) base signal magnitude is much higher than the base to emitter conduction-voltage drop. Assuming constant gain of the transistor over its working range, the load current can be seen to follow the applied base signal. Fig.5.20 (b) shows a typical load voltage and base signal waveforms. The other transistor will also be dissipating identical power during its conduction.

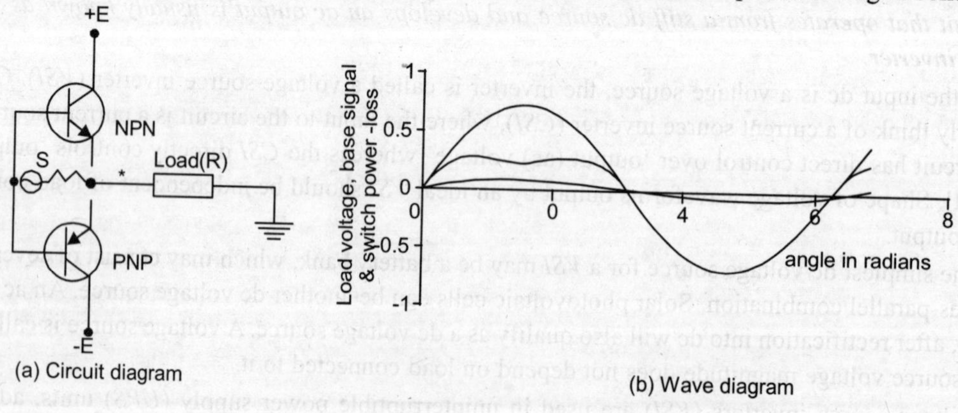

Fig.5.20: Transistors working in active (amplifier) mode

When the transistors work in switched-mode

The circuit for the transistors working in switched-mode is shown in Fig.5.21(a). The conducting switch remains fully on having negligible on-state voltage drop and the non-conducting switch remains fully off allowing no leakage current through it. The load voltage waveform output by switched-mode circuit of Fig.5.21(b) is rectangular with magnitude $+E$ when the *npn* transistor is on and $-E$ when *pnp* transistor is on. Such waveform is shown in Fig.5.21(b). The on and off durations of the two transistors are controlled so that (*i*) the resulting rectangular waveform has no dc component (*ii*) has a fundamental (sinusoidal) component of desired frequency and magnitude and (*iii*) the frequencies of unwanted harmonic voltages are much higher than that of the fundamental component so that those can be filtered. The fundamental sine wave is also shown in Fig.5.21 (b). This waveform is identical to the sinusoidal output voltage waveform of the system voltage.

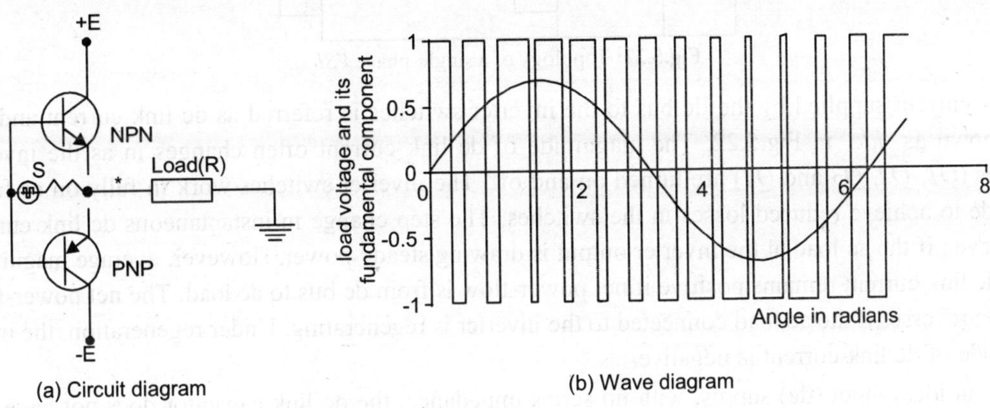

(a) Circuit diagram (b) Wave diagram

Fig.5.21: Transistors working in switch mode

Conclusion

Both amplifier mode and switched mode circuits are capable of producing ac voltages of controllable magnitude and frequency. However, the amplifier circuit is not acceptable in power-electronic applications due to high switching power loss. The switched mode circuit is acceptable although it generates significant amount of unwanted harmonic voltages along with the desired fundamental frequency voltage. These high frequency voltage harmonics can easily be blocked by using small size filters. Thus, a desired quality of load voltage can be obtained.

5.9 GENERAL STRUCTURE OF VOLTAGE SOURCE INVERTERS (*VSI*)

Single-phase voltage source inverter

A typical power-circuit topologies of a single-phase voltage source inverter is shown in Fig.5.22. For medium output power applications, only one dc source and the preferred devices are:

(i) n-channel *IGBTs* (*Q1, Q2, Q3* etc.- act as fast and controllable switches)

(ii) a large dc link capacitor (*Cdc*)

(iii) fast recovery diodes (*D1, D2, D3* etc.)

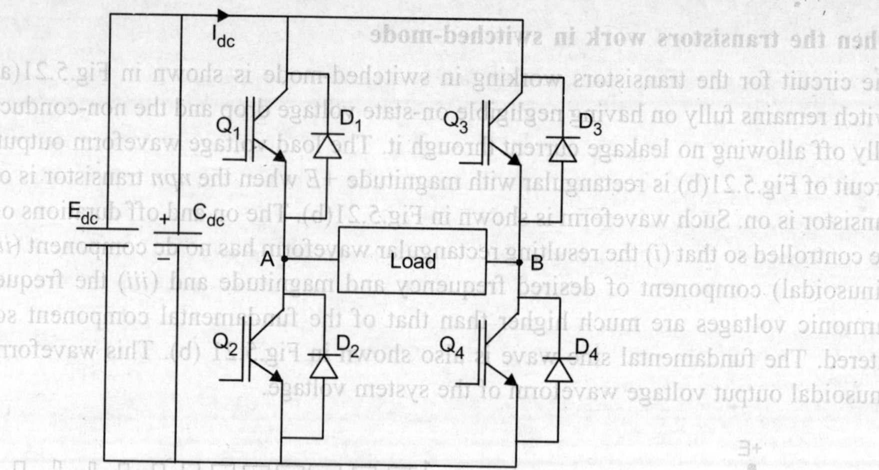

Fig.5.22: Topology of a single-phase *VSI*

The current supplied by the dc bus to the inverter switches is referred as dc link current and has been shown as *'idc'* in Fig.5.22. The magnitude of dc link current often changes in as the inverter switches (*Q1, Q2, Q3* and *Q4*) are turned on and off. The inverter switches work in fully-on or fully-off mode to achieve reduced losses in the switches. The step change in instantaneous dc link current occurs even if the ac load at the inverter output is drawing steady power. However, average magnitude of the dc link current remains positive if net power-flow is from dc bus to ac load. The net power-flow direction reverses if the ac load connected to the inverter is regenerating. Under regeneration, the mean magnitude of dc link current is negative.

For an ideal input (dc) supply, with no series impedance, the dc link capacitor does not have any role. However a practical voltage supply may have considerable amount of output impedance. The supply line impedance, if not bypassed by a sufficiently large dc link capacitor, may cause considerable voltage spike at the dc bus during inverter operation. This may result in deterioration of output voltage quality, it may also cause malfunction of the inverter switches as the bus voltage appears across the non-conducting switches of the inverter. Moreover, in the absence of dc link capacitor, the series inductance of the supply line will prevent quick build up or fall of current through it and the circuit behaves differently from the ideal *VSI* where the dc voltage supply is supposed to allow rise and fall in current as per the demand of the inverter circuit.

Therefore, dc link capacitor should be applied to remove the ripples. These should be put very close to the switches so that it provides a low impedance path to the high frequency component of the switching currents. The capacitor itself must be of good quality with very low equivalent series resistor and inductor.

When switches are to be connected in bridge fashion, using n-channel *IGBTs*, the gate (base) signals to the *VSI* switches need to be isolated.

While framing the circuit, the leads that interconnect switches and diodes to the dc bus must be of minimum length to avoid insertion of significant amount of stray inductances in the circuit. The overall layout of the power circuit has a significant effect over the performance of the inverter circuit.

A diode in anti-parallel with each controlled switch, like *IGBT*, is used in *VSI* to allow a non-unity power factor load at the output.

Essentially, it is a full bridge circuit as shown in Fig.5.22. The circuit has two legs of switches, each leg consisting of an upper switch and a lower switch. Junction point of the upper and lower switches is the output point of that particular leg. Voltage between output point of legs and the mid-potential of the dc bus is called as '**pole voltage**' referred to the mid potential of the dc bus. One may think of pole voltage referred to negative bus or referred to positive bus too but unless otherwise mentioned pole voltages are assumed to be referred to the mid-potential of the dc bus. The two pole voltages of the single-phase bridge inverter generally have same magnitude and frequency but their phases are 180° apart. Thus the load connected between these two pole outputs (between points '*A*' and '*B*') will have a voltage equal to twice the magnitude of the individual pole voltage.

Three-phase voltage source inverter

A typical power-circuit topologies of a three-phase voltage source inverter is shown in Fig.5.23. For medium output power applications, only one dc source and the preferred devices are:

(*i*) n-channel *IGBTs* ($Q_1, Q_2, Q_3, Q_4, Q_5, Q_6$ - act as fast and controllable switches)

(*ii*) a large dc link capacitor (C_{dc})

(*iii*) fast recovery diodes ($D_1, D_2, D_3, D_4, D_5, D_6$)

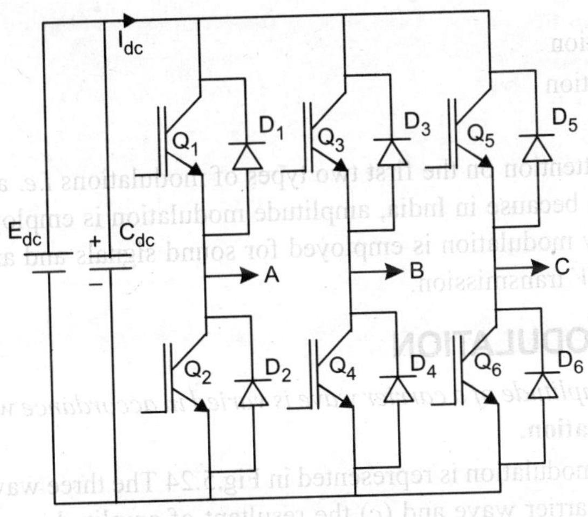

Fig.5.23: Topology of a three-phase VSI

The working of 3-phase inverter bridge, shown in Fig.5.23, is similar as that of 1-phase inverter. The only difference is that the pole voltages of the 3-phase inverter bridge are 120° phase apart across the terminals *A*, *B* and *C*.

5.10 MODULATION

*The process of varying some characteristic (e.g. amplitude, frequency or phase) of a carrier wave by the *modulating signal is called* **modulation.**

The audio (message) signals or waves have small power and cannot be transmitted to long distances. Therefore, these waves or signals are '*added*' to the high frequency (carrier) waves because these have sufficient power to travel through long distances.

* The modulating signal may be audio or vedio signal.

The carrier wave is usually represented by the expression:

$$v_c = V_c \sin(\phi_c t + \phi)$$

The three variables in this equation are amplitude V_c; frequency f_c (as $\omega_c = 2pf_c$) and phase angle ϕ. Therefore, the audio waves are *added* to the carrier waves in such a way that some characteristic (amplitude frequency or phase) of carrier varies in accordance with the signal. This process by which the carrier is *modified* in accordance with the signal is called ***modulation*.

5.11 NEED OF MODULATION

Modulation is extremely necessary in radio/TV communication, processing weak signals in instrumentation and in other electronic systems to:

(*i*) increases the operating range

(*ii*) transmit the signal without wireand

(*iii*) reduces the size of transmitting antenna.

5.12 TYPES OF MODULATION

As discussed earlier, modulation is a process of varying some characteristic *i.e.* amplitude, frequency or phase of the carrier wave in accordance with the message signal. Accordingly, the modulation is of three types, namely;

(*i*) amplitude modulation

(*ii*) frequency modulation

(*iii*) phase modulation.

We shall confine our attention on the first two types of modulations *i.e.* amplitude modulation and frequency modulation. It is because in India, amplitude modulation is employed in radio broadcasting system, whereas, frequency modulation is employed for sound signals and amplitude is employed for picture (video) signals in *TV* transmission.

5.13. AMPLITUDE MODULATION

The process by which the amplitude of a carrier wave is varied in accordance with the modulating signal is called **amplitude modulation.**

Principle of amplitude modulation is represented in Fig.5.24 The three waves shown in Fig.5.24 are (*a*) modulating signal, (*b*) carrier wave and (*c*) the resultant of amplitude modulated wave. It may be noted that when a signal of frequency f_s is modulated with carrier wave of frequency f_c, a resultant wave (modulated wave) is produced. It retains the frequency of carrier wave *i.e.* f_c, but its amplitude varies in accordance with the signal. The amplitude of the resultant wave varies in both the halves (positive and negative) simultaneously. It is because when the signal in increasing in the positive sense, the amplitude of the carrier wave also increases. On the other hand when the signal is changing in the negative sense, the amplitude of the carriers decreases. This modulation (mixing of two waves) is done with the help of an electron circuit using solid state devices or valves. The circuit is known as *modulator*.

The important points to be noted are:

(*i*) Amplitude of the modulated wave varies in accordance with message signal.

(*ii*) Frequency of the modulated wave is the same (f_c) as that of the carrier frequency.

** In fact, the name modulation has come from modification.

(*iii*) While doing modulation, the amplitude of carrier varies on both the halves simultaneously in accordance with the signal.

5.14 MODULATION FACTOR AND ITS SIGNIFICANCE

We have seen that in amplitude modulation, the amplitude of the carrier increases during positive half cycle of the signal, whereas it decreases during the negative half-cycle of the signal (See Fig.5.24). The extent to which the amplitude of carrier wave is changed by the signal is described by the factor called *modulation factor*. It may be defined as under:

The ratio of change in amplitude of carrier wave due to modulation to the original amplitude of the normal carrier wave is called **modulation factor.** It is generally represented by letter '*m*'.

Modulation factor, $$m = \frac{\text{Change in amplitude of carrier}}{\text{amplitude of normal carrier}}$$

The above relation clearly shows that the modulation factor depends upon the amplitude of both the waves; carrier and signal.

Significance of modulation factor: In amplitude modulation, the amplitude of the carrier wave varies as per the modulating signal wave. The degree of modulation factor shows the strength of message signal *i.e.* the greater the degree of modulation (*m*), the stronger and clear will be the message signal and vice-versa. Hence, for $m = 1$ or 100%, the message signal being transmitted will be the strongest. However, when the carrier is over modulated (*i.e.* $m > 1$ or 100%), the AM wave will be clipped off and a huge distortion will occur during reception. Therefore, degree of modulation should *never exceed* 100%.

5.15 ANALYSIS OF AMPLITUDE MODULATED WAVE

Let the carrier voltage wave be represented by the equation;

$$v_c = V_c \sin(\omega_c t + \phi)$$

For simplicity, consider phase angle ϕ to be zero, then

$$v_c = V_c \sin \omega_c t \qquad \ldots(i)$$

where, v_c = Instantaneous value of carrier

V_c = amplitude of carrier

$\omega_c = 2\pi f_c$ = angular velocity at carrier frequency f_c.

Let this carrier is modulated with the modulating signal represented by the equation;

$$v_s = V_s * \cos \omega_s t \qquad \ldots(ii)$$

where, v_s = instantaneous value of signal

V_s = amplitude of signal

$\omega_s = 2\pi f_s$ = angular velocity at signal frequency f_s

If *m* is the modulation factor, then

$$m = \frac{V_s}{V_c} \quad \text{or} \quad V_s = mV_c \qquad \ldots(iii)$$

Substituting the value of V_s in eqn. (*ii*), we get,

$$v_s = mV_c \cos \omega_s t \qquad \ldots(iv)$$

In amplitude modulation, the amplitude of the carrier wave (V_c) varies in accordance with the signal as shown in Fig.5.24. Hence, the amplitude of the resultant modulated wave varies at signal frequency. Therefore, the amplitude of *AM* wave is given as;

$$= V_c + mV_c \cos \omega_s t$$

$$= V_c (1 + m \cos \omega_s t) \quad \ldots(v)$$

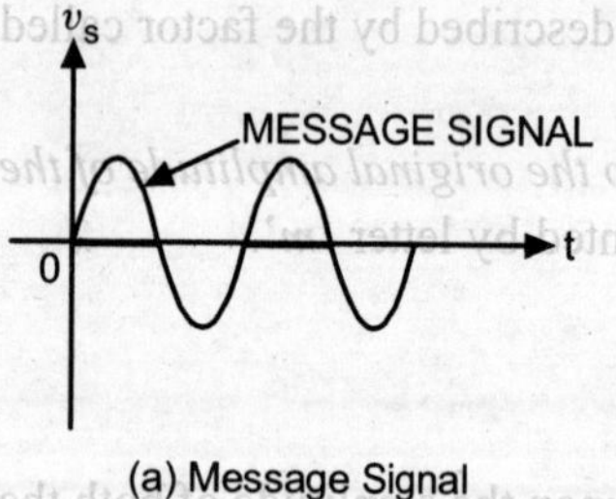

(a) Message Signal

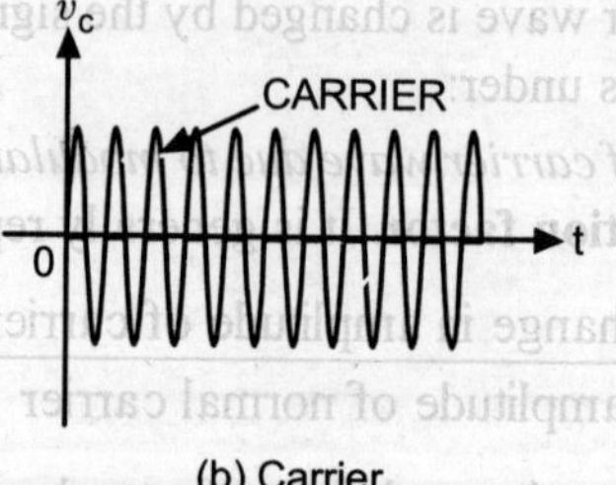

(b) Carrier

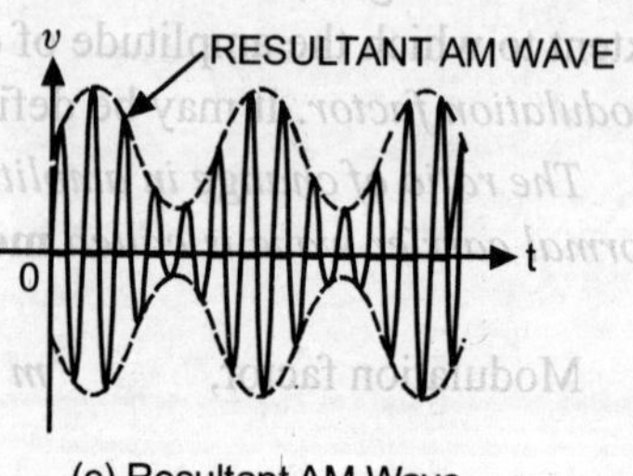

(c) Resultant AM Wave

Fig.5.24: Amplitude modulated wave.

Thus, the instantaneous value of the resultant modulated wave will be;

$$v = \text{Amplitude} \times \sin \omega_c t$$

$$= V_c (1 + m \cos \omega_s t) \sin \omega_c t$$

$$= V_c \sin \omega_c t + mV_c \sin \omega_c t \cos \omega_s t$$

$$= V_c \sin \omega_c t + \frac{mV_c}{2} (2 \sin \omega_c t \cos \omega_s t)$$

$$= V_c \sin \omega_c t + \frac{mV_c}{2} [\sin (\omega_c t - \omega_s t) + \sin (\omega_c t + \omega_s t)]$$

$$= V_c \sin \omega_c t + \frac{mV_c}{2} \sin (\omega_c - \omega_s)t + \frac{mV_c}{2} \sin (\omega_c + \omega_s)t \quad ..(vi)$$

The following points may be noted from the expression (*vi*) of modulated voltage wave;

(i) The *AM* waves is just a summation of the following three waves varying sinusoidally

(a) First having magnitude V_c and frequency **f_c

(b) Second having magnitude $mV_c/2$ and frequency $(f_c - f_s)$

(c) Third having magnitude $mV_c/2$ and frequency $(f_c + f_s)$.

(ii) The *AM* wave contains the fundamental frequency of carrier f_c and produces two more frequencies $(f_c - f_s)$ and $(f_c + f_s)$ known as side-band frequencies. Where, $(f_c - f_s)$ is called lower sideband frequency and $(f_c + f_s)$ is called the upper sideband frequency.

5.16 LIMITATIONS OF AMPLITUDE MODULATION

Amplitude modulation suffers from the following shortcomings:

* To make the calculation work simple, the signal is represented by consine of the angle. This only shows that the reckoning point is taken at positive maximum position.

** Here, $f_c = \frac{\omega_c}{2\pi}$; $(f_c - f_s) = \frac{(\omega_c - \omega_s)}{2\pi}$ and $(f_c + f_s) = \frac{(\omega_c + \omega_s)}{2\pi}$

1. **Poor efficiency:** In amplitude modulation, sidebands contain the signal, therefore, the power existing in the sidebands is only the useful power. The sideband power is only one-third (33·3%) of the total power of *AM* wave. Therefore, it is said that the amplitude modulation transmits signal at poor efficiency.
2. **Noisy reception:** In an *AM* wave, the amplitude variation in carrier contains the signal. Unfortunately, all the natural and man made noises consists of electrical disturbances. Moreover, radio receiver is such a device that it cannot distinguish between the amplitude variations that present due to desired signal or noise. Hence the reception is noisy. For instant, if we start a scooter or motor-cycle near the radio receiver, a noise. Hence the reception is noisy.
3. **Smaller operating range:** Owing to the poor efficiency of amplitude modulation, the transmitters employing this method have smaller range *i.e.* message (signal) cannot be transmitted to longer distances.
4. **Poor audio quality:** Because of more interference of the electrical amplitude disturbances, the audio quality obtained is poor when the signal a transmitter by amplitude modulation.

5.17 FREQUENCY MODULATION

The process by which the frequency of a carrier wave is varied in accordance with the modulating signal is called **frequency modulation.**

Principle of frequency modulation is represented in Fig.5.25. The three waves shown in the figure are (*a*) signal, (*b*) carrier and (*c*) the resultant frequency modulated wave. It may be noted that when a signal of frequency f_s is modulated with carrier wave of frequency f_c, a resultant modulated wave is produced.

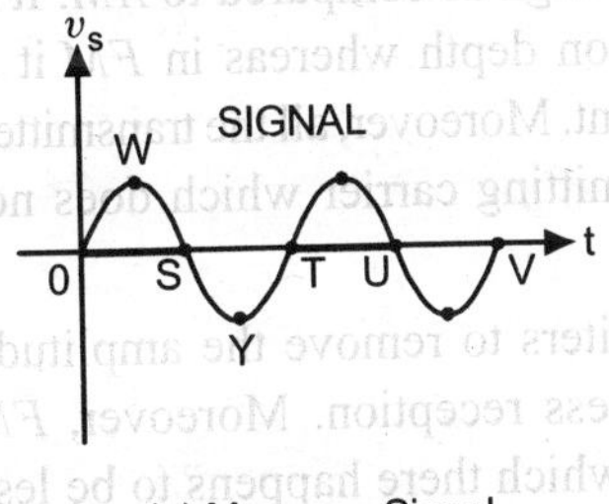

(a) Message Signal

(b) Carrier

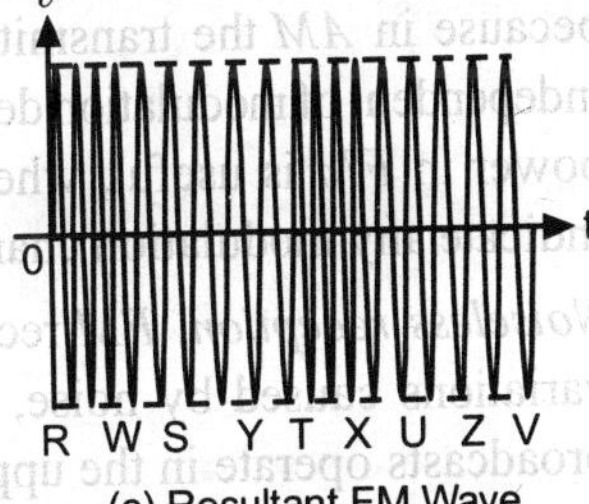

(c) Resultant FM Wave

Fig.5.25: Frequency modulated wave.

It retains the amplitude of the carrier wave (*i.e.* V_c) but its frequency varies in accordance with the instantaneous magnitude of the signal. Thus, when the signal voltage is zero, frequency of the carrier wave is not modulated and remains the same (see position *RSTU* and *V* in Fig.5.25-*c*). However, when the magnitude of signal is reaching its positive peak value, the frequency of carrier wave (after modulation) also increases as represented by closely spaced cycles (see position *W* and *X* in Fig.5.25-*c*). Whereas, during negative peak value, the frequency of carrier decreases as represented by widely spaced cycles (see position *Y* and *Z* in Fig.5.25-*c*).

Mathematically;

In the carrier voltage is given by the equation;

$$*v_c = V_c \sin \omega_c t$$

* In eqn. $v_c = V_c \sin(\omega_c t + \phi)$, if phase angle ϕ is neglected for simplicity, then $v_c = V_c \sin \omega_c t$.

and the modulating wave or signal is given by the equation;

$$v_s = V_s \cos \omega_s t$$

Thus, the instantaneous value of FM voltage is given as;

where V_c = amplitude of carrier wave

$\omega_c = 2\pi f_c$ = angular velocity at carrier frequency f_c

$\omega_s = 2\pi f_s$ = angular velocity at signal frequency f_s

δ = maximum frequency deviation

$= k V_s f_c$ where k is proportionality constant and V_s is amplitude of signal wave

Modulation index for *FM*, m_f is defined as:

$$m_f = \frac{\text{max. frequency deviation}}{\text{modulating signal frequency}} = \frac{\delta}{f_s} \quad ...(iv)$$

Hence, $v = V_c \sin (\omega_c t + m_f \omega_s f)$ ...(*v*)

The following points are worth noting:

(*i*) The amplitude of the modulated wave is the same as that of the carrier wave.

(*ii*) The frequency of the modulated wave varies in accordance with the message signal.

Advantages

(*i*) *High transmission efficiency*: Transmission efficiency of *FM* is very high as compared to *AM*. It is because in *AM* the transmitted power is governed by the modulation depth whereas in *FM* it is independent of modulation depth as the amplitude of *FM* wave is constant. Moreover, all the transmitted power in *FM* is useful, whereas in *AM* most of it is used in transmitting carrier which does not indicate any modulation change.

(*ii*) *Noiseless reception*: *FM* receivers can be fitted with amplitude limiters to remove the amplitude variations caused by noise, therefore, *FM* receivers provide noiseless reception. Moreover, *FM* broadcasts operate in the upper **VHF* and *UHF* frequency range at which there happens to be less noise than in the *MF* and *HF* rangers occupied by *AM* broadcasts.

(*iii*) *Better audio quality*: There is less adjacent-channel interference in *FM* than in *AM*. It is because **Standard frequency allocations provide a guard band between commercial *FM* stations.

Disadvantages

FM transmissions also suffer from some of the drawbacks, otherwise there would have been no *AM* transmission left. These drawbacks are listed below:

---

* *VHF* means Very High Frequency and
UHF means Ulta High Frequency
MF means Medium Frequency
HF means High Frequency

** Standard frequenices are allocated worldwide by the International Radio Consultative Committee (*CCIR*) of the *I.T.U.*

(*i*) *Wider channel is needed*: A much wider channel is required by *FM*, upto 10 times as large as that needed by *AM*. This is one of the major disadvantage.

(*ii*) *Equipment used is more complex and costly*: *FM* transmitting and receiving equipment particularly modulators and demodulators are more complex and costly.

(*iii*) *Smaller area of reception*: In this case reception is limited to line of sight, therefore, area of reception for *FM* is much smaller than for *AM*. This property may be considered as an advantages for co-channel allocation otherwise it is true that it reduces the area of reception.

5.18 PHASE MODULATION

The process by which the phase angle of a carrier wave is varied in accordance with the modulating signal is called **phase modulation.**

In phase modulation, the phase deviation is proportional to the amplitude of the modulating signal (V_s) but in independent of its frequency (f_s). In this case, the phasor of the modulated wave leads the reference position during positive half-cycle of the modulating signal and lags the reference position during the negative half-cycle of the modulating signal. The angular displacement of the phasor of the modulated wave (f_s) is restricted within the limits by the modulating signal.

Mathematically

If carrier voltage is given by the equation:

$$v_c = V_c \sin (\omega_c t + \phi) \quad ...(i)$$

and the modulating signal wave is given by the equation;

$$v_s = V_s \cos \omega_s t \quad ...(ii)$$

In phase modulation, the phase in eqn. (*i*) is varied in accordance with the instantaneous amplitude of the modulating voltage. The expression for the *PM* wave will be

$$v = V_c \sin (\omega_c t + \phi_s \sin \phi_s t) \quad ...(iii)$$

where f_s is the maximum value of phase change introduced by this particular modulating signal and is proportional to the maximum amplitude of this modulation. For the sake of uniformly, the above equation may be rewritten as;

$$v = V_c \sin (\omega_c t + m_p \sin \omega_s t) \quad ...(iv)$$

where $m_p = \phi_s$ = modulation index for phase modulation

SUMMARY

1. *Power converter:* A power converter is an electrical or electro-mechanical or electronic device that converts electrical energy to mechanical, mechanical energy electrical, *DC* to *AC, AC* to *DC, DC* to *DC* of different voltage level, *AC* to *AC* of different voltage level or different frequency.
2. *Simple linear voltage regulator:* It is just a potential-divider.
3. *Buck converter*: It is a *DC*-to-*DC* power converter which steps down voltage and steps up the current simultaneously such that input power supply remains the same as that of the output (load).

4. *Applications* of buck converters: These are very useful for converting a computer's main (bulk) supply voltage (often 12*V*) down to lower voltages needed by *USB, RAM* and the *CPU* (1.8*V* or less).

5. *Boost converter:* It is a *DC*-to-*DC* power converter which steps up voltage and steps down the current simultaneously such that input power supply remains the same as that of the output (load).

6. *Applications of boost converters:* Two battery-powered applications that use boost converters are used in hybrid electric vehicles (*HEV*) and lighting systems.

7. *Buck–boost converter:* It is a type of *DC*-to-*DC* power converter that has the ability to step-down or step-up the output voltage (magnitude) and steps-up or step-down the current simultaneously such that input power supply remains the same as that of the output (load).

8. *Duty cycle:* The period or duration in which a signal or system remains active is called duty cycle.

9. *Voltage source inverter:* A circuit that operates from a stiff dc source and develops an ac output is usually known as voltage source inverter.

10. *Modulation:* The process by which the characteristics of one waveform (carrier) is modified in accordance with the variations in another wave (modulating signal wave) is called modulation.

11. *Equation for carrier wave:* The carrier wave is given by the equation:

$$v_c = V_c \sin (\omega ct + \phi)$$

This wave has three variable characteristics i.e. amplitude V_c, frequency fc ($\omega c = 2\phi fc$) and phase angle ϕ.

12. *Necessity of modulation:*

(*i*) to increase operating range (*ii*) to permit transmission without wire

(*iii*) to reduce the size of transmitting antenna.

13. *Amplitude modulation:* When the amplitude of a carrier wave is varied in accordance with the modulating signal the process is called amplitude modulation.

14. *Modulation factor:* The ratio of change in amplitude of carrier wave due to modulation to the original of the normal carrier wave is called modulation factor.

$$m = \frac{V_s}{V_c}$$

15. *Limitations of amplitude modulation:* (*i*) poor efficiency (*ii*) noisy reception (*iii*) smaller operating range (*iv*) poor audio quality.

16. *Frequency modulation:* The process by which the frequency of a carrier wave is varied in accordance with the modulating signal is called frequency modulation.

17. *Advantages to FM*: (*i*) High transmission efficiency (*ii*) Noiseless reception (*iii*) Better audio quality.

Disadvantages of FM: (*i*) Wider channel is needed (*ii*) Equipment used is more complex and costly (*iv*) Smaller area of reception.

18. *Phase modulation:* The process by which the phase angle of a carrier wave is varied in accordance with the modulating signal is called phase modulation.

SHORT ANSWER QUESTIONS

1. What is a power converter?
2. How does a power converter work?
3. What is the difference between an inverter and a converter?
4. How do you convert *DC* to *AC*?
5. What is the difference between a rectifier and an inverter?
6. How does a potential divider used to get reduced voltage?
7. Why buck converter is preferred over potential divider?
8. How does a buck converter work?
9. How current can be increased in a circuit?
10. How does boost converter work?
11. What is the use of a boost converter?
12. What is a buck boost converter?
13. How does a buck boost converter work?
14. How does a *DC* to *AC* inverter work?
15. What is the duty ratio?
16. What does 60% duty cycle mean?
17. What is a 100% duty cycle?
18. How does a voltage source inverter convert dc into ac?
19. Name the major components used in a *VSI*, draw its circuit.
20. What is difference between a 1-phase and 3-phase *VSI*?
21. What is the difference between a *UPS* and an inverter?
22. What is a sinusoidal modulator?
23. What are the different types of modulation?
24. What is meant by Over-modulation?
25. What is meant by under-modulation?

26. Where amplitude modulation is used?

27. What do you mean by amplitude modulation?

28. Why *FM* is better than *AM*?

29. How amplitude modulation is work?

30. What is amplitude modulation index?

31. What is *FM* signal?

MULTI-CHOICE QUESTIONS

1. A diode in anti-parallel with the controlled switch, like *IGBT*, is used in *VSI* to:
(*a*) prevent reversal of dc link current.
(*b*) allow a non-unity power factor load at the output.
(*c*) protect the circuit against accidental reversal of dc bus polarity.
(*d*) none of the above.

2. A large capacitor, put across *dc* bus of a voltage source inverter, is intended to:
(*a*) to protect against switch failure.
(*b*) to minimize high frequency current ripple through the ideal dc source.
(*c*) to maintain a constant dc link current.
(*d*) allow a low impedance path to the high frequency component of dc link current.

3. Gate (base) signals to the *VSI* switches, using n-channel *IGBTs*, need to be isolated to allow:
(*a*) protection of switches against short at the inverter output terminals.
(*b*) lower losses in the gate drive circuit.
(*c*) switches to be connected in bridge fashion.
(*d*) a dc link voltage higher than the switch voltage rating.

4. The inverter switches work in fully-on or fully-off mode to achieve:
(*a*) reduced losses in the switches.
(*b*) minimum distortion in the output voltage waveform.
(*c*) easier gate control circuit for the switching devices.
(*d*) satisfactory operation for non-resistive load at the output.

5. *AM* is a process by which
(*a*) the frequency of the carrier is varied in accordance with the signal.
(*b*) the amplitude of the carrier is varied in accordance with the signal.
(*c*) the phase of the carrier is varied in accordance with the signal.
(*d*) the frequency and amplitude both are varied in accordance with the signal.

TEST QUESTIONS

1. What do you mean by power converter? How will you classify them?
2. Explain the working of a buck converter with the help of a circuit diagram.
3. Draw the input and output wave-shapes of a buck converter working in continuous mode.
4. How will you differentiate between continuous and discontinuous mode of operation of buck converter?
5. Explain the working of a boost converter with the help of a circuit diagram.
6. Draw the input and output wave-shapes of a boost converter working in discontinuous mode.
7. How will you differentiate between continuous and discontinuous mode of operation of boost converter?
8. Mention the characteristics and important applications of buck converter and boost converter.
9. Explain the working of a buck-boost converter with the help of a circuit diagram.
10. What is the significance of duty cycle?
11. How dc is converted into ac with the help of a voltage source inverter (*VSI*)?
12. Why switch mode *VSI* is preferred over amplifier mode *VSI*?
13. Explain the working of a single-phase *VSI* with the help of a circuit diagram.
14. What is modulation ? State the need of modulation in communication system.

    or

    What do you mean by modulation ? Why is essential ?
15. Define *AM* and *FM*. Use sketches to explain these definitions. Where are these employed ?
16. What do you understand by modulation factor ? What is its significance ?
17. In what respect *FM* differ from that of *AM* ? Discuss relative merits and demerits of the two methods of modulation.
18. Define amplitude modulation and derive an cxpression for a sinusoidal carrier voltage, amplitude modulated by another sinusoidal voltage?

ANSWER TO MULTI-CHOICE QUESTIONS

Multi-Choice questions:

1. (*b*) 2. (*d*) 3. (*c*) 4. (*a*) 5. (*b*)

ELECTRICAL INSTALLATIONS

LEARNING OBJECTIVES

- What is the meaning of switchgear?
- What are basic functions of switchgear?
- What is the meaning of *LV, MV* and *HV* switchgear?
- Which are the major devices or equipment employed in *LV* switchgear?
- How *LV* switchgear devices such as Switch Fuse Unit (*SFU*), *MCB, ELCB, MCCB* protect the system?
- What is a fuse and *an MCB?* What are their roles in electrical installations ?
- What is *an ELCB* or *RCCB* ?
- What is *MCCB* and where it is applied?
- What are different types of cables used in wiring installations?
- Which are different types of wiring systems used in domestic, commercial and industrial installations?
- What is the power rating of various domestic appliances?
- How to select a cable or insulated wire for domestic, commercial or industrial wiring installation?
- How domestic appliances are being earthed?
- What is earthing and its purpose ?
- What are different methods of earthing ?
- What is double earthing and its purpose ?
- What is a battery ?
- What are the constructional features of a lead-acid battery ?
- How to check the charged condition of a battery ?
- What are the important characteristics of lead-acid battery and a Nickle-iron battery?
- How to calculate the electrical energy consumption charges?

INTRODUCTION

The rapid industrial growth is mainly due to utilisation of electrical energy in various fields. To-day, the dependence on electrical energy is so much that it has become a part and pencil of our daily life.

Electricity is a good servant but a bad master if not handled properly. Therefore, while working with electrical installation and handling electrical equipment, one should always take care of his own as well as of other's safety. A little carelessness may result in an accident which may be fatal. Therefore certain safety measures must be observed before dealing with electricity.

To provide safety to the operators and equipment various safety devices have been developed In this chapter, we shall discuss various common safety devices, earthing etc.

6.1 SWITCHGEAR

In an electric power system, the combination of electrical switches, fuses or circuit breakers used to control, protect and isolate electrical equipment is called switchgear.

Switchgear is employed to improve the reliability of the electric supply.

Broadly, a switchgear has two types of components:

- Power conducting components, such as switches, circuit breakers, fuses, and lightning arrestors, that conduct or interrupt the flow of electrical power
- Controlling components, such as control panels, current transformers, potential transformers, protective relays, and associated circuitry, that monitor, control, and protect the power conducting components

Thus, the basic functions of switchgear is protection. It interrupts the short-circuit and overload fault currents while maintain the service to unaffected circuits. Switchgear also provides isolation of circuits from power supplies In power system, the electrical switchgears are categorized as HV, MV & LV switchgears.

HV Switchgears: The switchgear system that deals with voltage above 33 kV, is referred as high voltage switch gears. High voltage circuit breaker, is the main component of HV switchgear,

MV Switchgears: The switchgear system which is to handle voltages between 3 kV to 33 kV is categorized as medium voltage switchgear or MV switchgear. These switchgears are of many types.

LV Switchgears: Generally electrical switchgear rated upto 1 kV is termed as low voltage switchgear. The term LV Switchgear includes low voltage circuit breakers, switches, off load electrical isolators, HRC fuses, earth leakage circuit breaker, miniature circuit breakers (MCB) and molded case circuit breakers (MCCB) etc i.e. all the accessories required to protect the LV system.

6.2 LOW VOLTAGE SWITCHGEAR OR LV SWITCHGEAR

In LV switchgear system, electrical appliances are protected against short circuit and over load conditions by electrical fuses or electrical circuit breaker. However, the human operator is not adequately protected against the faults occurs inside the appliances. The problem can be overcome by using earth leakage circuit breaker. This operates on low leakage current. The earth leakage circuit breaker can detect leakage current as low as 100 mA and is capable of disconnecting the appliance in less than 100 m sec.

Let us deal with the most popular switchgear devices such as Switch fuse Unit (SFU), MCB, ELCB, MCCB.

6.3 FUSE

A short piece of metal wire, inserted in series with the circuit, which melts when predetermined value of current flows through it and breaks the circuit is called a ***fuse.**

A *fuse* is connected in series (see Fig.6.1) with the circuit to be protected and carries the load current without overheating itself under

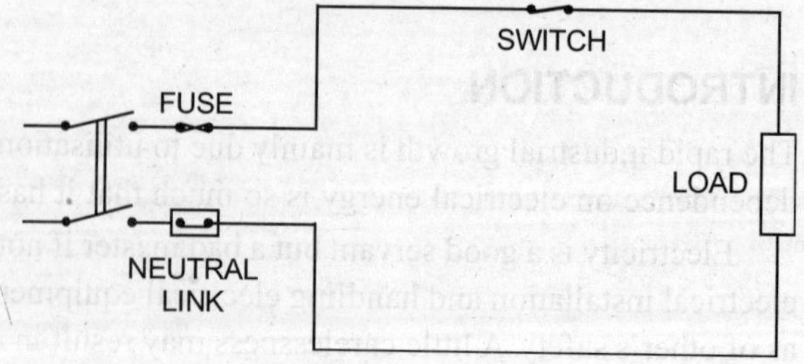

Fig.6.1: Circuit with fuse

normal conditions. However, when abnormal condition occurs, an excessive current (more or equal to the predetermined value for which the fuse is designed) flows through it. This raises the temperature of the fuse wire to the extent that it melts and opens the circuit. This protects the machines or apparatus from damage which can be caused by the excessive current.

Time-Current Characteristics: The time required to blow out a fuse depends upon the magnitude of excessive current. Larger the current, smaller is the time taken by the fuse to blow out. Hence a fuse has inverse time current characteristic as shown in Fig.6.2 which is desirable for a protective device.

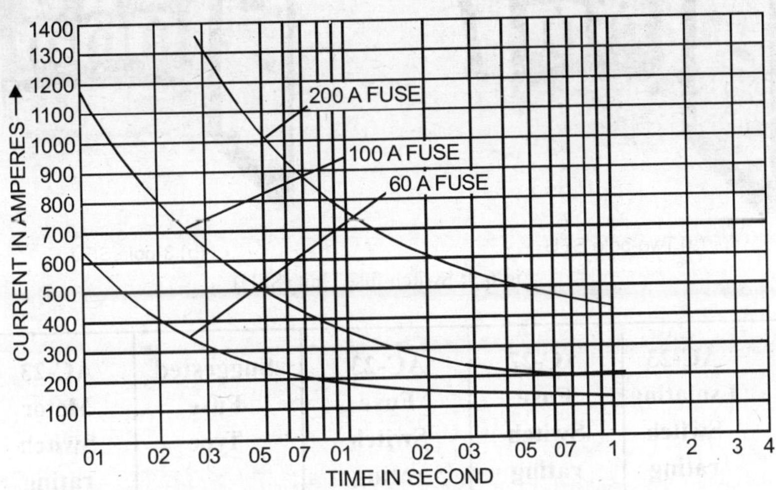

Fig.6.2: Time-current characteristics of various fuses

Advantages

(*i*) The cost of this protective device is very low.

(*ii*) It requires no maintenance.

(*iii*) It interrupts heavy current without noise or smoke.

(*iv*) The smaller size of fuse element imposes a current limiting effect under short circuit.

(*v*) The minimum time of operation can be predetermined by selecting proper material of the fuse wire.

(*vi*) The inverse time-current characteristic makes it suitable for overcurrent protection.

Disadvantages

(*i*) Considerable time is lost in re-wiring or replacing fuses after every operation.

(*ii*) On short-circuit, determination between fuses in series can only be obtained if there is considerable difference in the relative sizes of the fuse concerned.

6.4 SWITCH FUSE UNIT (SFU)

Switch Fuse: Switch fuse Unit comprises of various porcelain rewireable fuses or *HRC* fuse fittings complete with their conducting parts. The switch is fitted with sturdy side operating handle with quick-break type mechanism, as shown in fig.6.3.

Contacts are made of electrolytic copper silver-plated. The fixed contacts are provided with removable shield. Switch Fuse units are provided with rewireable fuse or *HRC* Fuse Links. All these parts are assembled in an enclosure.

The Enclosure is made of sheet steel duly phosphatised and powder-coated. They are provided with conduit knock-outs. Door inter-clock is provided to prevent opening when the switch is in '*ON*' condition.

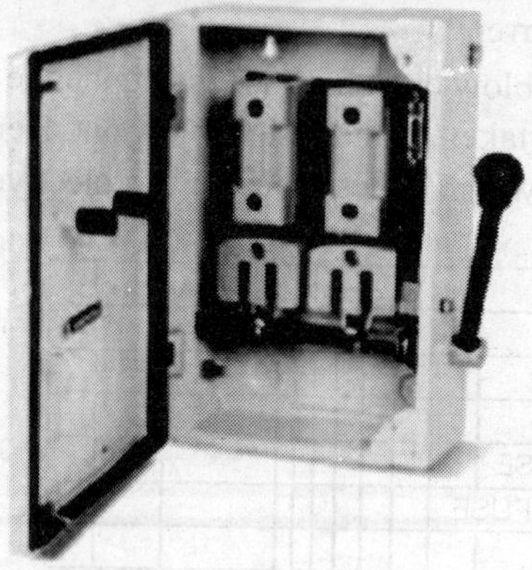

(a) Two-pole SFU

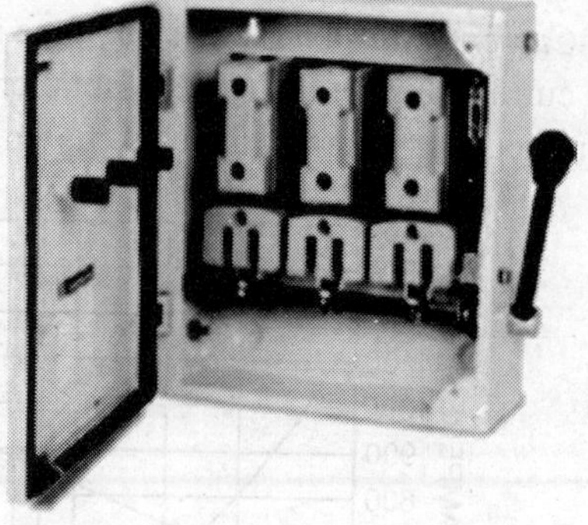

(b) 3-pole SFU

Fig.6.3: Switch fuse Unit (SFU)

Rating Thermal	AC-23 Lsolating Switch rating	AC-22 Fuse Switch rating	AC-23 Fuse Switch rating	Suggested Fuse Type	AC-23 Motor Switch rating	Mechanical Endurance (Operating Cycles)
63A	80A	80A	63A	H TS	28 KW	10,000
100A	125A	125A	100A	H TSD	45 KW	10,000
160A	200A	200A	160A	H TSF	80KW	8,000
200A	250A	250A	200A	H TSF	90 KW	8,000
320A	400A	320A	320A	H TSK	150 KW	5,000
400A	500A	460A	400A	H TSMF	185 KW	5,000
500A	630A	630A	500A	H TTM	225 KW	5,000
800A	800A	750A	630A	H TLM	300 KW	3,000

6.5 MINIATURE CIRCUIT BREAKER (MCB)

Miniature Circuit Breaker (*MCB*) is a device which ensures definite protection of wiring system and sophisticated equipment against over current and short circuit faults. The outer view and the internal details of a miniature circuit breaker are shown in Fig.6.4 and 6.5 respectively.

Construction

Construction of an *MCB* can be explained by considering the following main parts:

(*i*) *Outer body or housing:* The outer body or housing of an *MCB* is moulded from a special grade glass fibre reinforced polyster with the help of an injection moulding machine. The outer body and other polyster components of *MCB* are fire retardant, anti-tracking and non-hygroscopic. These polyster parts and housing have the ability to withstand high temperature and mechanical impacts.

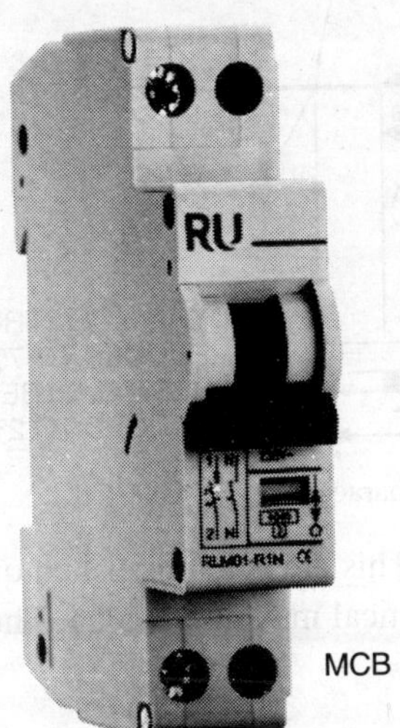

Fig.6.4: Outer view of an MCB

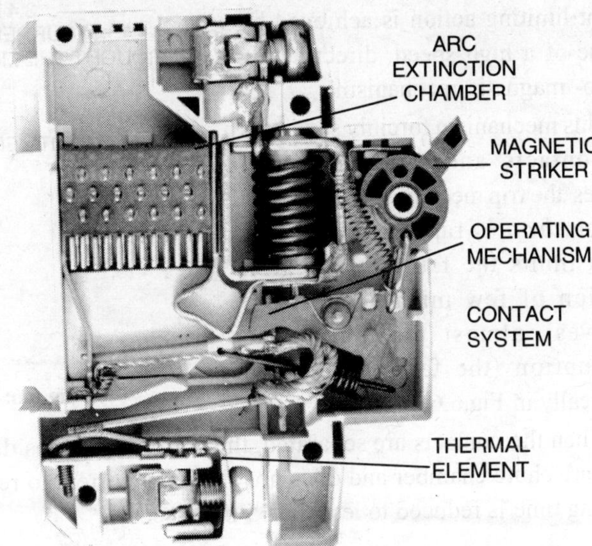

Fig.6.5: Internal structure of MCB

(*ii*) *Contacts:* The contacts of an *MCB* are made of pure silver. This provides definite advantages such as long contact life, low contact resistance, ensures quick arc removal and low heat generation.

(*iii*) *Operating mechanism:* All the components of the operating mechanism are made of special plastic that they are self lubricating which eliminates wear and tear, rust and corrosion. These components are very light in weight and have low inertia thereby ensure snap make the break ability. The reliability and ruggedness of the operating mechanism is thus maintained.

(*iv*) *Arc extinguishing Chamber*: The arc produced during breaking of circuit is extinguished abruptly by providing a special arc chute chamber.

(*v*) *Fixing arrangement:* The *MCB* mounting clip gets easily snapped on to the Din-bar and can be removed easily by a simple operation with a screw driver. This saves the time which would have been required for fixing it with screws.

(*vi*) *Mechanical interlocking of Multiple MCBs*: The levers of all the (3 or 4) multiple *MCB*s are connected internally. This ensures simultaneous tripping of all poles even if the fault develops in any one of the phases.

Working

MCB may operate under two different conditions:

(*i*) *Moderate Overload Condition:* Detection of moderate overload conditions is achieved by the use of a thermo-metal which deflects in response to the current passing through it. The thermo-metal moves against the trip lever releasing the trip mechanism.

(*ii*) *Short circuit conditions*: When the current flowing through the *MCB* reaches a pre-determined level (as per its setting or rating), it pushes the solenoid plunger which releases the trip mechanism and simultaneously separates the contacts.

Under short circuit conditions, the current limiting action is achieved by the use of a high speed, direct acting electro- magnetic mechanism.

This mechanism forcibly separates the contacts and simultaneously releases the trip mechanism. A high *arc voltage drop* is rapidly introduced which limits the fault current to a duration of few milliseconds and achieves almost instantaneous interruption (the facts are shown graphically in Fig.6.6).

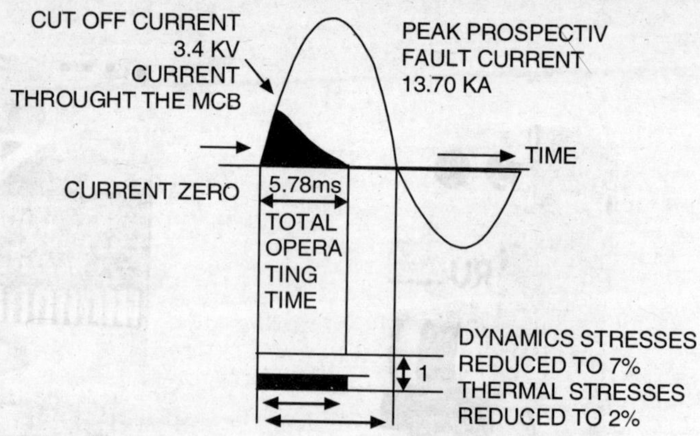

Fig.6.6: Operating characteristics of an *MCB*

When the contacts are separated, the current still rises due to arc. This arc is extinguished quickly in the arc chute chamber and does not allow the current to reach theoretical maximum value. The total breaking time is reduced to less than 5 millisecond.

Applications

Since MCBs are available with different current ratings of 0.5, 1.6, 2, 2.5, 3, 4, 5, 6, 7.5, 10, 16, 20, 25, 32, 35, 40 and 63 ampere and voltage ratings of 240/415 volt ac and upto 220V dc. Moreover, they have very small breaking time (5 millisecond), therefore, these are generally employed to protect the important and sophisticated appliances used commercially and for domestic purposes such as computers, air conditioners, compressors, refrigerators and many others.

6.6 EARTH LEAKAGE CIRCUIT BREAKER (ELCB)

In the industrial, commercial and domestic building some times (usually in rainy season) leakage to earth occurs. This leakage may cause electric shock or fire. Hence, the leakage to earth is very dangerous and needs protection.

ELCB is a device which provides protection against earth leakage faults.

Construction and internal circuit details

The enclosures of the *ELCB* is moulded from high quality insulating material. The materials are fire retardant, anti- tracking, non-hygroscopic, impact resistant and can withstand high temperatures. The body contains spring loaded mounting arrangement on din-channel which ensures snap fitting of *ELCB* into position. However, these also have the facility to screw-on directly to any surface with the help of two screws. A 2-pole *ELCB* is used for 1-phase supply and a 4-pole *ELCB* is used for 3-phase, 4 wire supply. A 4-pole *ELCB* is shown in Fig.6.7.

Internal wiring diagram of a 2-pole *ELCB* is shown in Fig.6.8. As shown in Fig.6.8 an *ELCB* contains a core balanced transformer (ferrite ring on which one or two turns of phase and neutral wire; and a few turns of operating coil of relay are wound), and a relay. A test button is placed between phase and neutral in series with a limiting resistor. The terminal designation and connection diagram for a 2-pole and 4-pole *ELCB* are shown in Fig. 6.9 and Fig.6.10 respectively.

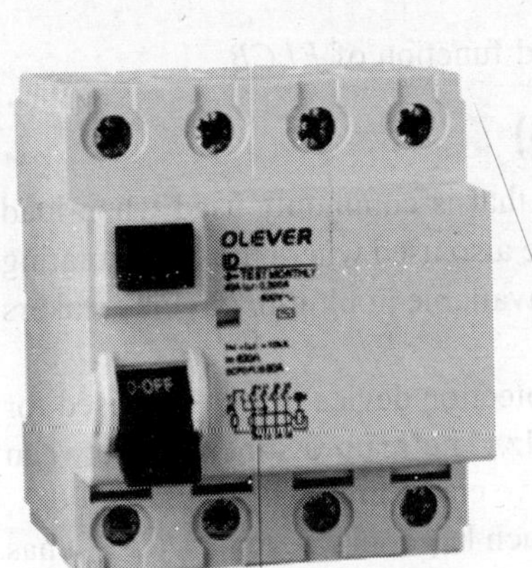

Fig.6.7: 4-pole ELCB

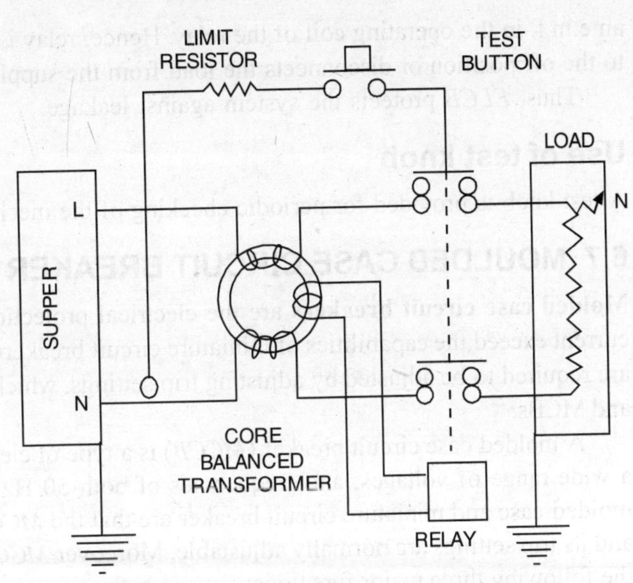

Fig.6.8: Internal wiring diagram of a 2-pole ELCB

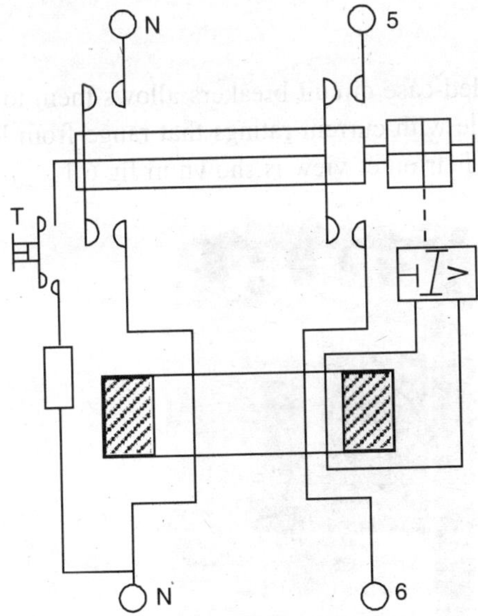

Fig.6.9: Circuit for 2-pole ELCB

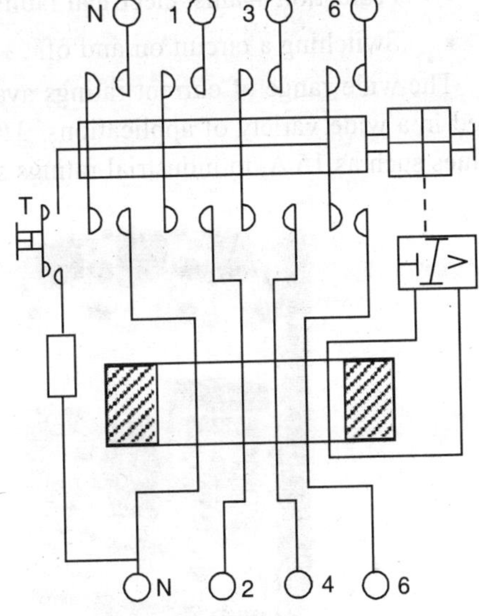

Fig.6.10: Circuit for 4-pole ELCB

Principle of Operation

Under normal conditions, the magnitude of currents flowing through the phase wire and neutral are the same and core of the core balanced transformer does not carry any flux (*i.e.* two windings neutralise the flux). Thus, no e.m.f. is induced in the operating coil of the relay wound on the same core. However, when the earth fault (earth leakage) occurs, the current in the phase wire becomes more than the neutral wire. This unbalancing sets up flux in the core of the core balanced transformer which in turn induces an e.m.f. in the operating coil of the relay. Hence, relay is energised and the plunger of the *ELCB* goes to the off position or disconnects the load from the supply.

Thus, *ELCB* protects the system against leakage.

Use of test knob

A test knob is provided for periodic checking of the mechanism and function of *ELCB*.

6.7 MOULDED CASE CIRCUIT BREAKER (MCCB)

Molded case circuit breakers are the electrical protection device that is commonly used when load current exceed the capabilities of miniature circuit breakers. They are also used where the current rating are required to be adjusted by adjusting trip settings, which are not available in plug-in circuit breakers and MCBs.

A molded case circuit breaker (*MCCB*) is a type of electrical protection device that can be used for a wide range of voltages, and frequencies of both 50 Hz and 60 Hz. The main distinctions between molded-case and miniature circuit breaker are that the *MCCB* can have current ratings of up to 2500 A, and its trip settings are normally adjustable. Moreover *MCCBs* are much larger in size than *MCBs*. It has the following three major functions:

- Protection against overload
- Protection against electrical faults
- Switching a circuit on and off

The wide range of current ratings available from molded-case circuit breakers allows them to be used in a wide variety of applications. *MCCBs* are available with current ratings that range from low values such as 15 A, to industrial ratings such as 2500 A. Their outer view is shown in fig 6.11.

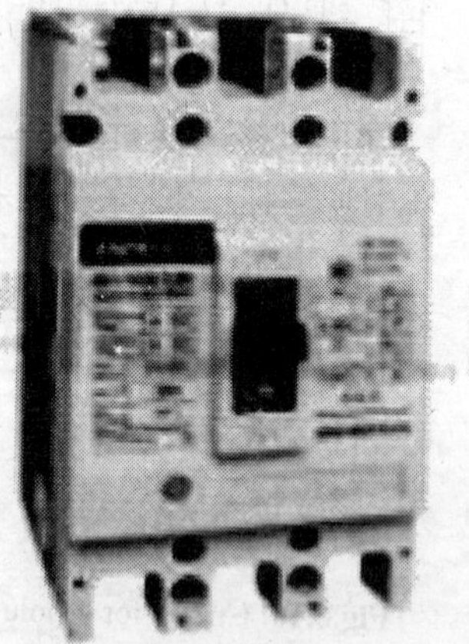

100 AMP MCCB Moulded Case Circuit Breakers

100 AMP MCCB 4 pole Electrical Circuit Breakers

Fig.6.11: Moulded case circuit breakers

Operating Mechanism

At its core, the protection mechanism employed by *MCCBs* is based on the same physical principles used by all types of thermal-magnetic circuit breakers.

Overload protection: It is accomplished by means of a thermal mechanism. *MCCBs* have a bimetallic contact what expands and contracts in response to changes in temperature. Under normal operating conditions, the contact allows electric current through the *MCCB*. However, as soon as the current exceeds the adjusted trip value, the contact will start to heat and expand until the circuit is interrupted.

Fault protection: On the other hand, fault protection is accomplished with electromagnetic induction, and the response is instant. Fault currents should be interrupted immediately, no matter if their duration is short or long. Whenever a fault occurs, the extremely high current induces a magnetic field in a solenoid coil located inside the breaker, this magnetic induction trips a contact and current is interrupted.

Manual operation: As with all types of circuit breakers, the *MCCB* includes a disconnection switch which is used to trip the breaker manually.

6.8 TYPES OF WIRES AND CABLES

Before considering the various types of wiring systems suitable for any installation, it is required to know about the various types of cables (insulated conductors) which are used for internal wiring systems.

A solid or stranded conductor covered with insulation is known as a **cable.**

The cable may be a single core or multi-core depending upon the number of conductors. Various types of insulating materials are employed for covering the conductors.

Accordingly, the cables (wiring conductors) may be classified as:

(*i*) Vulcanised Indian Rubber (*V.I.R.*) cables.

(*ii*) Poly-vinyl Chloride (*P.V.C.*) cables.

(*iii*) Tough Rubber sheathed (*T.R.S*) or Cab Tire Sheathed (*C.T.S.*) cables.

(*iv*) Lead sheathed cables.

(*v*) Weather proof cables.

6.8.1 Types of Wiring

The main types of wirings usually employed in residential buildings, commercial buildings and industries are:

1. Cleat Wiring; 2. Casing and Capping Wiring; 3. *C.T.S.* or *T.R.S.* Wiring;

4. Metal Sheathed Wiring; 5. Conduit Wiring.

1. **Cleat Wiring:** In this system of wiring, usually *V.I.R.* or *P.V.C.* conductors are employed. The conductors are supported in porcelain cleats which are placed at least 6 mm above the walls. The porcelain cleats are made of two parts, the lower one is known as base being having two or three grooves for the accommodation of conductors and the upper one is known as cap as shown in Fig.6.12. The conductors are run in the grooves, cap is placed over

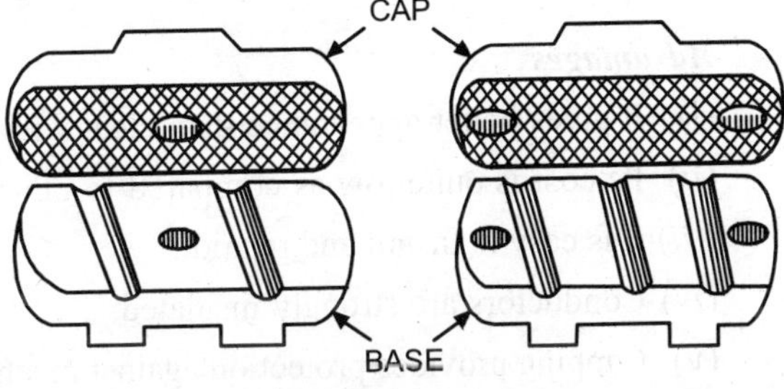

Fig.6.12: Porcelain cleats

the base and the whole assembly in fixed on to the wall with the help of wooden screws and gutties (wooden or *P.V.C.* plugs) already cemented in the wall. The screw not only fixes the cleats on the wall but also tightens the grip of the wires between the two halves of the cleat.

Advantages

(*i*) It is cheapest system to wiring.

(*ii*) A little skill is required to lay the wiring.

(*iii*) This wiring can be installed very quickly.

(*iv*) It is the most suitable system for temporary wiring.

(*v*) The wiring can be dismantled very quickly and whole of the material is recovered.

(*vi*) Inspection, alteration and additions can be made easily.

Disadvantages

(*i*) It gives a rubbish look.

(*ii*) It is rarely employed for permanent job.

(*iii*) At the time of white washing or distempering the lime falls over the wires which may damage insulation.

(*iv*) Oil and smoke are also injurious to *V.I.R.*

(*v*) Mechanical injuries may damage the conductor since there is no protecting cover.

2. **Casing and Capping Wiring:** In this system of wiring, generally *V.I.R.* wires are employed. The casing is just a base that consists of rectangular *P.V.C.* or wooden block of seasoned teak wood and has usually two grooves to accommodate wires. The casing is fixed on the wall with the help of wooden screws and gutties already cemented in the wall. The casing is usually placed 3 mm apart from the wall by means of porcelain discs in order to protect the casing from dampness. Then the wires of opposite polarity are laid in different grooves. After placing the wires in the grooves of casing at the top is covered by means of a rectangular strip of seasoned wood of the same width as the casing known as capping with the help of wooden screws. The assembled view of casing capping with the *V.I.R.* wires placed in the grooves is shown in Fig.6.13.

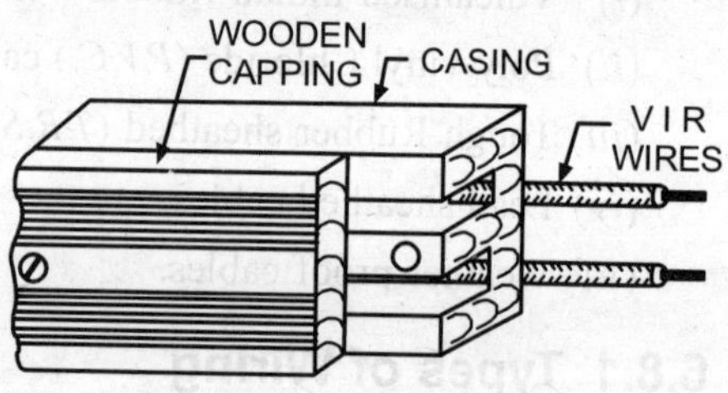

Fig.6.13: Casing – capping

Advantages

(*i*) It gives better appearance than cleat wiring.

(*ii*) Its cost is quite low as compared to other systems of wiring except cleat wiring

(*iii*) It is easy to install and repaid.

(*iv*) Conductors are strongly insulated.

(*v*) Capping provides protection against mechanical injury.

Disadvantages

(*i*) It is not suitable in damp situations.

(*ii*) There is a risk of fire.

(*iii*) To make the job good looking highly skilled labour is required.

3. **C.T.S. or T.R.S. Wiring:** In this system of wiring, generally *C.T.S.* (Cab Tire Sheathed) or *T.R.S.* (Tough Rubber Sheathed) conductors are employed. The conductors are run on well seasoned, perfectly straight and well varnished teak wood batten of thickness 13 mm. The width of the batten is chosen depending upon the number of wires to be run on it. While doing this type of wiring, the batten is fixed on to the wall by means of wooden screws and gutties already cemented in the wall. The wires are held on the batten with the help of clips already fixed on the batten with the help of nails as shown in Fig.6.14.

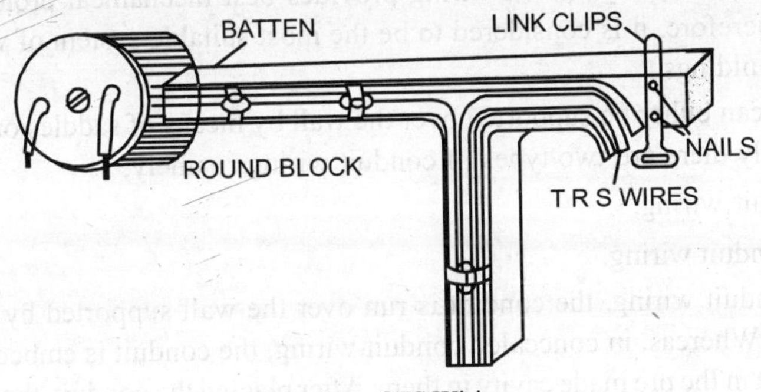

Fig.6.14: CTS Wiring

Advantages

(*i*) It is easy to install and repair.

(*ii*) It gives nice appearance.

(*iii*) The conductors have strong insulation, therefore, it has longer life.

(*iv*) It is fire-proof up to some extent.

(*v*) Chemicals do not affect the conductors insulation.

Disadvantages

(*i*) The conductors are open and liable to mechanical injury, therefore, this type of wiring cannot be used in workshops.

(*ii*) Its use in places open to sun and rain in restricted.

4. **Metal Sheathed Wiring:** This system of wiring is similar to *C.T.S.* or *T.R.S.* wiring. Only difference is that in this case *V.I.R.* conductors covered with lead alloy sheath (metal sheathed cable) are used. The metal sheathed cables are run on the wooden batten. The batten is fixed on the wall by means of screws and gutties already cemented into the wall. The cables are held on the batten with the help of link clips.

Advantages

(*i*) The conductors are protected against mechanical injury.

(*ii*) It can suitably employed under damp situations.

(*iii*) It gives batter appearance.

(*iv*) It has longer life.

(*v*) Conductors are protected against chemicals.

(*vi*) It can be installed in open space.

Disadvantages

(*i*) The metal sheathed cables are costlier than *C.T.S.* or *T.R.S.* wires.

(*ii*) In case of leakage, there is every risk of shock.

(*iii*) Skilled labour and proper supervision is required.

5. **Conduit Wiring:** In this system of wiring, *V.I.R.* or *T.R.S.* conductors are run in metallic or *P.V.C.* tubes called conduits. This system of wiring provides best mechanical protection, safety against fire and shock. Therefore, it is considered to be the most suitable system of wiring for workshops and commercial buildings.

The conduits can either be supported over the wall by means of saddles or can be buried under plaster. Accordingly there are two types of conduit wirings namely;

(*i*) Surface conduit wiring,

(*ii*) Concealed conduit wiring.

In surface conduit wiring, the conduit is run over the wall supported by means of saddles as shown in Fig.6.15 Whereas, in concealed conduit wiring, the conduit is embedded in the walls and ceiling by placing it in the pre made cavity in them. After placing the conduit, the insulated conductors (or cables) are drawn into them by means of *G.I.* wire known as pilot wire. Number of inspection boxes (conduit boxes) are provided along the run of conduit to facilitate the drawing of wires.

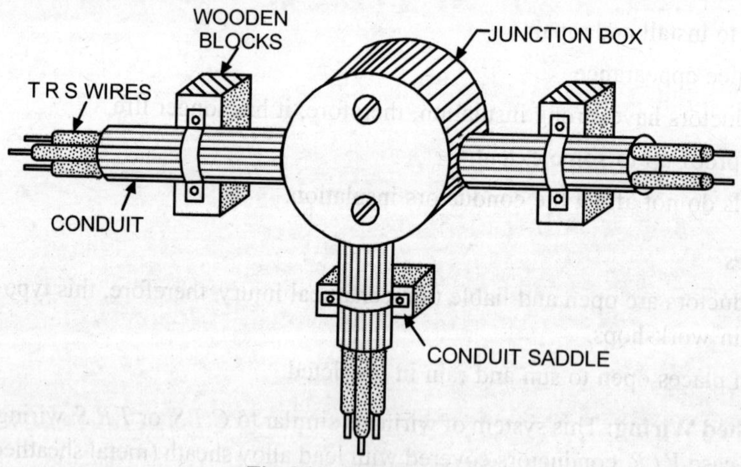

Fig.6.15: Conduit wiring

Advantages

(*i*) Conduit provides protection against mechanical injury and fire.

(*ii*) Conduit provides protection against chemicals.

(*iii*) Conductors are safely secured from moisture.

(*iv*) This wiring has far better look.

(*v*) It has a longer life.

Disadvantages

(*i*) It is costly system of wiring

(*ii*) It requires more time for erection.

(*iii*) It requires highly skilled labour.

6.8.2 Power rating of Basic house-hold equipment:

To calculate the rating of the main switch and MCB$_S$ / fuses to be used on the distribution board for a building and also for finalizing the size of the cable for service connection and internal wiring, it is necessary to assess the load of the building in ampere. For this purpose, following guidelines may be adopted after deciding the number and type of outlets installed in the building.

Incandescent light point (Bulb Point)	60 watt
Fluorescent tube 2 feet long (0.6 m)	20 watt
4 feet long (1.2 m)	40 watt
Ceiling fans	60 watt
Socket outlets (5A)	100 watt
15 ampere socket outlet (Power Socket)	1000 watt
Exhaust fan	100 watt
CFL/LED lamp	40 watt

Selection of cable or insulated wires used in wiring installations.

To find out the correct size of wire to be used at an installation, it is necessary to find out the load in ampere to be carried by the cables.

The following tables help us for selection of correct size of insulated wire based upon current and voltage drop of the conductors. Table shows only the copper wires used in the internal wiring system upto a current carrying capacity of 28 ampere, although, larger size wires are also available:

Table. Current Rating of Copper Conductor Single Core Cables

(VIR, PVC or Polythene insulated including tough rubber sheathed, PVC or lead sheathed cables)

Size of conductors		Two cables d.c. or single phase a.c.		Three or four cable balanced three phase a.c	
Nominal area mm^2	No. and dia of wire (mm)	Current Rating (A)	Approximate length of run for one volt drop (metre)	Current Rating (A)	Approximate length of run for one volt drop (metre)
1.0	1/1.12	5	2.9	3	2.8
1.5	3/0.737	10	3	10	3.7
2.5	3/1.06	15	3.4	13	4.3
4.0	7/0.737	20	3.7	15	4.8
6.0	7/1.06	28	4.0	25	5.2

Aluminium core cable or insulated wires can also be used as per the current rating and size given in the following table. But aluminium core cable are avoided in the internal wiring of the buildings since they have poor mechanical strength. However, these are commonly used in service connections.

Table. Current Rating of Aluminium Conductor Single Core Cables

(VIR, PVC or Polythene insulated including tough rubber sheathed, PVC or lead sheathed cables)

Size of conductor		Two cables d.c. or single phase a.c.		Three or four cables balanced three phase		Four cables d.c. or single phase a.c.	
Nominal area mm^2	No. and dia. of wire (mm)	Current Rating (A)	Approx run for one volt drop (m)	Current Rating (A)	Approx run for one volt drop (m)	Current Rating (A)	Approx run for volt drop (m)
1.5	1/1.40	10	2.3	9	2.9	9	2.5
2.5	1/1.80	15	2.5	12	3.6	11	3.4
4	1/2.24	20	2.9	17	3.9	15	4.1
6	1/2.80	27	3.4	24	4.3	21	4.3
10	1/3.55	34	4.3	31	5.4	27	5.4
16	7/1.70	43	5.4	38	7.0	35	6.8
25	7/2.24	59	6.8	54	8.5	48	8.5
35	7/2.50	69	7.2	62	9.8	55	9.0
50	7/3.00	91	7.9	82	10.1	69	10.0
50	19/1.80	91	7.9	82	10.1	69	10.0
70	19/2.24	134	8.0	131	9.5	–	–
95	19/2.50	153	8.8	152	10.0	–	

6.9 CLASSIFICATION OF CABLES

Underground cables are usually classified according to the voltage for which these are manufactured. Hence these are classified as:

(*i*) Low tension (L T) cables—for operating voltages upto 10.00V

(*ii*) High tension (H T) cables—for operating voltages upto 11 kV

(*iii*) Super tension (S T) cables—for operating voltages from 22 kV to 33 kV

(*iv*) Extra high tension (EHT) cables—for operating voltages from 33 kV to 66 kV

(*v*) Extra super voltage cables—for operating voltages from 66 kV to 132 kV and above.

Low Tension (LT) Cables

These cables are employed for system voltages upto 1000 volt. Generally these cables are used for distribution of electric power at 400 V. The cables are of two types *viz* single-core and multi-core cables. Due to economic reasons 4-core cable is mostly employed for distribution of electric power at low voltages.

The low voltage cables do not have special attention for their construction since in these cables the electrostatic stresses are small. Moreover, their thermal conductivity is quite sufficient to dissipate heat. Fig.6.16 represents the construction of a four-core cable. The conductors are insulated with rubber or polyvinyl chloride. Around all the four conductors a layer of fibrous compounded material is provided.

Above this layer a lead sheath is provided to restrict the moisture to enter the core of the cable. The lead sheath is covered with a compounded fibrous material. In this case, wire armoured cover may or may not be provided, however if it is provided, it is covered with a layer of serving.

6.10 EARTHING

The process of connecting metallic bodies of all the electrical apparatus and equipment to the huge mass of earth by a wire of negligible resistance is called **earthing.**

When a body is earthed, it is basically connected to the huge mass of earth by a wire having negligible resistance. Thus, the body attains zero potential *i.e.*, potential of earth. This ensures that whenever a live conductor comes in contact with the outer body, the charge is released to the earth immediately.

6.10.1 Purpose of Earthing

The basic purpose of earthing is to protect the human body (operator) from electric shock.

To illustrate the purpose of earthing consider an electrical circuit shown in Fig.6.17 where an electrical appliance of resistance *R* is connected to the supply through a fuse and a switch. When an operator touches the metallic body of the apparatus [see Fig.6.17(*a*)] having perfect insulation, the equivalent circuit is shown in Fig.6.17 (*b*). Where two parallel paths are formed. Since the insulation resistance R_i is very high as compared to appliance resistance *R*, whole of the current flows through appliance resistance and no current flows through human body (operator's body) resistance.

When earth fault occurs, the live (phase) wire directly comes in contact with the outer body and the insulation resistance reduces to zero as shown in Fig.6.17 (*c*). Now the body resistance is just in parallel with the appliance resistance. A heavy current flows through the human body and operator gets a severe shock.

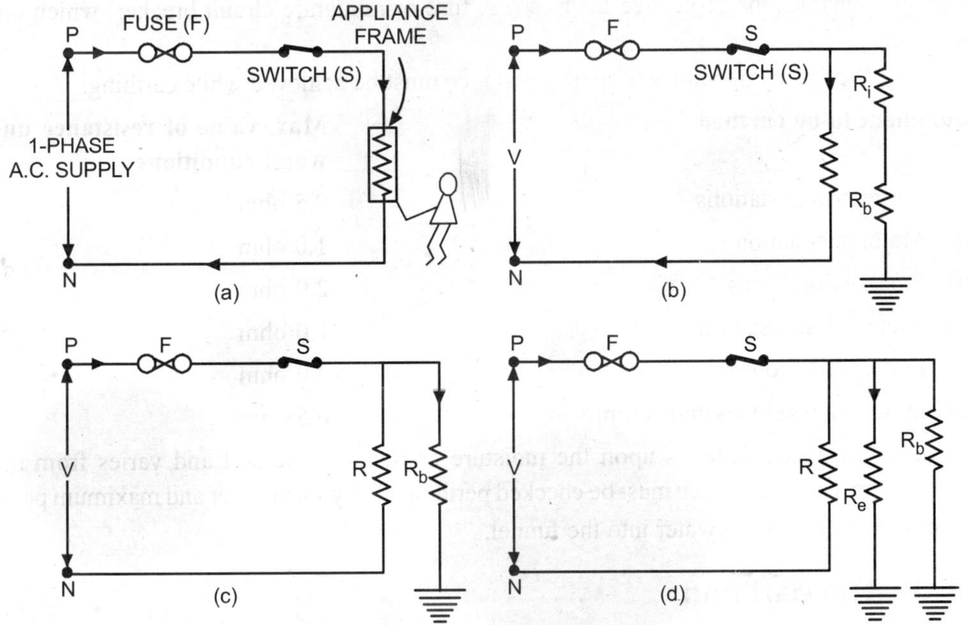

Fig.6.17: Earthing protects the operator

However, if the metallic body or outer frame of the appliance is properly earthed then under earth fault condition, the circuit will be as shown in Fig.6.17 (*d*). Where earth resistance R_e is just in parallel with the appliance resistance R and body resistance R_b. Since earth resistance is very small as compared to body resistance, almost whole of the fault current flows through the earth resistance and no current flows through the human body. *Thus, the operator is protected from electric shock.* Moreover, the fault current is much more than the full load current of the circuit which melts the fuse. Hence, the appliance is disconnected automatically from the supply mains.

6.10.2 Equipment Earthing

According to Rule 61 of Indian Electricity Rules 1956, it is obligatory to earth the following points and apparatus used in the power system, where the voltage is more than 125 V: —

(*i*) All the metal frames of motors, generators, transformers and controlling equipment.

(*ii*) The steel tower and steel tubular or rail poles carrying over head conductors.

(*iii*) The metal frames of portable electrical equipment such as heaters, table fans, electric iron, refrigerator, air conditioners, vacuum cleaners etc.

(*iv*) Other metal parts such as conduits, switch gear casings etc.

(v) Earth terminal of all the 3-pin outlet sockets.

(*vi*) In case of concentric cables, external conductor *i.e.* armouring of such cables.

(*vii*) Stay wires of overhead lines if stay insulator is not provided.

In case of insulation failure, the primary object of connecting all the above points and apparatus to earth is to release the charge accumulated on them immediately to earth so that the person coming in contact may not experience electric stock. The other object is that a heavy current when flows through the circuit that operates the protective devices (*i.e.* fuse or miniature circuit breaker) which opens the circuit.

Generally, the following values of earth resistance must be achieved while earthing:

Equipment to be earthed	Max. value of resistance under worst conditions
(*i*) Large power stations	0.5 ohm
(*ii*) Major sub-station	1.0 ohm
(*iii*) Small sub-stations	2.0 ohm
(*iv*) Factories sub-stations	1.0 ohm
(*v*) Lattice steel towers	3.0 ohm
(*vi*) Industrial machines and equipment	0.5 ohm

The earth resistance depends upon the moisture contents in the soil and varies from month to month. Therefore, earth resistance must be checked periodically by earth tester and maximum permissible value be obtained by pouring water into the funnel.

6.10.3 System Earthing

A proper system has to be adopted while earthing. In fact, all the heavy power equipment should be earthed by two separate distinct earth wires following the different routes. The two earth connections are applied to improve the reliability. If one of the earth wires breaks or fails to carry the fault current, the other carries that current and provides the required protection. Moreover, in factories and substations where more than one equipment is to be earthed, parallel connections should invariably be used. In no case series connections are done, as even a single bad contact or break in the earthing-lead will disconnect all the succeeding equipment from the earth.

Thus, *for proper earthing of heavy power equipment, double earthing system has to be adopted. Moreover, the number of apparatus must be connected in parallel to the earth.*

6.10.4 Methods of Earthing

As already discussed, *earthing* means to connect metallic bodies of the apparatus with the general mass of earth by a wire of negligible resistance. There are various methods of achieving this connection, some of them are given below:

1. **Strip Earthing:** This system of earthing employs the use of 5 SWG copper wire or strip of cross-section not less than 25 mm × 1.6 mm. The strips or wires are buried in horizontal trenches. This type of earthing is used where the earth bed has a rocky soil and excavation work is difficult.
2. **Earthing through water mains:** In this type of earthing, a stranded copper lead is used which is rounded on the pipe with the help of a steel binding wire and a properly designed earthing clip. The copper lead is soldered to make it solid. Before making connection to the water main, it should be ensured that G.I. pipe is used throughout.
3. **Rod Earthing:** it is the cheapest method of earthing and is employed in sandy areas. In this method, a copper rod is hammered directly into the ground and no excavation work is required. The earthing lead is joined to this rod with the help of nuts and bolts.
4. **Pipe Earthing:** Taking into consideration the factors such as initial cost, inspection, resistance measurement etc., G.I. pipe earthing is best form of ground connection. Iron is the cheapest material and remains serviceable even if put in salty mass of earth. The pipe used as earth electrode is galvanised and perforated. Its diameter is 38.1 mm and length is 2 metre. The length may be increased to 2.75 metre in case of dry soil. The diameter of pipe has very little effect on the resistance of the earth connection. To facilitate the driving in of the pipe into ground, it is provided with the tapered casting at the lower end. Another pipe of 19.05 mm diameter and of length 2.45 metre is connected at the top of the above said pipe. The connection between these pipes is made through a reducing socket, as shown in Fig.6.18.

   The earthing lead should be soldered and connected to the pipe. Alternate layers of charcoal and salt are provided around the G.I. pipe to keep the surroundings moist enough. The salt is poured at the bottom and thereafter alternate layers of charcoal and salt are arranged.
5. **Plate Earthing:** In this type of earthing, a copper or G.I. plate of dimensions not less than 60 cm × 60 cm × 3.18 mm or 60 cm × 60 cm × 6.35 mm is used as earth electrode instead of G.I. pipe. The plate is buried into ground in such a way that its face is vertical and the top is not less than 3 metres below the ground level. The G.I. wire is used for G.I. plate and copper wire for copper plate earthing. The size of wire is selected according to the installation and fault current. The earthing lead is suitably protected placing it under-ground in a pipe, as shown in Fig.6.19.

   Alternate layers of charcoal and salt are used around the plate. The layers of charcoal shall be placed immediately over the plate and thereafter successive layers of salt and charcoal are laid to keep the surroundings sufficiently moist.

Note: Pipe earthing and plate earthing are considered to be the best as they have resonably low value of earth resistance.

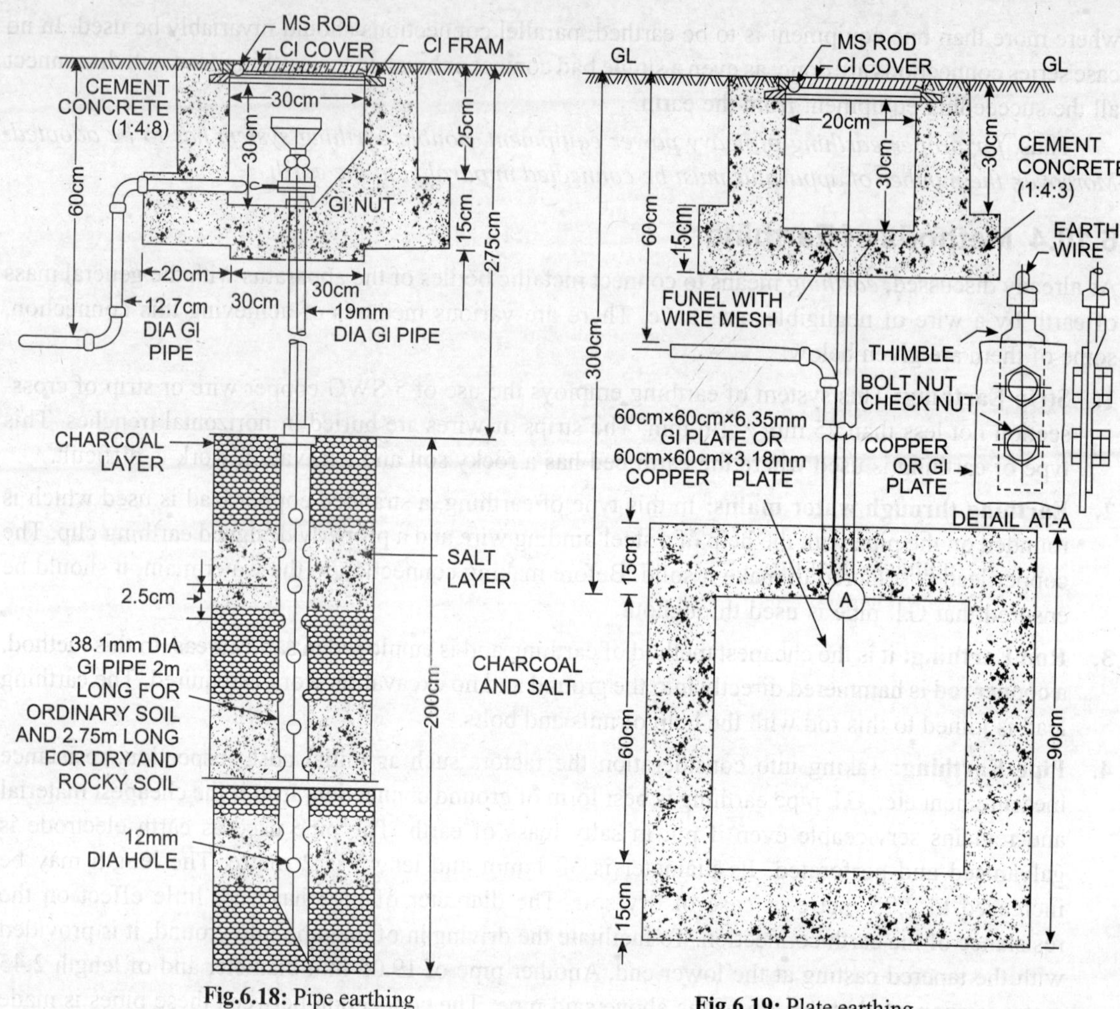

Fig.6.18: Pipe earthing

Fig.6.19: Plate earthing

6.10.5 Double Earthing

For providing better safety it is advisable to provide two separate earth wires, from two separate earth electrodes, connected to same metallic body of the equipment at two different points. This is known as double earthing. Double earthing is essential, as per Indian Electricity Rule, for metallic bodies of large rating equipment such as transformer, motors etc. working at 400 V and above.

Advantages of double earthing are:

1. Surety of safety, because if at any time one earthing is ineffective, then another will provide earth path to fault current.
2. As the two earth wires are in parallel so the effective resistance from equipment to earth electrode is reduced.

6.11 BATTERY

A series, parallel or series-parallel grouping of cells is called a **battery.**

Generally, a cell can deliver a small current at low voltage. For a circuit, if higher voltage is required – a battery containing number of cells connected in series is applied; if higher current in required a battery containing number of cells connected in parallel is applied; if larger current at higher voltage is required a battery containing number of cells in series further connected in parallel is applied.

Usually, a number of cells connected in series placed in a single container is called **a battery.**

6.11.1 lead-Acid Battery

Fig.6.20 shows the cut-away view of 6V commercial lead-acid battery. The following are the important parts of the battery.

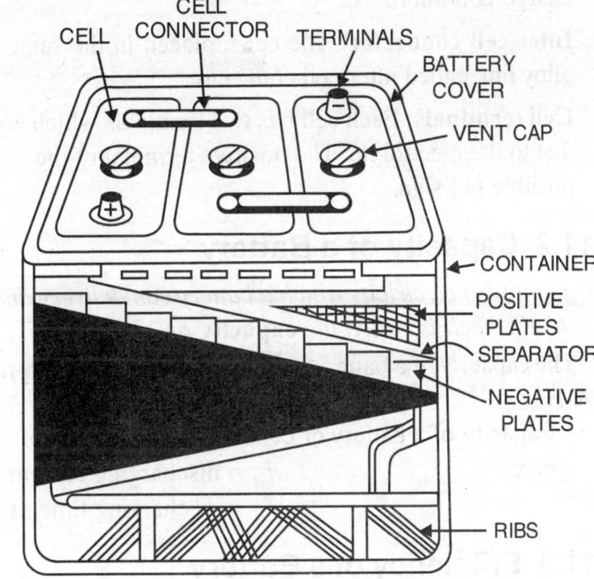

Fig.6.20: Lead-acid battery

1. **Container:** It is the outer body of the battery. It is made of a hard rubber or plastic material and is sealed at the top to prevent spilling of the electrolyte. A large space is left at the bottom of the container so that the sediments that drop from the plates are collected here and may not short circuit the positive and negative plates.
2. **Plates:** Generally, alloy of lead-antimony sheets covered with lead-peroxide and spongy lead forming positive and negative plates respectively are used as electrodes. To increase the capacity of the battery, we use a large number of plates in each cell instead of only two plates. The number of positive and negative plates (*i.e.* 11, 13, 15 or 17) of each cell are alternatively placed and sandwitched with an insulator called *separator* as shown in fig.6.21. One group of positive and negative plates forms a cell which develops an e.m.f. of 2.0 volt. A separate compartment is provided for each cell in the container of the battery.

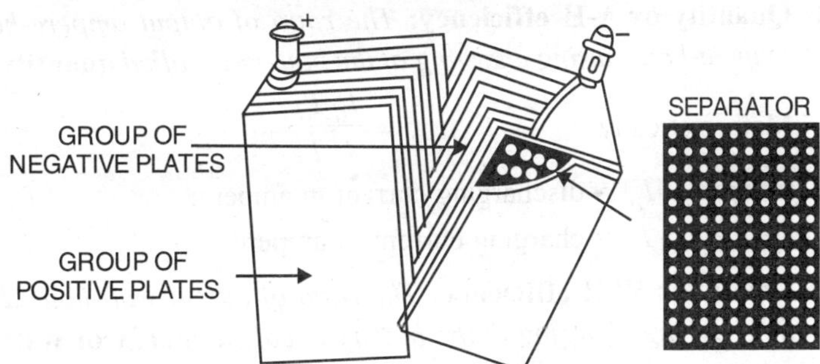

Fig.6.21: Battery plates and separator

3. **Separator:** To reduce the internal resistance of the cell and to save the space, the plates are placed very close to each other. To prevent the plates touching each other if they wrap or buckle, they are separated by a rubber sheet (non-conducting material) having large number of small holes called *separator* [see fig.6.21].
4. **Electrolyte:** Dilute sulphuric-acid (H_2SO_4) is used as an electrolyte in lead-acid batteries. Sulphuric-acid is added to water in such a proportion that with a fully charged battery, its specific gravity is about 1.28 to 1.29.

5. **Battery cover:** Each cell compartment is covered usually with a molded hard rubber and the joints between covers and container are sealed with an acid-resistance material. In each cell cover openings are provided — two for positive and negative terminals, and third for a vent. The whole container is fitted with a leak proof cover.
6. **Vent caps:** The vent-cap has a vent hole to allow free exit of the gases formed in the cell during charging. The vent caps can be easily removed for adding water. The vent cap is also removed to insert the nozzle of hydrometer for checking the specific gravity of electrolyte to check the battery charge condition.
7. **Inter-cell connector:** The cells, placed in the same container are connected in series with a lead alloy link called inter-cell connector.
8. **Cell terminals:** Each cell has two terminals which are generally made of lead as it does not corrode due to the electrolyte. The positive terminal of the battery is marked with a red colour or by a large positive (+) sign.

6.11.2 Capacity of a Battery

The quantity of electricity which a battery can delivery during single discharge until its terminal voltage falls to 1·8V/cell is called the **capacity** *of a battery.*

The capacity of a battery or cell is commercially expressed in ampere-hour and is generally denoted by A-H.

Capacity of a battery or cell = $I_d T_d$ ampere-hour

where, I_d = discharging current in ampere

T_d = discharging time of battery or cell in hour.

6.11.3 Efficiency of a Battery

The efficiency of a battery (or cell) can be defined in the following two ways:

(*i*) **Quantity or A-H efficiency:** *The ratio of output ampere-hour during discharging to the input ampere-hour during charging of the battery is called* **quantity or ampere-efficiency** *of the battery.*

Mathematically, $$\eta_{AH} = \frac{I_d T_d}{I_c T_c}$$

where, I_d = discharging current in ampere: T_d = discharging time in hour.

I_c = charging current in ampere, T_c = charging time in hour.

(*ii*) **Energy or W-H efficiency:** *The ratio of output watt-hour during discharging to the input watt-hour during charging of the battery is called* **energy or watt-hour efficiency** *of the battery.*

Mathematically, $$\eta_{WH} = \frac{I_d T_d V_d}{I_c T_c V_c}$$

where, V_d = Average terminal voltage during discharging,

V_c = Average terminal voltage during charging

Battery Back-up

The time (in hrs) for which a battery can deliver the desired current is called **battery back-up** *of the bank.*

Example 6.1. *An alkaline cell is discharged at a steady current of 4 A for 12 hour, the average terminal voltage being 1·2 V to restore it to its original state of voltage, a steady current of 3 A for 20 hour is required, the average terminal voltage being 1·44 V. Calculate the ampere-hour and watt-hour efficiencies in this particular case.*

Solution:

$$\text{Ampere-hour efficiency, } \eta_{AH} = \frac{I_d T_d}{I_c T_c} \times 100$$

where, $I_d = 4\text{A}$; $T_d = 12$ hour; $I_c = 3\text{A}$; $T_c = 20$ hour;

$$\eta_{AH} = \frac{4 \times 12}{3 \times 20} \times 100 = \mathbf{80\%} \textit{ (Ans.)}$$

$$\text{Watt-hour efficiency, } \eta_{WH} = \frac{I_d T_d V_d}{I_c T_c V_c} \times 100$$

where, $V_d = 1{\cdot}2$ V; $V_c = 1{\cdot}44$ V;

$$\eta_{WH} = \frac{4 \times 12 \times 1{\cdot}2}{3 \times 20 \times 1{\cdot}44} \times 100 = \mathbf{66{\cdot}67\ \%} \textit{ (Ans.)}$$

Example 6.2. *A discharged battery is put on charge at 5A for 4 hour at a mean charging voltage of 13.5 V. It is then discharged in 6 hour at a constant terminal voltage of 12 V through a resistance of R ohm. Calculate the value of R for an ampere-hour efficiency of 85%.*

Solution: Ampere-hour efficiency, $\eta_{AH} = \dfrac{I_d T_d}{I_c T_c} \times 100$

where, $\eta_{AH} = 85\%$; $T_d = 6$ hour; $I_c = 5$ A; $T_c = 3.5$ hour.

$$\therefore \quad 85 = \frac{I_d \times 6}{5 \times 3{\cdot}5} \times 100 \quad \text{or} \quad I_d = 2{\cdot}48 \text{ A} \quad \text{and} \quad R = \frac{V_d}{I_d} = \frac{12}{2{\cdot}48} = \mathbf{4{\cdot}84\ ohm} \textit{ (Ans.)}$$

6.11.4 Charge Indications of a lead-acid Battery or Cell

The charge condition of a battery is usually checked by checking the specific gravity of the electrolyte (H_2SO_4). For a fully charged battery, the specific gravity of H_2SO_4 is 1.28 to 1.29. However, when the specific gravity falls below 1.15, the battery is fully discharged. In fact, to increase the life of battery, it should be recharged when the specific gravity of the electrolyte is found to be less than 1.18. The values of specific gravity for different condition of charge are given below :

Specific gravity	***Condition***
1.280 to 1.290	100% charged
1.230 to 1.250	75% charged
1.190 to 1.200	50% charged
1.150 to 1.160	25% charged
below 1.130	fully discharged

To check the specific gravity of H_2SO_4, an instrument called *hydrometer* is used (see fig.6.22) which works on Archmedeies principle. In common use, the decimal point is omitted from the value of specific gravity *i.e.* 1.280 specific gravity is spoken as 1280 and so on.

However, the state of a battery can also be checked by checking.

(*i*) *Voltage:* When the terminal voltage of the battery on load is 2.1 to 2.5 V per cell, the battery is said to be fully charged. Whereas, when the voltage of battery falls to below 1.8 V per cell, the battery is considered to be fully discharged and it is immediately put on charging.

(*ii*) *Colour of plates*: When a lead-acid cell or battery is fully charged, its anode is PbO_2 which is *chocolate brown* in colour and cathode is of *Pb* which is *grey* in colour. However, when the battery is fully discharged, both the plates attain $PbSO_4$ as active material which is whitish in colour.

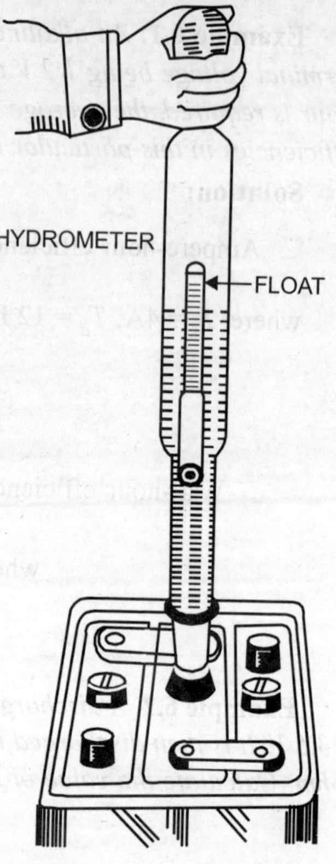

Fig.6.22: Checking of electrolyte

Characteristics of Lead-acid Battery

(i) The e.m.f. of a fully charged lead-acid cell is 2.2 V which decreases to 2.0 V rapidly.

However, the average e.m.f. of the cell is 2.0 V which decreases to 1.8 V when fully discharged.

(*ii*) The internal resistance of this cell is quite low.

- The A-H efficiency of this cell is nearly 80% whereas, the W-H efficiency is 60%.
- The specific gravity of electrolyte is 1280 to 1290 but to1150 when the battery is fully discharged.

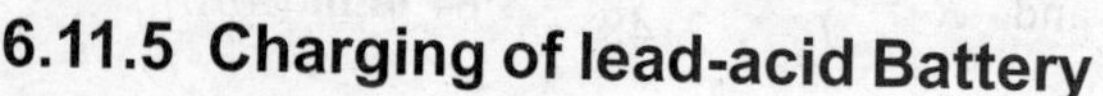

6.11.5 Charging of lead-acid Battery

Whenever, terminal voltage of a battery falls below 1.8V per cell, it is put under recharging. The following points must be kept in mind while charging a battery:

(*i*) Only a d.c. voltage source is applied for recharging.

6.11.6 Care and Maintenance of lead-acid Batteries

The average life of a lead-acid battery is two to four years depending upon its manufacturing qualities and technique. However, to obtain longer life and efficient service, the following points must be kept in view:

1. The battery should not be allowed to use when the e.m.f. of the battery falls to 1·8 V per cell. Otherwise, the lead sulphate of the plates partly changes to non-active lead sulphate and reduces the life of the battery.
2. The specific gravity to the electrolyte should not be allowed to falls below 1.15.
3. The battery should never be left standing in a discharged condition, otherwise sulphation will occur and the battery cells are permanently damaged.

4. When not in use, the battery must be fully charged and stored in a cool and dry place.
5. Great care should be taken that the acid used as electrolyte should not contain any substantial impurity. It should be colourless when viewed through a 12 cm column.
6. The electrodes must remain completely immersed in the electrolyte, preferably the level of electrolyte should always be about 10 mm above the electrodes.
7. Whenever the level of the electrolyte decreases due to evaporation or gassing, distilled water should be added so as to keep the same concentration of electrolyte.
8. The battery should be charged and discharged at low rate so that its temperature may not rise above 45 °C. The high temperature may buckle the plates and damage the separators and the battery may be totally damaged.
9. The battery terminals should never be short circuited.
10. While charging the polarity must be checked carefully.
11. The room where the batteries are charged should be well ventilated as the atmosphere near the batteries would be charged with corresponding acid fumes.
12. The flames must be kept away form the vent of the battery, otherwise hydrogen and oxyzen produced with in the battery cells may get fire.
13. The battery terminals should always be kept clean and periodically greased with vaseline to prevent corrosion.

6.11.7 Applications of lead-acid Batteries

Lead-acid batteries have innumerable commercial applications. Some of the important application are given below:

1. Used in automobiles for starting and lighting.
2. For lighting on steam and diesel railway trains.
3. Used at generating stations and sub-stations for operation of protective devices and for emergency lighting.
4. Used at telephone exchanges.
5. Used for emergency lighting at important places such as hospitals, theaters, banks etc.
6. Used for lighting purposes in remote rural areas.

6.12 NICKEL-IRON ALKALINE CELL

It is also known as Edison-cell as it was developed by an American scientist Thomson A Edison in 1909.

Construction

It contains two plates *i.e.* a positive plate (anode) and a negative plate (cathode). The active material of anode is $Ni(OH)_4$ and of cathode is iron (Fe) when fully charged. The two plates are immersed in the electrolyte, a solution of potassium hydroxide (KOH). A small quantity of lithium hydrate ($LiOH$) is also added to the electrolyte which increases the capacity and life of the cell. The specific gravity of the electrolyte is 1.2. In this case, the container is made of nickel-plated iron to which negative plates are connected. This cell is quit compact as small quantity of electrolyte is used.

Electrical characteristics

(*i*) The e.m.f. of a fully charged cell is 1.4 V which decreases to 1.3 V rapidly. However, the average e.m.f. of the cell is 1.2 V which decreases to 1.0 V when fully discharged.

(*ii*) The internal resistance of this cell is quite high nearly 5 times to that of a lead-acid cell.

(*iii*) The A-H efficiency of this cell is nearly 80% whereas, the W-H efficiency is 60%.

Advantages

It has the following advantages in comparison to that of a lead-acid cell.

(*i*) Longer life – about 5 year

(*ii*) Its electrolyte (*KOH*) is not harmful if spilled away.

(*iii*) The specific gravity of its electrolyte does not change when discharged, therefore, it can be left in a fully discharged condition for a considerable period of time without damage.

(*iv*) Lower weight – nearly half to that of lead-acid cell.

(*v*) It can be discharged and recharged at higher rate for longer period without damage.

(*vi*) It can withstand higher temperature.

(*vii*) It is more rugged and can withstand more mechanical and electrical stresses.

Disadvantages

(*i*) Higher cost-nearly double.

(ii) Higher cost-nearly double.

(iii) As the e.m.f. developed per cell is less (only 1.2V), more number of cells are required for a particular voltage.

(iv) Higher internal resistance-nearly 5 times. Therefore, it can not provide large current and is unsuitable for automobile starting.

(v) Lower efficiency.

6.13 CALCULATIONS FOR ENERGY CONSUMPTION

Electrical energy is supplied to a consumer by the supplier. To charge the electrical energy consumed by a consumer, an energy meter is installed to its quantity. The reading of the energy meter is taken every month. The difference between the fresh reading and the previous reading tell about the consumption of electrical energy in that month. This quantity of energy is multiplied by the rate (or tariff) fixed by the supplier to prepare an electricity bill. However, some other charges such as meter rent, GST, other taxes applicable etc. are also added in the bill.

Example 6.3. *A building has (i) 12 light points of 60 watt each burning 4 hour a day, (ii) a fan point of 75 watt each running 10 hour a day, (iii) a plug point for a 750 watt heater is used one hour a day. (iv) one radio 80 watt used 6 hour a day and (v) a 1/2 H.P. pump of 80% efficiency running 2 hour a day. Calculate the total connected load in kilowatt, maximum possible current, the daily consumption of energy and monthly bill. The supply is given at 230 volt and energy costs Rs. 5.15 per unit. The rent for a meter is Rs. 50 per month. Assume the month of 30 days.*

Solution:

Load Points	Connected Load	Energy Consumed/Day
(i) 12 lights of 60W each 4 hr/day	12 × 60 = 720 W	$\frac{720 \times 4}{1000} = 2{\cdot}88$ kWh
(ii) 4 fan points of 75W each – 10 hr/day	4 × 75 = 300 W	$\frac{300 \times 10}{1000} = 3{\cdot}00$ kWh
(iii) 1 plug point of 750W heater – 1 hr/day	1 × 750 = 750W	$\frac{750 \times 1}{100} = 0{\cdot}75$ kWh
(iv) 1 ratio of 80W – 6 hr/ day	1 × 80 = 80 W	$\frac{80 \times 6}{1000} = 0{\cdot}48$ kWh
(v) 1/2 H.P. Pump 80% eff. – 2 hr/day	$\frac{1}{2} \times \frac{735{\cdot}5}{80} \times 100 = 460$ W	$\frac{460 \times 2}{1000} = 0{\cdot}92$ kWh
Total	**2310 W**	**8·03 kWh**

Therefore connected load, P = 2310 W = **2·31 kW** (*Ans.*)
Maximum possible current, $I = P/V$ = 2310/230 = **10·043 A** (*Ans.*)
Energy consumption/day = **8·03 kWh** (*Ans.*)
Energy consumption/month = 8·03 × 30 = **240·9 kWh** (*Ans.*)
Rate of energy/month = Rs. 5.15
Energy cost/month = 5·15 × 240·9 = Rs. 1240·60
Meter rent/month = Rs. 50·00
Therefore monthly bill = 50 + 1240·60 = **Rs. 1290·60** (*Ans.*)

PRACTICE EXERCISE

Q.1 *An alkaline cell is discharged at a steady current of 4 A for 12 hour, the average terminal voltage being 1·2 V to restore it to its original state of voltage, a steady current of 3 A for 20 hour is required, the average terminal voltage being 1·44 V. Calculate the ampere-hour and watt-hour efficiencies in this particular case.* **(Ans.** *80%; 66.67%*)

Q.2 *A discharged battery is put on charge at 5A for 4 hour at a mean charging voltage of 13.5 V. It is then discharged in 6 hour at a constant terminal voltage of 12 V through a resistance of R ohm. Calculate the value of R for an ampere-hour efficiency of 85%.* **(Ans.** *4.84 ohm*)

Q.3 *A building has (i) 12 light points of 60 watt each burning 4 hour a day, (ii) a fan point of 75 watt each running 10 hour a day, (iii) a plug point for a 750 watt heater is used one hour a day. (iv) one radio 80 watt used 6 hour a day and (v) a 1/2 H.P. pump of 80% efficiency running 2 hour a day. Calculate the total connected load in kilowatt, maximum possible current, the daily consumption of energy and monthly bill. The supply is given at 230 volt and energy costs Rs. 5.15 per unit. The rent for a meter is Rs. 50 per month. Assume the month of 30 days.*

(Ans. *2·31 kW; 10.043 A; 240.9 kWh; 1290.60*)

SUMMARY

1. *Switchgear:* In an electric power system, the combination of electrical switches, fuses or circuit breakers used to control, protect and isolate electrical equipment is called switchgear.
2. *Power conducting components:* All the switches, circuit breakers, fuses, and lightning arrestors, that conduct or interrupt the flow of electrical power come under this category.
3. *Controlling components:* All the control panels, current transformers, potential transformers, protective relays, and associated circuitry, that monitor, control, and protect the power conducting components come under this category.
4. *HV, MV & LV switchgears:* In power system, the electrical switchgears are categorized as *HV, MV & LV* switchgears.
5. *HV Switchgears:* High voltage circuit breaker, is the main component of *HV switchgear,*
6. *MV Switchgears:* The switchgear system which is to handle voltages between 3 kV to 33 kV is categorized as medium voltage switchgear or *MV switchgear.* These switchgears are of many types.
7. *LV Switchgears:* Generally electrical switchgear rated upto 1 kV is termed as low voltage switchgear. The term *LV* Switchgear includes low voltage circuit breakers, switches, off load electrical isolators, *HRC* fuses, earth leakage circuit breaker, miniature circuit breakers (*MCB*) and molded case circuit breakers (*MCCB*) etc i.e. all the accessories required to protect the LV system.
8. *Switch Fuse Unit (SFU):* It comprises of a porcelain or *HRC* fuse and switch. The switch is operated manually with the help of a handle.
9. *Miniature Circuit Breaker (MCB):* It is a device which ensures definite protection of wiring system and sophisticated equipment against over current and short circuit faults.
10. *ELCB or RCCB:* Earth Leakage Circuit Breaker or Residue Current Circuit Breaker is a device which provides protection against earth leakage faults.
11. *Molded case circuit breakers:* Molded case circuit breaker is the electrical protection device that is commonly used when load current exceed the capabilities of miniature circuit breakers.
12. *Wires and Cables:* A solid or stranded conductor covered with insulation is known as a insulated wire or cable.
13. *Low Tension (LT) Cables:* These cables are employed for system voltages upto 1000 volt. Generally, these cables are used for distribution of electric power at 400 V. The cables are of two types viz single-core and multi-core cables.
14. *Earthing* : The process of connecting metallic bodies of all the electrical apparatus and equipment to the huge mass of earth by a wire of negligible resistance is called *earthing.*
15. *Purpose of earthing:* The basic purpose of earthing is to *protect the human body* (operator) from electric shock.
16. *Equipment earthing:* According to Indian Electricity Rules, it is obligatory to earth the metallic bodies of all the electrical equipment/apparatus/appliance which is operated at 125 V or more than this.

17. *System earthing:* All the electrical equipment/apparatus/appliance operated at 125 V or more must be properly earthed atleast at two places called double earthing.

18. *Methods of earthing*: The following method may be employed for earthing (i) strip earthing (ii) Earthing through water mains (iii) Rod earthing (iv) Pipe earthing (v) Plate earthing. Pipe earthing is considered the best method.

19. *Battery*: A number of cells connected in series, placed in single container is called a *battery*.

20. *Capacity of battery:* The quantity of electricity (in Ah) which a battery can deliver in single discharge is called its *capacity.*

21. *Efficiency of a cell or battery*:

$$\eta_{AH} = \frac{I_d\, T_d}{I_c\, T_c}; \ \eta_{WH} = \frac{I_d\, T_d\, V_d}{I_c\, T_c\, V_c}$$

22. *Charge indications of a lead-acid cell or battery:*

(*i*) Specific gravity of electrolyte: 1.280 to 1.290 – 100% charged

(*ii*) Voltage: more than 2V – fully charged; below 1·8 V – fully discharged.

(*iii*) Colour of plates: anode – chocolate brown and cathode – grey fully charged; both whitish – fully discharged.

23. *Nickle – Iron alkaline cell:* Anode – $Ni(OH)_4$; Cathode – *Fe*; Electrolyte – *KOH*

Electrical characteristics:

(*i*) E.M.F. of a fully charged cell is 1·4 V which decreases rapidly to 1·3 V. Average e.m.f. of the cell is 1·2 V which decreases to 1·0 V when fully discharged.

(*ii*) High internal resistance – 5 times to lead-acid cell.

(*iii*) η_{AH} – 80% (App); η_{WH} – 60% (App.)

24. *Calculations for energy consumption:* The quantity of electrical energy consumed by the consumer in a specified period is multiplied by the rate (or tariff) fixed by the supplier to prepare an electricity bill.

SHORT ANSWER QUESTIONS

1. What do you mean by switchgear?
2. What are the important elements of *LT* switchgear?
3. What is the function of *SFU*?
4. Why *MCB* is preferred over *SFU*?
5. What is the major function of *ELCB*?
6. How will you differentiate between *MCCB* and MCB?
7. What is *CTS* and *VIR* wire ?
8. Name the various types of wiring systems commonly used in residential and commercial buildings.
9. What do you mean by earthing ?

10. How earth resistance can be kept low in dry summer season ?

11. What do you mean by double earthing ?

12. Which type of earthing will you prefer pipe earthing or plate earthing ? Give reasons.

13. How the outer metallic bodies of electrical appliances are connected to earth?

14. What is the function of container of a lead-acid battery.

15. Why large number of holes are provided in the separator.

16. Mention five important applications of a lead-acid battery.

17. While checking the conditions of battery it is seen that the float dips at different levels at different specific gravity of electrolyte, why ?

18. What are the charge indications of a lead-acid battery ?

19. What do you mean by capacity of a battery ? What are its units ?

20. Define ampere-hour efficiency of a battery.

21. What do you mean by capacity of a batter, what are the factors on which it depends ?

22. How do we care a lead-acid battery ? Write five major points.

23. What are the applications of lead-acid batteries ?

24. Name the important components of a Nickel-iron alkaline cell.

25. What are the electrical characteristics of a nickel-iron cell ?

26. How will ypu calculate the charges of electrical energy used by a consumer?

27. Mention the charge indications of a lead-acid battery.

NUMERICAL FOR PRACTICE

1. An alkaline cell is discharged at a steady current of 5 A for 12 hour, the average terminal voltage being 1·2 V to restore it to its original state of voltage, a steady current of 4 A for 20 hour is required, the average terminal voltage being 1·44 V. Calculate the ampere-hour and watt-hour efficiencies in this particular case. (***Ans.*** *75%, 62.5%*)

2. A discharged battery is put on charge at 6A for 4 hour at a mean charging voltage of 13.5 V. It is then discharged in 6 hour at a constant terminal voltage of 12 V through a resistance of R ohm. Calculate the value of R for an ampere-hour efficiency of 85%. (***Ans.*** *3.53 Ω*)

3. Calculate the ampere-hour and watt-hour efficiencies of a secondary cell having 20 hour charge rate of 10 A and delivery rate of 5A for 36 hour with mean terminal voltage of 1.96 V. The terminal voltage during charging has a mean value of 2.35 V. (***Ans.*** *90%; 75.06%*)

4. A building has (i) 20 light points of 60 watt each lighting 5 hour a day, (ii) 10 fan point of 70 watt each running 18 hour a day, (iii) a plug point with a 750 watt heater is used two hour a day. (iv) two TVs each of 80 watt used 6 hour a day and (v) a 1/2 H.P. pump motor having 80% efficiency running 2 hour a day. Calculate the total connected load in kilowatt, maximum possible current, the daily consumption of energy and monthly bill. The supply is given at 230 volt and energy costs Rs. 6.00 per unit. The rent for a meter is Rs. 80 per month. Assume the month of 30 days. (***Ans.*** *3.285 kW, 14.28 A, 22.01 kWh, Rs. 3961.80*)

MULTI-CHOICE QUESTIONS

1. *In motor wiring installations, double earting is required.*
   (a) Single
   (b) double
   (c) tripple
   (d) all of these
2. *The material used for fuse element is*
   (a) copper
   (b) Aluminium
   (c) Tin-lead alloy
   (d) any of a, b or c
3. *The material used for fuse must have*
   (a) low melting point and low specific resistance
   (b) low melting point and high specific resistance.
   (c) high melting point and low specific resistance
   (d) low melting point with any specific resistance.
4. *Fuse is always connected*
   (a) in series with the circuit to be protected
   (b) in parallel with the circuit to be protected
   (c) either (a) or (b)
   (d) none of these
5. *Fuse is always connected in*
   (a) neutral
   (b) earth
   (c) phase
   (d) any (a), (b) or (c).
6. *The basic purpose of earthing is that*
   (a) it avoids faults
   (b) it allows the current to flow in the circuit
   (c) it protects the operator from electric shock
   (d) it stops current to flow in the circuit.
7. *For proper earthing, according to I.E. Rules, of heavy power equipment:*
   (a) single earthing is sufficient
   (b) double earthing system has to be adopted
   (c) half earthing is sufficient
   (d) any one of above.

8. *Earth resistance should be very low*

(a) very low

(b) 10 ohm

(c) 100 ohm

(d) very high

9. The capacity of a cell depends upon

(*a*) nature of plates material and electrolyte

(*b*) size of plates and quantity of electrolyte

(*c*) both *a* and *b*

(*d*) none of above

10. A 12 V lead-acid battery used in a car contains

(*a*) 10 cells connected is series

(*b*) 10 cells connected in parallel

(*c*) 6 cells connected is parallel

(*d*) 6 cells connected in series.

11. To check the specific gravity of an electrolyte, the name of the instrument used is

(*a*) hydrometer

(*b*) lactometer

(*c*) barometer

(*d*) voltameter

12. The electrolyte used in a Nickel-iron battery is

(*a*) H_2SO_4

(*b*) $K(OH)_2$

(c) *NaCl*

(*d*) none of above

13. When lead-acid battery is fully charged.

(*a*) The colour of positive plates will be chocolate brown

(*b*) The colour of negative plates will be whitish

(*c*) the specific gravity of electrolyte will be 1.15.

(*d*) All above

14. When a nickle-iron cell is fully charged

(a) the active material of positive and negative plates will be $Ni\ (OH)_4$ and *Fe* respectively.

(b) the p.d. across the terminal of the cell be 1.4 V

(c) The specific gravity of the electrolyte (*KOH*) will be 1.2.

(d) All above.

TEST QUESTIONS

1. What do you mean by electrical switchgear? How will you categorise it?
2. What do you mean by LV switchgear? What are the various a SFU.
3. Explain the construction and working of an MCB.
4. How an MCB is different to fuse?
5. What do you mean by MCCB? How it is different to MCB?
6. Explain the construction and working of an ELCB.
7. How does a fuse protect and maintain the life of an electrical equipment ?
8. Classify the cables (or wires) use in domestic installations
9. Explain with the help of a neat sketch the construction of 4-core LV cable.
10. What is earthing. What is its purpose ?
11. Explain system earthing and equipment earthing.
12. What are the various methods of earthing ? Explain pipe earthing.
13. What do understand by double earthing ?
14. Explain the construction and working of a lead-acid storage battery.
15. How hydrometer is used to determine the charged condition of a lead acid cell ? Explain.
16. Write a note on care and maintenance of a lead-acid battery.
17. Explain the important characteristics of a Lead-acid battery and NIckle-iron battery.
18. State the precautions to be observed in the use of a lead-acid cell.
19. Explain the constructional feature of a typical nickel-iron alkaline cell.
20. Give atleast five important application of a lead-acid battery.
21. How will you calculate the electricity bill of a consumer?

ANSWER TO MULTI-CHOICE QUESTIONS

1. (b)	**2.** (d)	**3.** (a)	**4.** (a)	**5.** (c)
6. (c)	**7.** (b)	**8.** (a)	**9.** (*b*)	**10.** (*d*)
11. (*a*)	**12.** (*d*)	**13.** (*a*)	**14.** (*d*)	

BASIC ELECTRICAL ENGINEERING

RITU SAHDEV

FEEDBACK FORM

1. Which topics from your examination point of view are inadequately or not covered in this book?

___

___

2. Is there any misprints/ mistakes/ factual inaccuracy in this book? If so, please specify.

___

___

3. What is your assessment of this book as regards the presentation of the subject-matter, expression, precision and price in relation to the other recommended books available on this subject?

___

___

4. Which competing books you regard as better than this? Please mention their author and publishers.

___

___

5. Any further suggestions/comments you would like to make for the improvement of the book?

___

___

PS: If needed, please attach a separate sheet.

Name : ___

College : ___

Address : ___

Email : ___

Mobile : ___

Please mail the feedback form at:

KHANNA BOOK PUBLISHING CO. (P) LTD.

4C/4344, Ansari Road, Daryaganj, New Delhi-110002

You can also email this feedback form at contact@khannabooks.com or Whatsapp at +91 – 99109 09320

BEST FEEDBACK WILL BE REWARDED SUITABLY